Text Book on
SURVEYING AND GEOMATICS

Text Book on

SURVEYING AND GEOMATICS

AUTHOR

P.K. GARG

Professor,

Civil Engineering Department

Indian Institute of Technology Roorkee, Roorkee

&

Former, Vice Chancellor

Uttarakhand Technical University, Dehardun

REVIEWER

Raaj Ramsankaran

Associate Professor,

Civil Engineering Department

Indian Institute of Technology Bombay

Mumbai

KHANNA BOOK
PUBLISHING CO. (P) Ltd.

4C/4344, Ansari Road, Darya Ganj,
New Delhi, India

All India Council for
Technical Education

Nelson Mandela Marg, Vasant Kunj,
New Delhi, India

Dear Readers,

To prevent the piracy, this book is secured with HIGH SECURITY HOLOGRAM on the front title cover. In case you don't find the hologram on the front cover title, please write us to at contact@khannabooks.com or whatsapp us at +91-99109 09320 and avail special gift voucher for yourself.

Specimen of Hologram on front Cover title:

Moreover, there is a SPECIAL DISCOUNT COUPON for you with EVERY HOLOGRAM.

How to avail this SPECIAL DISCOUNT:

Step 1: Scratch the hologram
Step 2: Under the scratch area, your "coupon code" is available
Step 3: Logon to www.khannabooks.com
Step 4: Use your "coupon code" in the shopping cart and get your copy at a special discount
Step 5: Enjoy your reading!

Thanks,
Khanna Publishing House

ISBN: 978-93-5538-700-4

Book Code: UG421EN

TEXT BOOK ON SURVEYING AND GEOMATICS
by **P.K. Garg**

First Edition: 2026

To view complete list of books, please scan the QR Code:

Published by:
Khanna Book Publishing Co. (P) Ltd.
CIN: U22110DL1998PTC095547

Visit us at: www.khannabooks.com
Write us at: contact@khannabooks.com

Printed in Bharat

सत्यमेव जयते

प्रो. टी. जी. सीताराम
अध्यक्ष
Prof. T. G. Sitharam
Chairman

अखिल भारतीय तकनीकी शिक्षा परिषद्
(भारत सरकार का एक सांविधिक निकाय)
(शिक्षा मंत्रालय, भारत सरकार)
नेल्सन मंडेला मार्ग, वसंत कुंज, नई दिल्ली-110070
दूरभाष : 011-26131498
ई-मेल : chairman@aicte-india.org

ALL INDIA COUNCIL FOR TECHNICAL EDUCATION
(A STATUTORY BODY OF THE GOVT. OF INDIA)
(Ministry of Education, Govt. of India)
Nelson Mandela Marg, Vasant Kunj, New Delhi-110070
Phone : 011-26131498
E-mail : chairman@aicte-india.org

FOREWORD

Engineers are the backbone of any modern society. They are the ones responsible for the marvels as well as the improved quality of life across the world. Engineers have driven humanity towards greater heights in a more evolved and unprecedented manner.

The All India Council for Technical Education (AICTE), have spared no efforts towards the strengthening of the technical education in the country. AICTE is always committed towards promoting quality Technical Education to make India a modern developed nation emphasizing on the overall welfare of mankind.

An array of initiatives has been taken by AICTE in last decade which have been accelerated now by the National Education Policy (NEP) 2020. The implementation of NEP under the visionary leadership of Hon'ble Prime Minister of India envisages the provision for education in regional languages to all, thereby ensuring that every graduate becomes competent enough and is in a position to contribute towards the national growth and development through innovation & entrepreneurship.

One of the spheres where AICTE had been relentlessly working since past couple of years is providing high quality original technical contents at Under Graduate & Diploma level prepared and translated by eminent educators in various Indian languages to its aspirants. For students pursuing 2nd year of their Engineering education, AICTE has identified 88 books, which shall be translated into 12 Indian languages - Hindi, Tamil, Gujarati, Odia, Bengali, Kannada, Urdu, Punjabi, Telugu, Marathi, Assamese & Malayalam. In addition to the English medium, books in different Indian Languages are going to support the students to understand the concepts in their respective mother tongue.

On behalf of AICTE, I express sincere gratitude to all distinguished authors, reviewers and translators from the renowned institutions of high repute for their admirable contribution in a record span of time.

AICTE is confident that these outcomes based original contents shall help aspirants to master the subject with comprehension and greater ease.

(Prof. T. G. Sitharam)

Preface

This present book is an outcome of the fast developments which are taking place in Surveying and Geomatics in terms of their applications in wide range of disciplines. The objective of writing this book is to make the engineering students aware about the basics and applications of various tools and techniques used in surveying and geomatics. The book provides a wide coverage of essential topics as recommended by the AICTE. Efforts are made to explain the fundamentals of the surveying approaches in a simple way. Presently, there is an acute shortage of books which cover all the topics in a single book, and therefore the students have to consult many books to understand all the topics. This book presents a comprehensive matter on surveying and geomatics tools and techniques so that students are able to understand many topics from a single book.

While preparing the manuscript, emphasis has been laid on the basic principles and field procedure to use various surveying equipment. To provide better understanding, theory has been presented in a very logical and systematic manner with several illustrations. The book covers medium and advanced level numerical problems to test the understanding of the students. It is hoped that the book will inspire the students of civil engineering, geomatics engineering, geology and geography, image processig, to learn and discuss the new ideas in surveying and geomatics, and will contribute to the development of a solid foundation of the subject, and make their career in surveying profession. The author will be thankful to all the constructive comments and suggestions for further improvement of the book.

P K Garg

Acknowledgement

The author is grateful to the authorities of AICTE, particularly Prof. M. Jagadesh Kumar, Chairman; Prof. M. P. Poonia, Vice-Chairman; Prof. Rajive Kumar, Member-Secretary and Dr Amit Kumar Srivastava, Director, Faculty Development Cell for their planning to publish the books on (Surveying and Geomatics). We sincerely acknowledge the valuable contributions of the reviewer of the book Prof. RAAJ Ramsankaran, Associate Professor, Civil Engineering Department, IIT Bombay, Mumbai, to make the contents and subject matter more meaningful. It is hoped that this book will cover the AICTE Model Curriculum and the guidelines of National Education Policy (NEP) -2020. I am extremely grateful to my family members; Mrs Seema Garg, Dr Anurag Garg, Dr Garima Garg, Mr Hansraj Aggrawal, Ms Pooja Aggrawal and Master Avyukt Garg, and all relatives & friends for their understanding, continuous encouragement, moral support and well-wishes. Above all, I express my gratitude to Almighty God for offering all blessings and giving me enough strength to work hard to complete the Book on time, as planned.

This book is an outcome of various suggestions of AICTE members, experts and authors who shared their opinion and thought to further develop the engineering education in our country. Acknowledgements are due to the contributors and different workers in this field whose published books, review articles, papers, photographs, footnotes, references and other valuable information enriched us while writing this book. Finally, I like to express our sincere thanks to the publisher, M/s. XXXXXXXXXXXXXXXXXXXXX whose have cooperated to publish this book in a timely manner.

P K Garg

Outcome Based Education

For the implementation of an outcome based education the first requirement is to develop an outcome based curriculum and incorporate an outcome based assessment in the education system. By going through outcome based assessments evaluators will be able to evaluate whether the students have achieved the outlined standard, specific and measurable outcomes. With the proper incorporation of outcome based education there will be a definite commitment to achieve a minimum standard for all learners without giving up at any level. At the end of the programme running with the aid of outcome based education, a students will be able to arrive at the following outcomes:

PO1. **Engineering knowledge**: Apply the knowledge of mathematics, science, engineering fundamentals, and an engineering specialization to the solution of complex engineering problems.

PO2. **Problem analysis**: Identify, formulate, review research literature, and analyse complex engineering problems reaching substantiated conclusions using first principles of mathematics, natural sciences, and engineering sciences.

PO3. **Design / development of solutions**: Design solutions for complex engineering problems and design system components or processes that meet the specified needs with appropriate consideration for the public health and safety, and the cultural, societal, and environmental considerations.

PO4. **Conduct investigations of complex problems**: Use research-based knowledge and research methods including design of experiments, analysis and interpretation of data, and synthesis of the information to provide valid conclusions.

PO5. **Modern tool usage**: Create, select, and apply appropriate techniques, resources, and modern engineering and IT tools including prediction and modeling to complex engineering activities with an understanding of the limitations.

PO6. **The engineer and society**: Apply reasoning informed by the contextual knowledge to assess societal, health, safety, legal and cultural issues, and the consequent responsibilities relevant to the professional engineering practice.

PO7. **Environment and sustainability**: Understand the impact of the professional engineering solutions in societal and environmental contexts, and demonstrate the knowledge of, and need for sustainable development.

PO8. **Ethics**: Apply ethical principles and commit to professional ethics and responsibilities and norms of the engineering practice.

PO9. **Individual and teamwork**: Function effectively as an individual, and as a member or leader in diverse teams, and in multidisciplinary settings.

PO10. **Communication**: Communicate effectively on complex engineering activities with the engineering community and with society at large, such as, being able to comprehend and write effective reports and design documentation, make effective presentations, and give and receive clear instructions.

PO11. Project management and finance: Demonstrate knowledge and understanding of the engineering and management principles and apply these to one's own work, as a member and leader in a team, to manage projects and in multidisciplinary environments.

PO12. Life-long learning: Recognize the need for and have the preparation and ability to engage in independent and life-long learning in the broadest context of technological change.

Course Outcomes

After completion of the course the students will be able to:

CO-1: Describe various classifications of surveying, purpose and the equipment used for survey work.

CO-2: Carry out linear, height and angular measurements from various surveying instruments and compute the coordinates, including plotting of ground details to prepare maps.

CO-3: Apply computations for setting out horizontal and vertical curves on the ground using various methods.

CO-4: Use of various modern surveying equipment for digital field data collection and analyisis with greater speed and accuracy.

CO-5: Apply the concepts of Photogrammetry for scale, relief displacement and height determination as well as providing horizontal and vertical controls for engineering projects.

CO-6: Analyze different types of optical remote sening images for creation of accurate thematic maps.

Course Outcomes	Expected Mapping with Programme Outcomes (1- Weak Correlation; 2- Medium correlation; 3- Strong Correlation)											
	PO-1	PO-2	PO-3	PO-4	PO-5	PO-6	PO-7	PO-8	PO-9	PO-10	PO-11	PO-12
CO-1	3	2	2	2	1	1	1	1	1	1	1	1
CO-2	3	2	2	2	1	2	1	1	2	2	2	1
CO-3	3	3	2	2	1	2	1	1	2	2	2	1
CO-4	3	3	3	3	3	2	1	1	2	2	2	1
CO-5	3	3	2	2	2	2	2	1	2	2	2	1
CO-6	3	3	3	3	3	2	2	1	2	2	2	1

Guidelines for Teachers

To implement Outcome Based Education (OBE) knowledge level and skill set of the students should be enhanced. Teachers should take a major responsibility for the proper implementation of OBE. Some of the responsibilities (not limited to) for the teachers in OBE system may be as follows:

- Within reasonable constraint, they should manoeuvre time to the best advantage of all students.
- They should assess the students only upon certain defined criterion without considering any other potential ineligibility to discriminate them.
- They should try to grow the learning abilities of the students to a certain level before they leave the institute.
- They should try to ensure that all the students are equipped with the quality knowledge as well as competence after they finish their education.
- They should always encourage the students to develop their ultimate performance capabilities.
- They should facilitate and encourage group work and team work to consolidate newer approach.
- They should follow Blooms taxonomy in every part of the assessment.

Bloom's Taxonomy

Level	Teacher should Check	Student should be able to	Possible Mode of Assessment
Create	Students' ability to create	Design or Create	Mini project
Evaluate	Students' ability to justify	Argue or Defend	Assignment
Analyse	Students' ability to distinguish	Differentiate or Distinguish	Project/Lab Methodology
Apply	Students' ability to use information	Operate or Demonstrate	Technical Presentation/ Demonstration
Understand	Students' ability to explain the ideas	Explain or Classify	Presentation/Seminar
Remember	Students' ability to recall (or remember)	Define or Recall	Quiz

Guidelines for Students

Students should take equal responsibility for implementing the OBE. Some of the responsibilities (not limited to) for the students in OBE system are as follows:

- Students should be well aware of each UO before the start of a unit in each and every course.
- Students should be well aware of each CO before the start of the course.
- Students should be well aware of each PO before the start of the programme.
- Students should think critically and reasonably with proper reflection and action.
- Learning of the students should be connected and integrated with practical and real life consequences.
- Students should be well aware of their competency at every level of OBE.

Symbols and Abbreviations

LIST OF ABBREVIATIONS

Abbreviations	Full Form	Abbreviations	Full form
AI	Artificial Intelligence	LISS	Linear Imaging Self Scanning Sensor
ALISS	Advanced Linear Imaging Scanning Sensor	NavIC	Navigation with Indian Constellation
ALS	Airborne Laser Scanner	NAVSTAR	Navigation Satellite Timing and Ranging
ANN	Artificial Neural Networks	NDVI	Normalized Difference Vegetation Index
AS	Anti-Spoofing mode	NIR	Near Infrared
ASTER	Advanced Spaceborne Thermal Emission and Reflection Radiometer	NOAA	National Oceanic and Atmospheric Administration
AVHRR	Advanced Very High Resolution Radiometer	NRSC	National Remote Sensing Centre
AVIRIS	Airborne Visible/infrared Imaging Spectrometer	OCM	Ocean Colour Monitor
AWiFS	Advanced Wide Field Sensor	OLI	Operational Land Imager
BEIDOU	Beidou Navigation Satellite System	PAN	Panchromatic
BIM	Building Information Modelling	PC	Point of Curvature
BM	Bench Mark	PI	Point of Intersection
BS	Back Sight	POC	Point on Curve
C/A code	Course/Acquisition-code	POT	Point on Tangent
CP	Change Point	PP	Principal Point
DEM	Digital Elevation Model	PT	Point of Tangency
DGPS	Differential GPS	RADAR	RAdio Detection And Ranging
DN	Digital Number	RBV	Return Beam Videcon
DoD	Department of Defence	RGB	Red, Green, Blue
DOP	Dilution of Precision	RL	Reduced Levels
DPW	Digital Photogrammetric Workstation	rms	Root Mean Square

Abbreviations	Full Form	Abbreviations	Full form
DSM	Digital Surface Model	MIR	Middle Infrared
DSS	Decision Support Systems	MODIS	Moderate Resolution Imaging Spectroradiometer
DTM	Digital Terrain Model	MSL	Mean Sea Level
ERTS	Earth Resources Technology Satellite	MSS	Multispectral Scanner
EDM	Electronic Distance Measurement	NDVI	Normalised Difference Vegetation Index
EMR	Electro-magnetic Radiations	OTF-AR	On-The-Fly–Ambiguity Resolution
EMS	Electro-magnetic Spectrum	PCA	Principal Component Analysis
ETM	Enhanced Thematic Mapper	P-Code	Precision-code
ETM+	Enhanced Thematic Mapper Plus	PPS	Precision Positioning Signals
FCC	False Colour Composite	PRN	Pseudorandom Noise
FS	Fore Sight	QZSS	Quasi-Zenith Satellite System
GCP	Ground Control Points	RINEX	Receiver Independent Exchange Format
GDOP	Geometric Dilution of Precision	RTK	Real-Time Kinematic
GIS	Geographical Information System	SA	Selective Availability
GMS	Geostationary Meteorological Satellite	SAR	Synthetic Aperture Radar
GOES	Geostationary Operational Environmental Satellite	SoI	Survey of India
GPR	Ground Penetrating Radar	SPOT	Système Pour l'Observation de la Terre
GPS	Global Positioning System	SPS	Standard Positioning Signals
HE	Histogram Equalization	SRTM	Shuttle Radar Topography Mission
HRG	High Resolution Geometrical	SST	Sea Surface Temperature
HRV	High Resolution Visible	SWIR	Short Wave Infrared
HR-VIR	Visible & Infrared High-Resolution	TCT	Tasselled Cap Transformations
IFOV	Instantaneous Field of View	TIR	Thermal Infrared
ILWIS	Integrated Land and Water Information Management	TLS	Terrestrial Laser Scanner
IoT	Internet of Things	TM	Thematic Mapper

Abbreviations	Full Form	Abbreviations	Full form
IR	Infrared	UAV	Unmanned Aerial Vehicle
IRNSS	Indian Regional Navigation Satellite System	UTM	Universal Transverse Mercator
IRS	Indian Remote Sensing Satellites	UV	Ultraviolet
IS	Intermediate Sight	VI	Vegetation Indices
ISODATA	Iterative Self-Organizing Data Analysis Technique	WAAS	Wide Area Augmentation System
GLONASS	Global Navigation Satellite System	WGS84-	World Geodetic System1984-
GNSS	Global Navigation Satellite System	WiFS	Wide Field Sensor
LANDSAT	Land Satellites		
LiDAR	Light Detection and Ranging		

List of Figures

Unit 1: Surveying

Unit 2: Curves

Unit 3: Modern Field Survey Systems

Unit 4: Photogrammetry Surveying

Unit 5: Remote Sensing

Contents

01 SURVEYING

UNIT SPECIFICS

Through this unit we have discussed the following aspects:

- Principle of Surveying and historical background
- Various types of surveying
- Various maps and their characteristics
- Linear, angular and graphical methods of surveying and their utility
- Types of levels, including auto levels and digital levels, and their salient features
- Levelling observation methods
- Contour mapping and utility
- Magnetic bearing and compasses
- Measurements of magnetic bearing
- Various types of theodolites and their salient features
- Theodolite observation methods for taking horizontal and vertical angles
- Plane table survey
- Traverse, Triangulation and Trilateration
- Traversing with theodolites
- Error adjustments
- Types of triangulations and associated processes
- Numerical problems

In addition to the basic principle of surveying, the working of levels, compass and theodolites has been explained. The practical utility of these surveying equipment is presented for field data collection required for creating maps, as well as making these observations error free. Once the data is collected and corrected, topographic maps can be created by plane table survey methods in the field. Thenatic maps may be created from high resolution images, cutting down the requirements of large manpower and funds. Questions of short and long answer types are given following lower and higher order of Bloom's taxonomy, and a list of references and suggested readings is given in the unit so that the students can go through them for acquiring additional in-depth knowledge.

RATIONALE

This unit provides details of various types of surveying, maps, levels, compass and theodolites. Each of these equipments are used for a specific purpose of data collection required for creating maps. It explains various components of level, compass, plane table, and theodolite. Working of these equipments using various methods has been explained so that the students make use of these equipment in the field. Various methods used in the field and errors associated are also given. In levelling, angular observations, and traversing, errors and their minimisation are also discussed so that the users can minimise the errors from the field data. The computed coordinates of the surveyed points provide good horizontal and vertical control for civil engineering projects. Plane table survey will further enhance the understanding of map making processes.

PRE-REQUISITE

Mathematics: geometry and trigonometry, Earth surface.

UNIT OUTCOMES

List of outcomes of this unit is as follows:

U1-O1: Describe various types of Land survey, Maps, Levels, Compasses, and Theodolites

U1-O2: Explain the essential components and characteristics of Maps, Levels, Compasses, and Theodolites

U1-O3: Realize the role of Maps, Levels, Compasses, and Theodolites for field data collection

U1-O4: Describe various methods of data collection using, Levels, Compasses, and Theodolites, and apply corrections to observations

U1-O5: Apply the parameters collected in the field for providing horizontal and vertical controls, and creating the maps.

Unit-1 Outcomes	Expected Mapping with Programme Outcomes *(1- Weak correlation; 2- Medium correlation; 3- Strong correlation)*					
	CO-1	CO-2	CO-3	CO-4	CO-5	CO-6
U1-01	2	2	3	2	1	2
U1-02	3	1	2	1	2	-
U1-03	2	3	2	3	-	3
U1-04	2	2	3	2	2	-
U1-05	2	3	2	1	-	2

1.0 INTRODUCTION

Surveying is a core subject of civil engineering and it has an important role to play. It is the starting point of many projects, such as roads and railways, buildings, bridges, pipelines, transmission lines, dams, and many more (Schofield, 1984). Surveying is "*the technique of accurately determining the relative position of natural and man-made features above or below the surface of the Earth, by means of direct or indirect elevation, distance and angular measurements*". According to the American Congress on Surveying and Mapping (ACSM), '*Surveying is the science and art of making all essential measurements to determine the relative position of points or physical and cultural details above, on, or beneath the surface of the Earth, and to depict them in a usable form, or to establish the position of points or details'.*

Surveying is a means of making accurate measurements of the Earth's surfaces, including the interpretation of data so as to make it usable, and establishment of relative position and size (Survey Manual, 2014). It involves largely the field work which is done to capture and storage of field data using instruments and techniques specific for the type of survey work. Surveying also includes the technique of establishing the points by pre-determined angular and linear measurements. Thus, surveying has two distinct functions: (i) to determine the relative horizontal and vertical positions of the objects/features for the process of mapping, and (ii) to demarcate the land boundaries, establish the points to exactly layout the project on the ground, and control the construction work of a project.

The word 'Map' originates from the Latin word '*mappa*', means a tablecloth or napkin where 3-dimensional Earth features are represented on 2-dimensional cloth or paper. A map represents the 2D projection of the 3D terrain surveyed, which could be utilised to draw plans and sections to compute the area of land, or volume of a land mass or layout of an engineering structure. In land surveying, data are collected by using field surveying equipment and represented graphically on a piece of paper, called 'Map'. As the surveying technology grew, advanced materials, electronics, sensors and software are introduced in data collection, and analysis (Garg, 2021). These measurements may be used for the representation of different features in different forms. These features/details may be represented in analogue form as a topographic map with contours, plan or chart, or in digital form, such as a Digital Terrain Model (DTM).

As surveying allows us to acquire data on the relative positions, horizontal distances, and elevations of points, the objectives of surveying can be stated as follows (Schofiled, 1984; Basak, 2017):

1. Collect and record data about the relative positions of points/objects on the surface of the Earth.
2. Establish horizontal and vertical controls required for accurate mapping and subsequently for construction.
3. Prepare maps required for various civil engineering projects.
4. Compute areas and volumes of earthwork, required in various projects.
5. Layout of various engineering works on the ground using survey data.

1.1 IMPORTANCE OF LAND SURVEYING

Ever since the mankind acquired the sense of possessing the land and property, the art of surveying and mapping came into existence (Survey Manual, 2014). In the early days, the demarcation of land and defining the boundaries were extremely difficult tasks, and done as rough representation using conventional devices. With the growth of knowledge, skill and technology, development of surveying instruments and techniques of data collection and analysis improved considerably. Today, surveying is of vital importance as accurate planning and design of all civil engineering projects (Garg, 2021), such as railways, highways, tunneling, irrigation canal, dams, reservoirs, sewerage works, airports, seaports, building complex, etc. are based upon the quality of surveying measurements.

The knowledge of surveying is advantageous in many phases of civil engineering. The first task in surveying is to prepare a detailed topographical map of the area, and then either draw the sections of the best possible alignment, or compute the amount of earthwork or plan the structures on the map, depending upon the nature of the project (Chandra, 2007). The geographic and economic feasibility of the engineering project cannot be properly ascertained without undertaking a detailed survey work. Thus, the knowledge of surveying is fundamental and very useful to civil engineering professionals.

Surveying helps demarcating the land boundaries accurately on the ground. Surveying is essential to fix the national and state boundaries, map rivers and lakes, prepare maps of coastlines, etc. It is important to know the exact shape and size of the land boundary for acquisition of land or paying compensation to the land owners, and planning the construction. Many land properties have noticeable problems, mainly due to improper surveys, miscalculations in past surveys, titles of the land, error in measurements, etc. Also many land properties are created from multiple divisions of a larger piece of land over the years; and with every additional division, the risk of miscalculation/error in demarcation increases. The dispute in land measurements may lead to court cases to decide about the land ownerships. Many such problem can thus be detected on time through proper survey, and actions are taken on the basis of facts and figures provided by surveying measurements.

Surveying is also important in archaeology, geology, geophysics, landscape architecture, astronomy, meteorology, and seismology, including military engineering. Civil engineers must know the accuracy achieved in design and layout processes, as well as in construction. In addition to understanding the limits of accuracy, surveyors and engineers must know at what precision field data is to be collected to justify the accuracy. In particular, civil engineers must have a thorough understanding of the methods and instruments used, including their capabilities and limitations. This knowledge is best obtained by making observations with the kinds of modern equipment generally used in practice.

Civil engineers play an integral role in land development from planning and designing to the final construction of roads, buildings, bridges, dams, canals, stadium, buildings and utilities, and subsequently their maintenance and/or upgradation. The first and foremost requirement therefore is to have an accurate mapping of the land, as surveyors are the first to work at any construction site. For example, in a multipurpose building project (smart city), designing the project, planning the structures accurately and safely, and ensuring the buildings to be constructed as per specifications; all are the responsibilities of civil engineers. Since surveying measurements are used for providing horizontal and vertical controls on the ground, these are required not only for layout of structure accurately on the ground for construction purpose, but also useful to control the accuracy of entire construction of the structures (Duggal, 2004). These controls can also be used to make the actual estimate of the construction of structures while making payment to the contractor.

The role of surveying as a professional activity is expanding to include other techniques and skills, under one umbrella, known as Geospatial technology. The surveying professional are slowly adapting the modern approaches of data collection, data analysis and mapping in digital environment. These surveying approaches require a good working knowledge of photogrammetry, remote sensing, laser scanners, Global Positioning Systems (GPS), Unmanned Aerial Vehicles (UAVs), computer cartography, Geographical Information Systems (GIS), and associated software to analyse the data, generate detailed maps, visualize the terrain from satellite images, or integrate multiple geo-referenced databases (Garg, 2019).

1.2 BASIC PRINCIPLES OF SURVEYING

There are two basic principles of surveying which are to be followed for accurate and systematic survey measurements on the Earth surface. These are given below.

1.2.1 Working from whole to part

This is a fundamental and most important principle of surveying. Almost all survey works are required to follow this principle, particularly larger (whole) areas. In working from whole to part, a large (whole) area is divided into smaller parts by providing horizontal controls throughout the area. The smallest part of the area will consist of a triangle. If surveying is done without dividing into smaller parts, any error occurred in a part gets magnified at the end of entire survey work, and the error becomes large which can't be accepted for a good work. Whereas, on the other hand, any error occurred in smaller parts (triangle) is adjusted independently, and at the end of survey no error is left. Thus, the basic objective of this principle is to adjust the error locally within each small figure (triangle) independently and preventing the accumulation of errors.

In India, for large area survey works, Survey of India (SoI), Dehradun, established the control points very accurately at large distances by using triangulation method. From these control points, smaller areas are surveyed by setting out a network of triangles. Such sub-division continues with reference to the previously established systems of points, till all the details are surveyed.

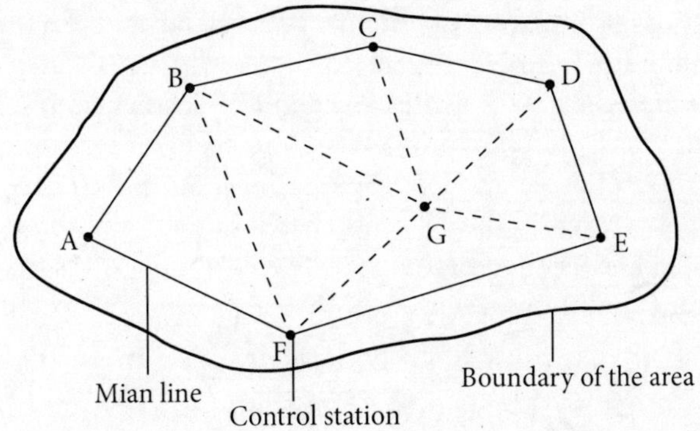

Fig.1.1: First principle of surveying: working from whole to part

1.2.2 Establishing a point by at least two independent measurements

Horizontal control points in surveying are located by linear and/or angular measurements. If two control points are established by surveying measurements, a new point (third point) can be established with the help of these two known control points by taking two linear or two angular measurements, or by one linear and one angular measurement. In other words, indirectly the location of the new point is established using the geometry or trigonometry of the triangle formed by these three points.

In Figure 1.2a, let P and Q be two control points, and a new point R is to be established by means of observations from points P and Q. Point R can be established using the distances PR, and R'R. In Fig. 1.2(b), R is established using the distances PR and QR. In Fig. 1.2(c), R is established by the angle RPQ and distances PR or by establishing angle RPQ and distance QR. In Fig. 1.2(d), R is established using the angles PQR and QPR.

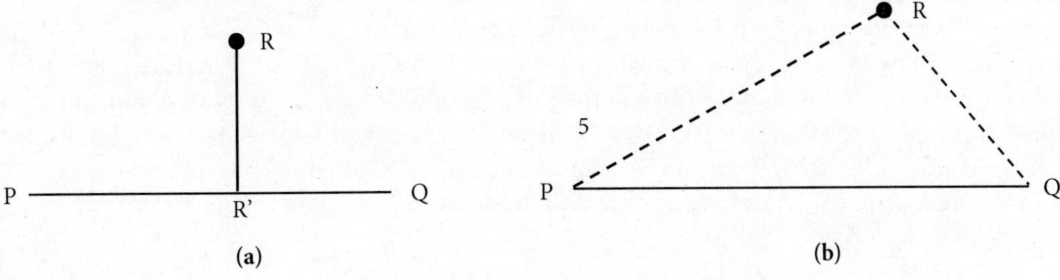

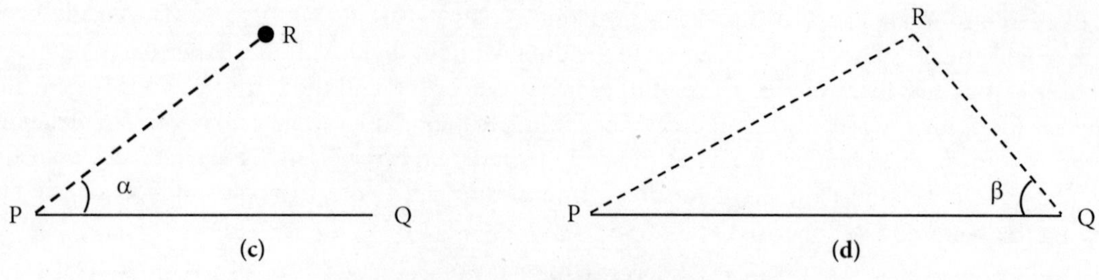

Fig.1.2: Second principle of surveying

1.3 HISTORY OF MAPPING: INDIAN PERSPECTIVE

As early as 1400 BCE, evidence of some form of boundary surveying has been found in the fertile valleys and plains of the Tigris, Euphrates, and Nile rivers. In India, evidence of map making skills has been found in the *Brahmand Purana* as far back as the 5th century (Chadha, 1991). Knowledge of land was presented in graphical form which described the extent and shape of territories. The art of surveying and techniques of mensuration of areas are described in *Sulva Sutra* (science of mensuration) and in the *Arth Shastra* of Chanakya, written in the 3rd century BC. In the 5th century, Aryabhatta, who wrote *Surya Siddhant*, calculated the Earth's circumference to be 25,080 miles- less than 200 miles off modern measurements of the equator (Phillimore, 1945).

Official surveying and mapping has been in practice in India, since way back 16th century. Raja Todar Mal during Akbar's and Sher Shah Suri regime introduced agriculture land measurements, termed as *cadastral survey*, which was done with foot and iron chains (Satyaprakash, 2010). Drawing used to be prepared cartographically with free-hand and scales. In India, the revenue maps were earlier prepared on a piece of cloth, called *'Khasra'* maps. These maps were earlier prepared by a method of distance measurement, known as pacing, which gave an idea of boundary and dimensions of land and property. Later, iron chains of 20 ft and 30 ft long were used to improve the distance measurement accuracy. *Khasra* maps and chains are used even today by many *Patwaris* and Village Development Officers for demarcation of property and collection of revenue.

The Indian terrain was completely mapped by the painstaking efforts of distinguished surveyors, such as Mr. Lambton and Sir George Everest (Phillimore, 1945). The angle-measuring instruments, called Theodolites, were developed to study the astronomy that were based on arcs of large radii, making such instruments too large for field use. The 36 inch theodolite used in the Indian triangulation is shown in Figure 1.3.

A 16 inch diameter Transit Theodolite with magnifiers, which has been used for measuring horizontal and vertical angles for the alignment, layout and construction of Ganga Canal, was designed and manufactured by the Canal Foundry (now known as Canal Workshop), Roorkee. (Late) Prof. H. Williams in 1937 carried out modifications in the Theodolite which was used for teaching and training to Engineers as well as for taking observations. Surveyor Compass with 1° least count was developed and manufactured at Roorkee to find out the bearing and direction required for the construction of Ganga Canal. In this Compass, numerals have been shown in English as well as Arbi languages.

Fig.1.3: The 36 inch theodolite used in the Indian triangulation (Keay, 1983)

The East India Company, after the Agra famine in 1937-38 where about one million people died, felt the need to build an irrigation system in the Doab region (Meerut to Allahabad zone) with support from skilled hands. Colonel Cautley was given the charge of constructing the Ganga canal. He suggested to James Thomason, the then Lieutenant-Governor of the North-West provinces, about the need to train locals in civil engineering for completing their ongoing projects. The students were to be trained systematically in surveying, leveling and drawing with proper infrastructure. This led to the foundation of the country's first-ever engineering college at Roorkee on September 23, 1847.

The Survey of India (SoI), under the Department of Science & Technology, is the oldest scientific department of the Govt. of India, which was set up in 1767 to help consolidate the territories of the British East India Company (Chadha, 1991). The SoI is the principal mapping agency of the country. The great trigonometric series spanning the country from north to south and east to west are some of the best geodetic control series available in the world. The foremost topographical survey in the south was the one carried out by Colin Mackenzie in Kanara and Mysore. The SoI is fully equipped with digital devices and software to ensure that the country's domain is explored and mapped suitably using the latest technology. It has oriented the technology to meet the needs of defense forces, planners and scientists in the field of geo-sciences, land and resource management, and many scientific programs of the country.

From the 19th century, the use of trigonometric method marked a historic moment in the process of surveying. This was the first time that a purely geometric method was utilised for making geographical calculations. The method of measurement through triangulation was conceived by the EIC officer William Lambdon and later under his successor George Everest, which became the responsibility of the SoI. One of the greatest accomplishments of the trigonometric survey was the measurement of Mount Everest, K2 and Kanchenjunga.

As human brain is always engaged in development of new procedures to overcome the shortcomings, a rapid growth has taken place since then. Digital maps became the authentic tools to claim the ownership. Over the past, several changes have occurred in surveying and mapping field, resulting in the techniques today which are fully automated and software based from data collection to analysis to mapping.

1.4 TYPES OF SURVEYING

The two broad types of land survey used are plane surveying and geodetic surveying, depending upon whether the spherical shape of the Earth is taken into account or not (Punmia et al., 2016). These are briefly described below.

1.4.1 Plane surveying

In plane surveying, the Earth surface is considered as a plane surface, and its spheroidal shape is neglected. For small projects covering area less than 250 sq.km, Earth curvature is not accounted for distance measurements. The line joining any two stations is considered to be straight, and three points will make a plain triangle. Surveys for engineering projects fall under this category. Plane surveying uses normal instruments, like measuring tape, theodolite, level, etc., and the survey accuracy is low.

1.4.2 Geodetic surveying

This type of surveying takes into account the true shape of the Earth. Earth curvature correction is applied to the observations. The triangle formed by any three points is considered as spherical. Geodetic surveys are carried out for areas greater than 250 sq.km. For large areas, degree of accuracy of measurements is high, and therefore geodetic survey requires sophisticated and high precision equipment, like the GPS, Total Station. Geodetic survey is used to provide control points to which small surveys can be connected.

1.5 CLASSIFICATION OF SURVEY

Surveys can be further classified into several categories depending on the purpose, instruments, techniques used, etc. These classifications are shown in Table 1.1.

Table 1.1: Classification of surveys based on

Instruments	Place	Methods	Purpose
Chain or tape	Land	Levelling	Topographic
Level	Water	Trigonometrical levelling	Geodetic
Theodolite	Underground	Traversing	Engineering
Compass	Aerial	Tachometry	Route
Tachometer		Triangulation	Cadastral
EDM		Trilateration	City
Total Station			Mine
Plane Table			Geology
GPS			Underground utility
GPR			Hydrographic
Laser scanners			
Aerial			
Satellite			
UAV/Drone			

1.5.1 Types of survey based on instruments

Earlier, measurements of distances and directions were mainly used to prepare maps of the area, and surveys were classified based upon the instruments used. Chains, tapes, levels, magnetic compasses, and theodolites were popularly used, either single or in combination with other equipment. With the advancement in electronics and digital technology, many modern instruments are now available for data collection and mapping, which are capable to collect the data alone, and no other instrument is generally required, for example Total station, Laser scanners and GPS. Figure 1.4 shows the images of some of these equipment.

a. Chains

In chain survey, a metallic chain is used to measure the linear distances. The chain traversing is done by dividing the small area to be surveyed into a number of triangles. Using the sides of the triangles by chain surveying, other details are worked out. Chains meant for survey work are available in length of 20 or 30 m. Later, chains due to their inaccuracy, difficulty to use, and wear & tear in the field, were replaced by the tapes. In addition, many corrections are applied while making the long distance measurements with chain.

b. Tapes

Many different varieties of measuring tapes are available, with different materials, lengths and designs. The most common type of tape is made of fibre glass reinforced plastic tapes and those with stainless steel. Steel tapes stretch very little, and are little expensive but gets kinks easily if not handled properly. Steel tapes also get rust very quickly. Measurements over 30 m are not recommended as that length of tape is difficult to manage and the sag in the tape makes the measurements inaccurate.

c. Levels

A level instrument is used to find the difference in elevation between the points. It is used along with a graduated rod, known as the leveling staff. If we link the observations to a known point, the elevation of other unknown points can be determined.

d. Theodolites

A theodolite is used to measure horizontal and vertical angles. Theodolites are frequently used to carry out traversing where angles and distances are measured simultaneously. These angle are required to establish horizontal and vertical controls in the area. Theodolites can also be used for prolongation of a line, alignment work, and levelling work.

e. Compass

A magnetic compass works on the principle that a freely suspended needle points in the magnetic north-south direction (reference direction). This instrument gives the direction of a line with respect to reference direction. A compass, together with a chain or tape or Total Station, can be used to survey a given area using methods, such as traversing.

f. Techeometer

A tacheometer is similar to a theodolite but has an anallactic lens and a stadia diaphragm having three horizontal hairs. The readings taken on a levelling staff against all the three cross hairs enable the horizontal distances to be computed. The central wire reading is linked to some Bench Mark (BM) to determine the elevation.

g. EDM

The Electronic Distance Measuring (EDM) device measures the slope distance between EDM instrument and a prism (reflector). The EDM generates infrared (IR) or microwave which travels through the atmosphere and strikes the reflector kept at an unknown point. The instrument measures the travel time of return beam, and computes the distance (distance= velocity x time). The phase difference between the instrument station and the reflector station is also used to compute the slope distance. For details of EDM refer to Unit-3.

(a) Chain (b) Tape (c) Level and Staff

(d) Theodolite (e) Compass (g) EDM

(h) Total Station (i) Plane Table (j) GNSS

(k) GPR (l) Laser Scanner (m) UAV/Drone

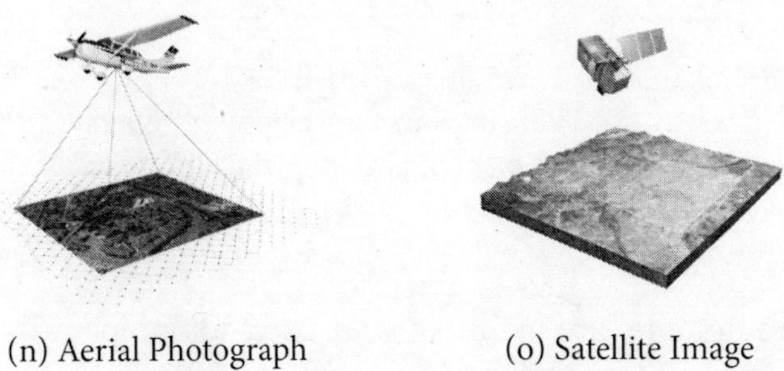

(n) Aerial Photograph (o) Satellite Image

Fig.1.4: Surveying based on equipment and tools

h. Total Station

A total station is a combination of an electronic theodolite and an EDM. Horizontal distances and horizontal & vertical angles are determined using a total station. Total station equipment is used all forms of surveys requiring a very high level of precession. A total station can display and store the field data, and transfer the data to a computer for further processing. For details of Total Stations refer to Unit-3.

i. Plane Table

A plane table is like a drawing board, which is used in the field to prepare a map of the area at desired scale. In plane table survey, fieldwork and plotting work proceed simultaneously. It is used along with an alidade which provides the direction towards the object to be plotted, and with a chain or tape, and tacheometer or a Total Station to measure the distance and elevation of the objects/features.

j. GNSS

The Global Navigation Satellite System (GNSS) device (receiver) is very useful in surveying and mapping. Depending on the type of receiver, it shall receive signals from the various GNSS constellations such as Global Positioning System (GPS), GLONASS, GALILEO satellites and provides 3D coordinates of the point on the Earth surface, anytime, where observations have been taken (Garg, 2019). These 3D coordinates can be used in software to process and generate a map as well creation of digital elevation model (DEM). For details of GNSS refer to Unit-3.

k. GPR

The Ground Penetrating Radar (GPR) device emits radar beams that penetrate into the ground to a certain depth, depending on the wavelength and frequency of antenna of radar beam (Garg, 2021). It will display the return signals from the objects which are below the Earth surface. Any underground utility, buried structures & objects, ground water table, etc., can be mapped with the GPR. Mapping all the underground utility is necessary in all modern townships, including smart cities. For details of GPS refer to Garg (2021).

l. Laser Scanners

Laser scanners emit laser beams that strike to the objects and return back to the instrument. The instrument provides 3D coordinates of the objects which could be used to generate the DEM of the area (Garg, 2021). There are two types of laser scanners; terrestrial (ground based) and aerial laser scanners, and depending on the use, 3D models and maps or profiles can be generated for the area. For details of Laser Scanners refer to Garg (2021).

m. UAVs/Drones

The Unmanned Aerial Vehicle (UAV) or drone are low flying aircrafts used for collecting very high-resolution images of laser point cloud data (Garg, 2020). These data are very helpful to generate DEM of the area. Today, UAVs have large number of applications in mapping, monitoring and management of civil engineering and other projects. For details of UAVs/Drones refer to Garg (2020).

n. Aircrafts

In aerial photogrammetry, aerial photographs are acquired and used for preparation of various thematic maps at different scales. Manual photogrammetric techniques require simple equipment, such as stereoscopes, parallax bar, analytical plotting devices, whereas the digital photogrammetric methods require sophisticated devices (workstations) and photogrammetric software to create various kind of maps (Garg, 2019). For details of photogrammetry refer to Unit-4.

o. Remote Sensing Satellites

Large number of satellite images are available from various satellites, which can be processed through image processing software or in GIS environment to generate maps at different scales (Garg, 2019). Using different time images in GIS, a change scenario can be generated which is very helpful in planning, designing and monitoring various civil engineering projects. In addition, GIS can be used to make a query form the exhaustive database which otherwise is not possible. For details of remote sensing refer to Unit-5.

1.5.2 Types of survey based on purpose

a. Control surveys

These are carried out to establish a network of horizontal and vertical controls that serve as a reference framework for completing the survey work with desired accuracy. Many control surveys performed today are done using Total Station and GPS instruments.

b. Topographic surveys

They are carried out to determine the locations of natural and man-made features, and their elevations and representation on a map.

c. Land boundary and cadastral surveys

These are used to establish property lines and corners of the property. The term cadastral is applied to surveys of the public lands systems.

d. Hydrographic surveys

They are used to define shorelines and depths of lakes, streams, oceans, reservoirs, and other bodies of water.

e. Alignment surveys

These are conducted to plan, design, and construct highways, railroads, pipelines, and other linear projects.

f. Construction surveys

They provide line, grade, control elevations, horizontal positions, dimensions, and configurations for construction operations. They are also used for computing the bill and quantities of construction.

g. Mine surveys

These are performed above and below the ground to guide tunnelling and other operations associated with mining. These surveys are carried out for mineral and energy resource exploration.

1.6 MAPS

One of the basic objectives of surveying is to finally prepare various types of maps useful for engineering works (Survey Manual, 2014). Maps are graphical representation of Earth surface features to some scale. Two broad types of maps; planimetric and topographic, are prepared as a result of surveys. The former, called plan map, depicts the natural and cultural features in their plan (x-y) views only, while the latter includes planimetric features, and the elevation (z) of ground surface. The topography represents the shape, configuration, relief, slope, roughness, or three-dimensional characteristics of the Earth's surface. Both types of maps are used by a larger community, such as engineers and planners to determine the most desirable and economical locations of projects, such as highways, rails, roads, canals, pipelines, transmission lines, reservoirs, and other facilities; by architects in housing and landscape design; by geologists to investigate the mineral, oil, water, stable slopes, and other resources; by foresters to locate wildlife, forest fire-control routes, identifying new sites for plantations; and by archeologists, geographers, and scientists in numerous fields.

Maps depict the locations as well a relief of natural and cultural features on the Earth's surface (Garg, 2021). Natural features normally shown on maps include forests, vegetation, rivers, lakes, snow, oceans, etc., while cultural features are the man-made features and include roads, rails, buildings, bridges, tower, canal etc. Traditionally, engineering maps have been prepared using manual drafting methods and plane table survey. With the availability of digital data from total station, GPS, laser scanners and digital images form photogrammetry and remote sensing, now the majority of maps are produced using computers, computer-aided drafting (CAD), and GIS software.

The natural and cultural features on the maps are depicted graphically by points, lines and polygons, with different symbols and colours. For example, the relief of the Earth includes its hills, valleys, plains, slopes and other surface irregularities, which are represented in the form of contour lines. A map will also have the names of places, rivers and legends to identify the different objects. The standard symbols and colours used for various natural and man-made objects, are shown in Table 1.2. For example, contours as burnt sienna, buildings as light red, water as persian blue, trees and vegetation as green, agricultural land as yellow, etc.

1.7 MAP SCALE

The distance between any two points on a map, measured along a straight line, is called the map distance, while the distance between the same two places on the ground, measured along a straight line, is called the ground distance. The ratio between the map distance and the ground distance is called the scale of map. Map scales are represented in three ways; (i) by ratio or representative fraction (RF) or ratio, such as 1:1200 or 1/1200, (ii) by an equivalence, for example, 1 in.= 100 ft. (1200 in.), and (iii) graphically using a linear scale.

In defining scale by ratio or RF, same units are used for map distance and corresponding object distance, and thus 1:1200 could mean 1 unit on the map is equivalent to 1200 unit on the ground. An equivalence scale of 1 inch= l00 ft. indicates that 1 inch on the map is equivalent to 100 ft. or 1200 in. on the ground. It is possible to convert from a ratio to an equivalence scale and vise-versa. As an example, 1 inch= 100 ft. is converted to a RF by multiplying 100 ft by 12 which converts it to inches and gives a ratio of 1:1200.

Table 1.2: Colours and symbols of various features on topographic maps (https://geokult.com/2011/09/14/map-symbolisation/)

TOPOGRAPHIC MAP SYMBOLS

VARIATIONS WILL BE FOUND ON OLDER MAPS

Primary highway, hard surface
Secondary highway, hard surface
Light-duty road, hard or improved surface
Unimproved road
Road under construction, alinement known
Proposed road
Dual highway, dividing strip 25 feet or less
Dual highway, dividing strip exceeding 25 feet
Trail

Railroad: single track and multiple track
Railroads in juxtaposition
Narrow gage: single track and multiple track
Railroad in street and carline
Bridge: road and railroad
Drawbridge: road and railroad
Footbridge
Tunnel: road and railroad
Overpass and underpass
Small masonry or concrete dam
Dam with lock
Dam with road
Canal with lock

Buildings (dwelling, place of employment, etc.)
School, church, and cemetery — Cem
Buildings (barn, warehouse, etc.)
Power transmission line with located metal tower
Telephone line, pipeline, etc. (labeled as to type)
Wells other than water (labeled as to type) — Oil ..... Gas
Tanks: oil, water, etc. (labeled only if water) — Water
Located or landmark object; windmill
Open pit, mine, or quarry; prospect
Shaft and tunnel entrance

Horizontal and vertical control station:
Tablet, spirit level elevation — BM △ 5653
Other recoverable mark, spirit level elevation — △ 5455
Horizontal control station: tablet, vertical angle elevation — VABM △ 9519
Any recoverable mark, vertical angle or checked elevation — △ 3775
Vertical control station: tablet, spirit level elevation — BM × 957
Other recoverable mark, spirit level elevation — × 954
Spot elevation — × 7369 × 7369
Water elevation — 670 670

Boundaries: National
State
County, parish, municipio
Civil township, precinct, town, barrio
Incorporated city, village, town, hamlet
Reservation, National or State
Small park, cemetery, airport, etc.
Land grant
Township or range line, United States land survey
Township or range line, approximate location
Section line, United States land survey
Section line, approximate location
Township line, not United States land survey
Section line, not United States land survey
Found corner: section and closing
Boundary monument: land grant and other
Fence or field line

Index contour — Intermediate contour
Supplementary contour — Depression contours
Fill — Cut
Levee — Levee with road
Mine dump — Wash
Tailings — Tailings pond
Shifting sand or dunes — Intricate surface
Sand area — Gravel beach

Perennial streams — Intermittent streams
Elevated aqueduct — Aqueduct tunnel
Water well and spring — Glacier
Small rapids — Small falls
Large rapids — Large falls
Intermittent lake — Dry lake bed
Foreshore flat — Rock or coral reef
Sounding, depth curve (10) — Piling or dolphin
Exposed wreck — Sunken wreck
Rock, bare or awash; dangerous to navigation

Marsh (swamp) — Submerged marsh
Wooded marsh — Mangrove
Woods or brushwood — Orchard
Vineyard — Scrub
Land subject to controlled inundation — Urban area

A linear scale is useful to measure the distance directly from the map. It allows an accurate measurement on the map, as direct measurement on a paper map using numerical scale may not be very accurate due to little contraction or expansion of paper with the climatic conditions. But a linear scale drawn on a map will yield correct result, even after enlargement or reduction of a map. In the graphic or linear scale, map distance is shown using a straight line. The length of the line of a linear scale depends on the size of the map, but is usually between 12 cm and 20 cm that is divided into parts known as primary divisions. The primary units to the left of zero must be exactly the same size or length as the primary units to the right of zero. The first primary division on the left is further sub-divided into smaller parts known as secondary divisions. The starting or zero point of the linear scale should be after the first primary division from the left. The primary divisions are to the right of zero, while the secondary divisions are to the left of zero. The example in Figure 1.5 shows a linear scale where 1 cm on the map equals to 1 km on the ground. Secondary divisions are divided into five equal parts (it may also be divided into 10), and each part represents 200 m on the ground. Thus, the ground distance can be computed using linear scale by measuring that distance on map

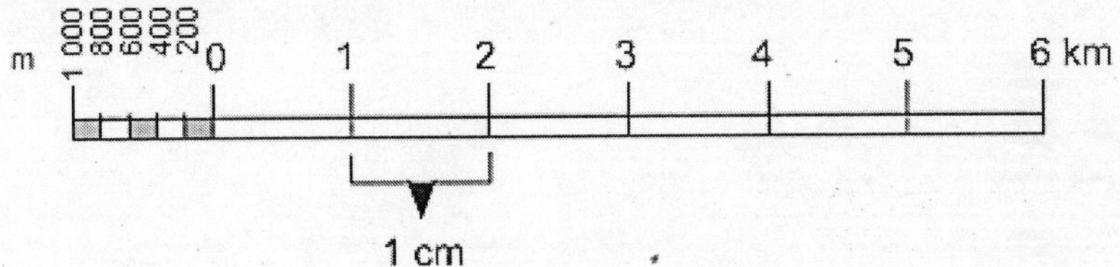

Fig.1.5: A typical linear scale

The Survey of India prepares various maps based on actual survey of land for the whole country. These maps which show the details about landforms, drainage patterns, land use, settlement patterns, transport, forest and cultural features. The topographical maps or topo sheets, prepared by Survey of India, are available at three scales; 1: 25,000, 1: 50,000 and 1: 2,50,000. The topographical maps also show a network of parallels (lines of Latitudes) and meridians (lines of Longitudes), forming grids on the map. These grids are useful to find the location of a place (Latitude, Longitude) on the Earth. Figure 1.6 shows a sample of toposheet (B&W) of Himachal Pradesh region at 1: 50,000 scale, and various associated features.

Map scales may be classified as large, medium, and small scale. Large scale maps are created where relatively high accuracy is needed over small areas, for example design of engineering projects, like dams, airports, stadium, and water & sewage systems. Medium scale maps are often required for projects where large area is involved but moderate accuracy is needed, such as township, proposed transportation systems, and existing facilities. Small scale maps are commonly created for mapping very large areas where a lower order of accuracy will be needed, such as district level maps. Choice of mapping scale would depend on the purpose, size of area, topography and required precision of the finished map, and of course, the time required to complete the mapping work. The scales of various maps normally used in survey work are given in Table 1.3. The greater the number in the denominator the smaller is the scale and vice versa.

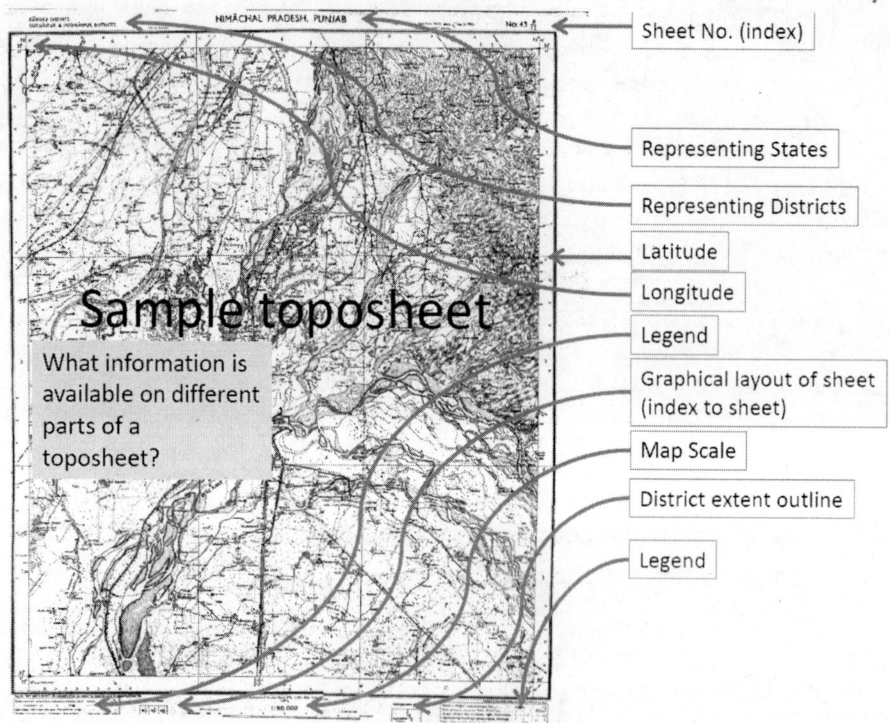

Fig.1.6: A typical toposheet

Table 1.3: Various engineering maps and their scales (Garg, 2021)

Type of purpose of survey	Scale	R.F.
(a) Topographic survey		
1. Building sites	1 cm = 10 m or less	$\frac{1}{1000}$ or less
2. Town planning schemes, reservoirs etc.	1 cm = 50 m or 100 m	$\frac{1}{5000}$ to $\frac{1}{10000}$
3. Location surveys	1 cm = 50 m or 200 m	$\frac{1}{5000}$ to $\frac{1}{20000}$
4. Small scale topographic maps	1 cm = 0.25 km or 2.5 km	$\frac{1}{25000}$ to $\frac{1}{250000}$
(b) Cadastral maps	1 cm = 5 m or 0.5 km	$\frac{1}{500}$ to $\frac{1}{5000}$
(c) Geographical maps	1 cm = 5 km to 160 km	$\frac{1}{500000}$ to $\frac{1}{16000000}$

(d) Longitudinal sections		
1. Horizontal scale	1 cm = 10 m to 200 m	$\frac{1}{1000}$ to $\frac{1}{20000}$
2. Vertical scale	1 cm = Im to 2 m	$\frac{1}{100}$ to $\frac{1}{200}$
(e) Cross-sections (Both horizontal and vertical scales equaly)	1 cm = Im to 2 m	$\frac{1}{100}$ to $\frac{1}{200}$

The plotting accuracy of a map can be determined with a simple concept. Cartographically, any line on a map is drawn with a thickness of 0.25 mm, which is considered to be the least dimension of a smallest dot. It means the thickness of the line or dot on the ground can be computed by multiplying the least dimension of 0.25 mm with the scale of map.

Plotting accuracy = 0.25 mm x scale of mapping (1.1)

At 1:1,200 scale, the plotting accuracy will be 0.25x1200= 300 mm= 30 cm. So, 30 cm is the maximum allowable error which could be presented in 1: 1,200 scale maps. A feature smaller than 30 cm on the ground can't be theoretically shown on this map.

1.8 SURVEY STATIONS

To survey the boundary of an area, the stations taken along the boundary of an area as controlling points are known as *survey stations*. Figure 1.7 shows the survey stations (P, Q, R and S) marked by black dot enclosed by circle. These stations are definite points on the Earth whose locations have been determined by surveying methods. Their location is marked on the ground over which survey instruments are kept to take the observations. Angular, linear, bearing and height observations may be taken at these stations, depending upon the purpose of survey. These stations are selected by the surveyors at commanding positions in an optimal manner, keeping in view the following criteria:

i. Survey stations should be visible from at least two or more other survey stations.
ii. As far as possible, survey stations must be situated so as to cover minimum elevation and maximum elevation of the area.
iii. The criteria for selecting the survey stations is that lines joining them should not make obtuse angle (>120°) or acute angle (<30°) with each other.
iv. Survey stations should be as few as possible, as more the stations more will be the observational work.
v. Survey stations should be avoided at busy locations as it might affect the smoot measurement of survey observations.

The purpose of survey stations is to provide horizontal control (planimetry, X and Y) and vertical control (Z) during mapping work. In surveying, it is important to know the absolute position of a point/object with respect to a given origin, or its relative position with respect to a known point/object. The position information can be expressed as (X, Y and Z) in linear coordinate system or as (Ø, λ and Z) in polar coordinate system. When expressed as X, Y, generally it is referred to as the grid coordinates or Easting/Northing, and Z represents the elevation of the point above a known or fixed datum (i.e., mean sea level). When a point is expressed in terms of (Ø, λ), it refers to latitude and longitude of that point.

The horizontal control is provided by two or more stations on the ground which are precisely fixed in position by distance and direction. It provides the basis for controlling the scale of the surveyed map as well as locating the various topographic features present in the area. The most common methods used to locate a point/feature in the field are by measuring: one angle and the adjacent distance, or two distances, or two angles. For small areas, horizontal control for topographic work is usually established by a traverse, but sometimes for a very small area mapping, two survey stations making a single line may also be used. Vertical control is provided by the Bench Marks (BMs) in or near the area to be surveyed. It is required to correctly represent elevations on a surveyed map, and is usually established by lines of levels, starting and closing on BMs.

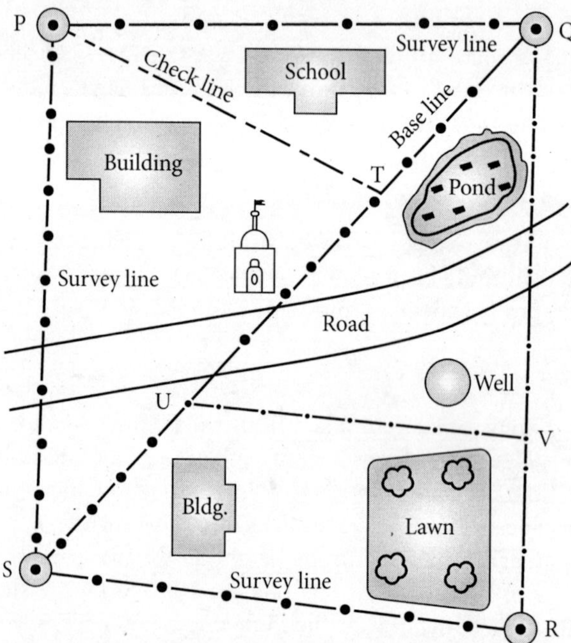

Fig.1.7: Survey stations and survey lines

1.9 SURVEY LINES

The lines joining the survey stations are called survey lines, as shown in Figure 1.7 (lines PQ, QR, RS, SP). The longest survey line (QS) is called as the base line whose distance is measured very accurately, and used for computational work as well as cross checking the measurements. It is the most important line in surveying, as it fixes up the directions of all other lines. Since the accuracy of whole survey work depends upon the accuracy of base line, it should be laid-off on a fairly level ground, if possible. As far as possible, a base line should be taken roughly through the middle of the area on which the framework of triangles covering the major portion of the area may be designed. There could be more than one base line depending upon the size of the area to be surveyed; one is used for computation and the other is used as check line (PT). The survey lines should be as few as possible such that the framework may be plotted. The survey lines should be measured in an order so as to unnecessary avoid walking between the stations to and fro.

1.10 SAFETY IN SURVEYING

Civil Engineers are generally involved both in field and office work. The field work demands making observations with various types of instruments either; (a) to determine the relative locations of points, or (b) to set out stakes as per plan to guide the construction operations. The office work involves computing and processing the surveyed data from field, and preparing maps, plots, charts, reports, etc. Sometimes, the fieldwork is performed in difficult or unsafe environment (e.g., sites close to landslides, volcanoes, heavy snow cover, forest fire, highways and railroads), so it is important that safety precautions are taken. Construction sites where heavy machinery operate are also hazardous and dangerous. The danger is often enhanced by poor hearing conditions due to excessive noise at the construction site as well as poor visibility caused by the dust, haze and obstructions.

In such situations, wherever possible, survey work should be carried out away from the dangerous sites or using indirect methods of measurements. If it is not possible to avoid, the work must be done with certain safety precautions. To ensure safety, vests of fluorescent yellow color should always be worn, and flagging materials of the same color can be attached to the surveying equipment to make them more visible from a distance. Depending on the work at the site, proper signage may be placed before the work areas so as to inform the working of a survey team ahead. Proper, barricades can also be placed at some sites to divert the traffic around surveying work, or manually people with flags can ask drivers of vehicles to slow or even stop, if necessary.

The *Occupational Safety and Health Administration* (OSHA) of the US Department of Labor (https://www.osha.gov), has developed safety standards and guidelines for various conditions and situations in the field. Depending on the location of survey site and the time of year, weather-related harms, such as frostbite and over-exposure to the sun, which can cause skin cancers, sunburns, and heat stroke, can also affect the field surveys. To overcome these problems, fluids at regular interval should be consumed, large hats can be worn and sunscreen applied. On extremely hot days, surveying should commence at dawn and stop around midday. Other hazards that could be present during field surveys, include wild animals, poisonous snakes, bees, insects, spiders, etc. Surveyors should be knowledgeable about the types of hazards that can be expected in any local area, and always be alert while working at the site. To prevent injury from these sources, protective boots and clothing should be worn and insect spray is used.

Certain tools can also be dangerous, such as saws and axes that are sometimes used for clearing the line of sight. These must always be handled with great care. Also, care must be taken while handling certain surveying instruments, such as long-range poles and levelling rods, especially when working around overhead live wires, to prevent accidental shocks.

Thus, it is important that surveyors always take precautions in the field, and follow the safety guidelines and standards. A first-aid kit should be carried in the field, which include all of the necessary antiseptics, ointments, bandage materials, and other accessories needed to provide first aid during minor accidents/injuries. The survey party should also carry important telephone numbers to call in emergencies/serious situations.

1.11 UNITS OF MEASUREMENTS

Measurements/observations are assigned specific units. In surveying, most commonly employed units are for length, area, volume, and angle. Two different systems are in use for specifying the units of observed quantities; the *English* (FPS) *system* and *Metric system* (MKS). Because of its widespread adoption, the metric system is also called the *International System of Units* (SI). According to Standards of Weights and

Measurements Act, India in 1956 decided to replace English (FPS) system used earlier with the MKS. In 1960, SI unit was approved by the conference of weights and measures. The major difference between MKS and SI is in the use of unit of force. In MKS unit of force is kg-wt (which is commonly called as kg only) while in SI it is newton.

The English system (FPS) has been the officially adopted standard for measurements in the United States for long time, where the linear units in feet and inches are most commonly used. Even today, FPS is used in construction industry. Because civil engineers perform all types of surveys and provide measurements for developing construction plans and guiding building operations, they must understand both the systems of units and be familiar of making conversions between them. The observations must be recorded in their proper units and then conversions are made correctly.

In FPS unit, the length is measured in *foot and inches.*

1 foot = 12 inches

1 yard = 3 feet (ft.)

1 inch = 2.54 cm (basis of international foot)

1 meter (m) = 39.37 inches (basis of U.S. survey foot)

1 rod = 1 pole = 1 perch = 16.5 feet

1 Gunter's chain = 66 feet = 100 links = 4 rods

1 mile = 5280 feet = 80 Gunter's chains

1 nautical mile = 6076.10 feet (nominal length of a minute of latitude, or of longitude at the equator)

In SI unit, the length is measured in *centimeter, meter, and kilometer.*

1 m = 10 dm (*decimeter*)

1 m = 100 cm

1 m = 1000 mm (*millimeter*)

1000 m = 1 km (*kilometer*)

1 km = five eighths of a mile

In the English system, areas are given in *square feet* or *square yards*. The most common unit for large areas is the *acre*.

1 acre = 10 square chains (Gunter's)

1 acre = 43,560 ft^2 (10 x 66^2)

SI units used for finding the areas are *square meter, square kilometre (km)* and *hectare (ha)*. Large tracts of land, for example, are given in *hectares*. Relations among them are:

1 ha = 100 m×100 m

1 ha = 1×10^4 m^2

1 ha = 2.471 acres

1 square km = 1000 m×1000 m = 10^6 m^2

1 square km = 100 ha

Volumes in the English system can be given in *cubic feet* or *cubic yards*. For very large volumes, for example, the quantity of water in a reservoir is designated in *acre-foot*.

1 acre-foot = 43,560 ft^3 (area of an acre having a depth of 1ft)

In the SI system, *cubic meter* (m^3) is used for volume.

The unit of angle used in surveying is the *degree*, defined as 1/360 of a circle. Degrees, minutes, and seconds, or the radian, are accepted SI units for angles. For measuring angles sexagesimal system is used, where:

1 circumference = 360°

1 degree = 60′ (minutes of arc)

1 minute = 60″ (seconds of arc)

Other methods are also used to subdivide a circle. A *radian* is the angle subtended by an arc of a circle having a length equal to the radius of the circle.

2π rad = 360°

1 rad ≈ 57°17'44.8' ≈ 57.2958°

0.01745 rad ≈ 1°

1 rad ≈ 206,264.8″

The recommended multipliers in SI units are given below:

Giga unit = 1×10^9 units

Mega unit = 1×10^6 units

Kilo unit = 1×10^3 units

Milli unit = 1×10^{-3} unit

Micro unit = 1×10^{-6} unit

1.12 VARIOUS ERRORS IN MEASUREMENTS

All measurements are subject to errors, irrespective of the instrument and method used. The 'true value' of a measured quantity is thus never known. The angular, linear and elevation measurements might have three basic errors present (Punmia et al., 206) ; (i) natural errors, (ii) instrument errors, and (iii) personal errors.

Broadly, these errors will fall under two categories: (a) Systematic or cumulative, and (b) Accidental, random or compensating. Systematic errors can always be identified and corrected because their magnitude and nature (sign) can both be determined. For example, a measuring tape is designed for its standard length under a particular pull and temperature, but if the pull or temperature changes during the field work, its effect on the length (increase or decrease in length) of the tape can be computed. Accidental, random or compensating errors are subject to chance, and hence follow the laws of probability. The magnitude and sign of errors are not known, as they are sometimes positive and sometimes negative, sometimes of small magnitude, and sometimes of large magnitude, and hence can't be determined or eliminated. To minimise it, we take a large number of observations to make an estimate of magnitude of such error that is likely to occur. In fact, there is one more error, i.e., mistake or blunder, but that cannot be classified under any category of error as these are mainly due to the carelessness of the observer. Mistakes can be corrected only if discovered, and comparison of several measurements of the same quantity could be used in isolating the mistakes.

There are two more terms involved when we deal with the errors in measurements; (i) accuracy, and (ii) precision. *Accuracy* is the closeness or nearness of the measurements to the 'true' or 'actual' value of the quantity being measured. The term *precision* (or repeatability) refers to the closeness with which the measurements agree with each other. Statistically, precision can be measured by means of a quantity σ, known as *standard deviation* or *standard error,* and is given by-

$$\sigma = \sqrt{\frac{\sum V^2}{n - 1}} \tag{1.2}$$

Where v^2 is the sum of the squares of the residuals, n is the number of measurements.

The smaller the value of σ becomes, the greater is the precision.

1.13 MEASUREMENT OF DISTANCES

One of the most important activity in surveying is the measurement or computation of horizontal distance between two points. If the points are at different elevations, the distance measured is the slope distance, and a vertical angle of the line joining the point is also required to compute the horizontal distance. Depending on the accuracy desired in measurement, there are two broad methods of measuring the horizontal distances:

a. Direct methods, and
b. Indirect methods

The direct methods can be employed using either of the following equipment.

Pacing: A distance between two points can approximately be determined by counting the number of paces and multiplying it with the average length of the pace. When instruments are not there, this is the method to get an approximate distance.

Passometer and Pedometer: Passometer is a small instrument which counts the number of paces. Pedometer directly gives the distance by multiplying the number of paces with the average pace length of the person carrying the instrument.

Odometer: It is a simple device which can be attached to the wheel of any device, two-wheelers, three-wheelers or four-wheelers or any such devices. It measures the distance by counting the number of revolutions made by the wheel to which it is attached and multiplying this number with the perimeter of the wheel.

Survey Chains: Chains (Figure 1.8a) are used to measure distances when great precision is not needed. Metallic chains of fixed lengths, 50 feet or 100 feet, having indicators on every 10 feet may be used. These chains are also used along with Optical Square equipment for establishing the right angle to the chain line and measuring the distance. In India, old cadastral maps/property maps have been prepared using chain survey. The chains give errors in measurements for larger distances, due to sag and pull and other constraints. In addition, it is laborious to take measurements from chains as these are heavy and subject to wear & tear. In India link type surveying chains of 30 m lengths with 100 links are frequently used in land measurement. There are tallies fixed at every 3 m to facilitate convenience in reading. Along with chains, small accessories used are; (i) Arrows, (ii) Pegs, (iii) Ranging rods, (iv) Plumb bob, (v) Hammer, etc. Now-a-days chains

Tapes: Tapes (Figure 1.8b) are used to measure the distances much accurately than the chains. These tapes are available in different types and lengths. Some are described below:

Fig.1.8: (a) Survey chain, and (b) Measuring tape

i. ***Linen tapes***: Linen tapes have painted strip of woven linen, attached to a spindle in a case. The tape is wound in leather, plastic or metal case when not in use. These tapes are light and handy but not very accurate for large distances as they are subject to serious variations in length.

ii. ***Metallic tapes***: These are ribbons of water proof fabric interwoven with thin brass or bronze wires to provide additional strength and to prevent stretching/breaking of tape. These tapes are available in 10, 20, 30 or 50 m length. Like linen tapes, these are also attached to a spindle in a leather, plastic or metal case. Non-metallic glass fibre tapes which are quite flexible, strong and non-conductive, are also available, and are used near power lines or electrical equipment.

iii. ***Steel tapes***: These tapes are superior to metal tape, and is more accurately graduated. It can't, however, withstand the rough usage. In case the tape gets wet, it is required to wipe it with a dry cloth immediately.

iv. ***Invar tapes***: For high precision distance measurements, such as base line measurements in triangulation/traverse survey, the tapes made of the alloy invar are used. Invar tape is made of 35% nickel and 65% steel. The tapes have a very low coefficient of thermal expansion, and hence are unaffected by temperature differences in the field.

1.13.1 Measurement of distance by chain or tape

The distance could be measured on level ground or sloping ground with a chain or a tape in a similar manner. On the ground, the line to be measured is marked at both ends and also at intermediate points, wherever necessary, in the same alignment direction. The chain or tape is laid on the ground in the alignment direction, and a peg is inserted at the other end. The chain or tape is again brought where the peg was inserted and full chain/tape length is again measured and another peg is inserted. The process continues till end point of the line is reached. The total distance is thus computed by adding up all the distances.

On uneven or sloping ground, the distance is directly measured in small stretches by keeping the chain/tape horizontal (by eye judgement only). This distance is measured and the chain/tape point at other end is projected on the ground with the help of a plumb bob, and marked with a peg. The starting end of chain/tape is brought to this peg and again a small horizontal distance is measured by keeping the chain/tape horizontal. Again, the chain/tape point is transferred on the ground with the help of a plumb bob, and peg inserted. This process is repeated till end point of the line is reached. The total distance is thus computed by adding up all the distances. Since, this approach provides rough measurements, now-a-days this method is not used on sloping ground.

If the length to be measured is very large, the direction of measurement is determined by a process, called *ranging* so that all the measurements fall in a straight line. However, in modern surveying, the EDM device and Total Station are used for measurement of long distances. These equipment have the advantages as these can provide measurement of longer distances in one single step, electronically. The details of EDM and Total Station are given in Unit 3 as well as Garg (2021).

1.13.2 Ranging of survey lines

Ranging is a process to establish intermediate points between these two stations in the same line so that the measurements are made along a straight line, following the principle of short distance, and not in a zig-zag manner. To measure the long distance between two stations on the ground, ranging rods (survey flags) are used (Figure 1.9) so that the movement is along a straight line. The ranging is carried out as (a) Direct ranging, and (b) Indirect ranging.

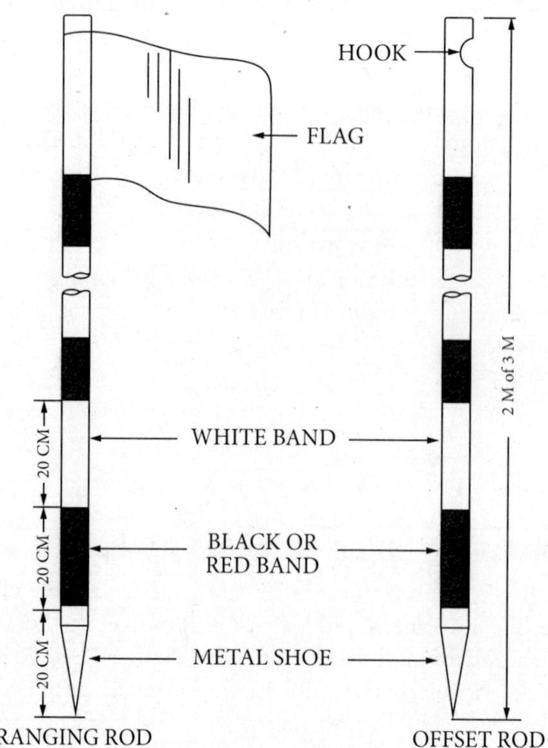

Fig.1.9: Specifications of a ranging rod

a. Direct Ranging

It is also called ranging by eye estimation method. Suppose ranging is to be carried between two stations A and B (Figure 1.10), the process involved is-

1. Firstly, a ranging rod (rod-1) or survey flag is established at a known station (point) A and fix on the ground firmly till the completion of work.
2. Second ranging rod (rod-2) is established at the farthest station (point) B and fix it firmly on the ground till the completion of work.

3. Third ranging rod (rod-3) is established by eye judgement at any convenient point P such that A, P, B stations are in one straight line and AP survey line could easily be measured with a tape. The position of observer's eye must be before point A, at least more than a meter away to help establishing the point P. Distance AP is now measured.
4. Fourth ranging rod (rod-4) is established at another convenient point Q such that P, Q, B stations are in one straight line and PQ survey line could easily be measured with a tape. Here, the position of observer's eye must be before point P, at least more than a meter away to help establishing point Q. Distance PQ is now measured.
5. In this way, process may be repeated for longer distances till you reach close to last station B. The corresponding distances and the last distance to station B are measured.
6. Finally, all the distances are added together (In this case, AP+PQ+QR) to get distance AB.

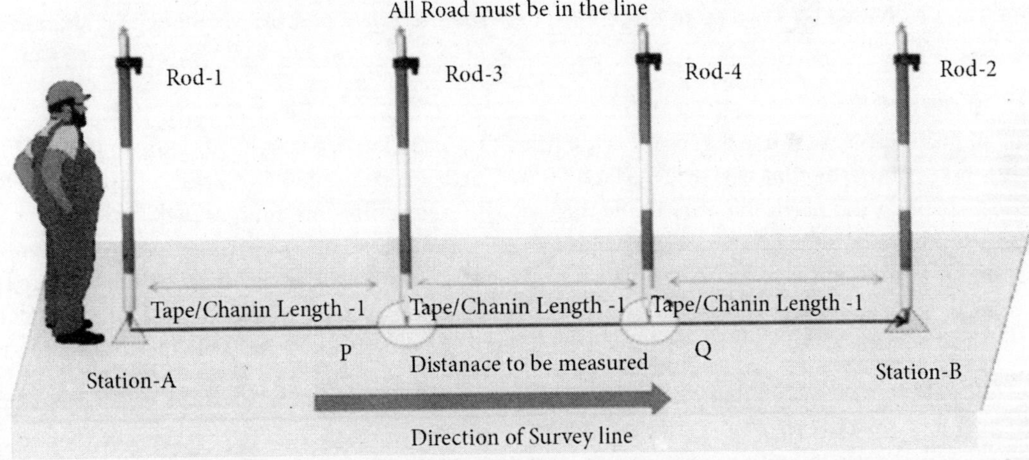

Fig.1.10: Direct ranging (https://dreamcivil.com/ranging-in-surveying/)

b. Indirect or reciprocal ranging

Due to intervening ground, if stations A and B are not intervisible, reciprocal ranging is used. Figure 1.11 shows this scheme of ranging. It finally establishes two intermediate points M and N indirectly in between the line AB. It involves two persons to occupy arbitrarily M_1 and N_1 points by each. Point M_1 and point N_1 are selected such that both stations A and B are visible from these. In addition, one surveyor is needed at station A and another surveyor at station B. To start the process, one surveyor near station A ranges the survey flag (ranging rod) near M_1 to position M_2 such that AM_2N_1 are in a straight line. Then surveyor near station B directs the person at N_1 to move the ranging rod to N_2 such that BN_2M_2 are in a straight line. Again, surveyor at station A will range the ranging rod at M_2 to move it to M_3 till AM_3N_2 are in a straight line. The process is repeated till intermediate points M and N are finally established such that AMNB are in a line.

M N

A B

(a) Sectional view

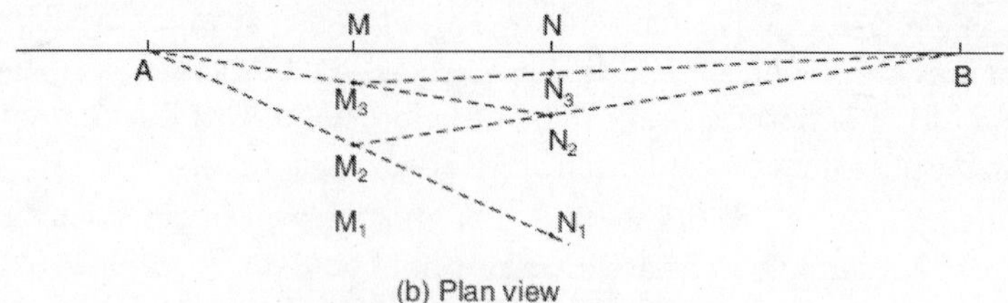

(b) Plan view

Fig.1.11: Reciprocal ranging

1.14 MEASUREMENT OF BEARINGS

Bearing is the direction of each survey line with respect to a reference meridian direction. Bearing of a line can be defined as the angle measured in clockwise direction from a given reference. Meridian direction can be divided into two main directions; (a) True meridian, and (b) Magnetic meridian. Both are explained below.

a. True meridian

The true or geographic meridian at a point is the line of intersection of plane passing through the north and south poles and the point with the surface of the Earth. Since the Earth represents approximately a spherical shape of the Earth, the meridians through different points are not parallel, and meet at the north and south poles. In case of small surveys, however, the meridians can be assumed to be parallel without making any significant error. The true meridian at a place can be established accurately through astronomical observations (Sun or Stars). Since, the direction of true meridian at a place remains unchanged, so bearing taken with respect to the true meridian does not change. True meridians are used from maps prepared by defence and army.

b. Magnetic meridian

The magnetic meridian through a point on the ground is the direction taken by a freely suspended magnetic needle at the point. The magnetic needle indicates the direction of magnetic north and south poles, and is not the same as true meridian. Since the magnetic meridian is observed with a magnetic needle, the magnetic meridian can be affected by magnetic substances if these are present near the area. A magnetic compass is used to measure the magnetic bearing of a line.

1.14.1 Types of bearings

Bearings are measured in two ways:

1. Whole circle bearing (WCB)
2. Reduced bearing (RB) or Quadrantal bearing (QB)

The *whole circle bearing* (WCB) of a survey line is the angle made by a survey line with the magnetic north direction, always measured in clock-wise direction, as shown in Figure 1.12. It can have a value between 0° and 360°. Since for trigonometrical calculations, we need RB angles, the whole circle bearing is reduced (converted) to RB. Figure 1.13 shows the relationship between the WCB of four lines with their corresponding RB in all the four quadrants (NE, SE, SW, and NW).

The *reduced bearing* (RB) or *quadrantal bearing* (QB) of a line is the angle made by the line with the either magnetic north or south direction, whichever is a smaller angle. The north-south direction and the east-west direction divide the horizontal plane into four quadrants, and any point on the ground will lie in one of the quadrants. Figure 1.13 shows these four quadrants. Thus, in quadrants I and IV, the reference direction is north and the angle is measured either to the east or west. In quadrants II and III, the angle is measured from the south, either to the west or east. Since the angle is observed from north or south towards the east or west, it is always less than 90°, as shown in Figure 1.13. RB is designated with the letters N (north) or S (south) and the direction in which the angle is measured; toward the east (E) or west (W). For example, the RB can be designated as N 30° 15'E; S 47° 45'E; S 40° 23'W; and N 57° 44'W.

Quadrant in which bearing lies	Coversion relation
NE	$\alpha = \theta$
SE	$\alpha = 180° - \theta$
SW	$\alpha = \theta - 180°$
NW	$\alpha = 360° - \theta$

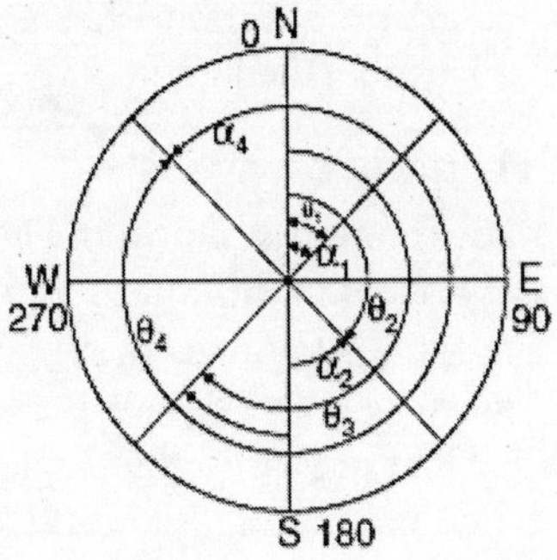

Fig.1.12: Representation of whole circle bearing

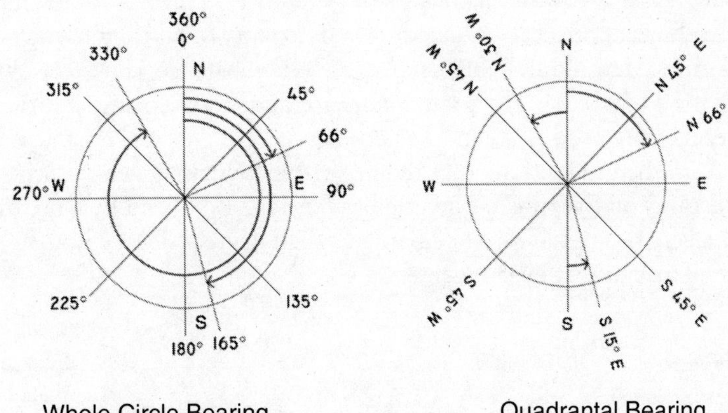

Fig.1.13: Representation of whole circle bearing into reduced bearing

1.14.2 Magnetic compasses

The magnetic compasses used in surveying may be classified as:

1. Prismatic Compass
2. Surveyor's Compass
3. Trough Compass
4. Tubular Compass

The compass works on the principle that a freely suspended magnetic needle takes the direction of the magnetic force at a place. It provides a reference direction with respect to which all angles can be measured. A magnetic needle is generally made perfectly symmetrical and is supported on a hard, pointed pivot so that it rotates freely. The north pole of a magnetic needle freely suspended gives the direction of the magnetic lines of force, and is used as a reference direction in compass surveying.

The Prismatic compass is very popular for survey work, and the Surveyor's compass is rarely used. However, the principle of operation of both the compasses is the same, but they are made differently. In Surveyor's compass, the graduations are done in quadrantal system, there is no prism attachment and the reading are taken with respect to north end. Trough and tubular compasses are used as add-ons to other instruments. They are not complete surveying instruments by themselves but are used only to indicate magnetic meridian direction to set the instrument in order to derive the magnetic bearing of the lines. A trough compass is generally used as an accessory to a plane table or theodolite, while a tubular compass is used only with a theodolite.

a. Prismatic compass

A prismatic compass is shown in Figure 1.14. The prismatic compass consists of an eye vane from where sighting is done is to the other station. An object vane is attached to metal frame, diametrically opposite to the eye vane. It is hinged at the bottom for folding over the glass cover when the compass is not in use. A fine silk thread or hair is fitted vertically in the centre of object vane frame, which is used to bisect the other objects/points. It has a perfectly symmetrical magnetic needle, balanced on a hard, steel pivot. When not in use, the needle can be lifted off the pivot, by folding the objective vane. It ensures that the magnetic needle is not rotating all the time on pivot, and pivot tip is not subject to undue wear. The needle is sensitive and takes up the north-south directly rapidly. A metal circular box supports an aluminum graduated ring, and is covered with glass so as to provide protection to aluminum ring with dust and moisture. The aluminum ring, graduated from 0° to 360°, is attached to the needle on its top a diametrical arm of the ring. Aluminum, being a non-magnetic substance, is used to ensure that the ring does not affect the rotation of needle. The graduations are done in a clockwise direction with 0°/360° marked on the south end of the needle, 90° marked on the west, 180° on the north, and 270° on the east directions. It is clear from the graduations that the prismatic compass gives the whole circle bearings of the lines. The graduations on the ring are inverted as they are to be read by a prism arrangement by creating an inverted image. The graduations are marked to half degrees (least count 30'), but reading can be taken to one-fourth of a degree by eye judgement.

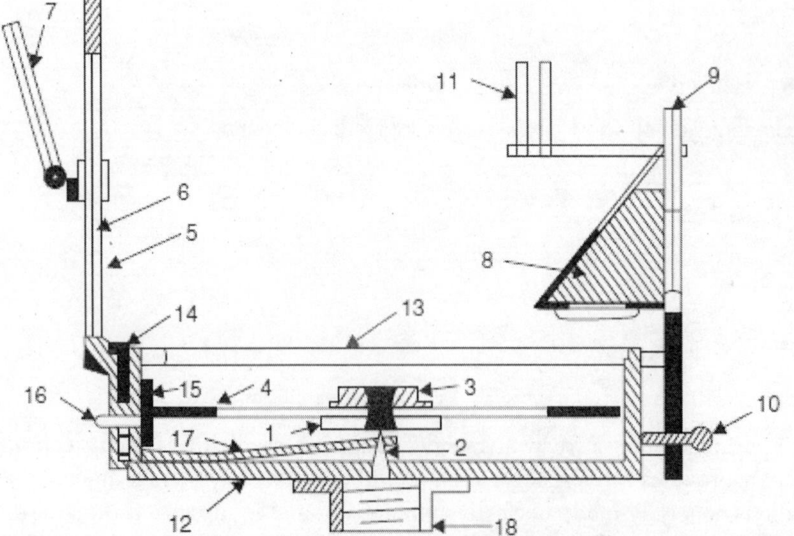

1. Needle
2. Pivot
3. Agate cap
4. Graduated disc
5. Slit metal frame
6. Horse hair
7. Mirror
8. Reflecting prism with cap
9. Eye vane
10. Focussing stud
11. Dark sunglasses
12. Box
13. Glass cover
14. Lifting pin
15. Light spring
16. Brake pin or knob
17. Lifting lever
18. Support to fit on tripod

Fig.1.14: Cross-sectional diagram of a prismatic compass (Duggal, 2017)

To measure an angle, the line of sight from compass station is established. The eye vane is made up of a rectangular frame with holes in the centre along its length. One of these holes is used to view the other object/stations by keeping observer's eye near the hole. A right-angle prism is hinged at the bottom of eye vane. A narrow slit in the prism holder enables to read the angles at the graduated ring. The prism also provides magnified view of the graduations on the ring. A metal cover is used to cover the graduated circle and the prism. The prism can be raised or lowered on the metal frame with a screw, as per the vision of the observer. Two dark coloured glasses are also provided on the metal frame, which can be brought in the view of line of sight while sighting towards bright objects or sun in the background so as to reduce the incoming glare. The metal box through a ball and socket arrangement can be levelled before using the compass. The bottom of metal plate has a screwed end that can be used to attach the compass to a tripod.

Steps to use prismatic compass

The following steps are required to use the Prismatic compass.

1. *Setting up and centering*: Fix the Prismatic compass on the tripod, and place the tripod over the station. Since, there is no centering device in the compass, it is centred over the station either with a plumb bob or by dropping a small piece of stone from the centre of the tripod. Tripod legs are moved to carry out proper centring over the station.
2. *Levelling*: Level the compass by approximation with the help of ball and socket arrangement so that the magnetic needle can move freely in a horizontal plane, after unfolding the objective vane and eye vane. There is no bubble tube attached to the compass for accurate levelling, hence it is carefully done with eye judgement only.
3. *Sighting the object:* Direct the object vane towards the object/station whose bearing is to be measured. Bisect the object/station with the vertical hair on object vane while looking through the hole of the eye vane. The metal circular plate of compass is rotated clockwise or anticlockwise so that the line of sight is passing through the object/station sighted and the cross hair of the object vane and also the view hole of eye vane.
4. *Taking readings*: The prism near the eye vane has to be adjusted with the help of a screw for a clear vision of the graduated ring readings. Once the magnetic needle comes to rest, record the reading at the point on the ring corresponding to the vertical hair seen directly through the slit in the prism holder. The prismatic compass will give the whole circle bearings of the lines. Disturb the line of sight, again align the line of slight on the same object/station, and take second reading. Repeat it at least one more time to take third reading. Take the average of at least three readings to improve the accuracy of observations.
5. Steps 1 to 4 are to be repeated at other compass stations.

1.14.3 Fore bearing and back bearing

Normally, bearings are used for traverse lines. Figure 1.15 illustrates how the bearing of a line is taken. Suppose fore bearing and back bearings of a line AB are to be measured. Set the compass at station A, bisect station B. The needle points to the north always and the reading is taken from the south end, in case of a Prismatic compass. The graduations made in the clockwise direction from the south end give the whole circle bearing of line AB, as can be seen from Figure 1.15. Measure the magnetic bearing towards B point (say 260°). This is called *fore bearing* of line AB. Now set the compass at station B, bisect station A, and bearing of line BA is measured from point B towards point A (say 80°) clockwise from magnetic north. This bearing is called *back bearing* of line AB or fore bearing of line BA. Both the bearings must differ by 180°, which is clear from the Figure also. This difference can also be used as a check while taking the measurements.

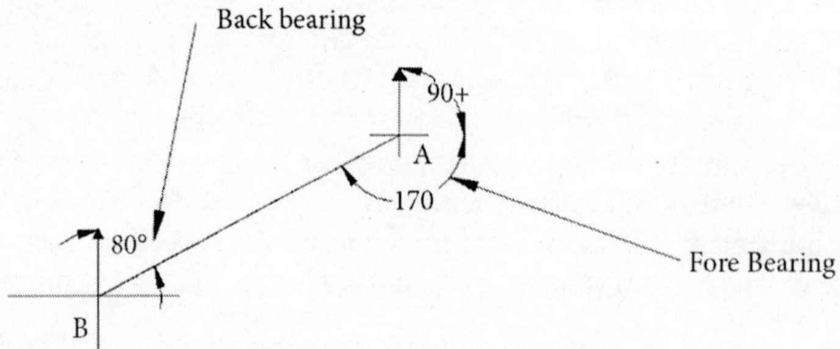

Fig.1.15: Fore bearing and back bearing

1.14.4 Magnetic declination

The magnetic meridian and the true meridian at any place are not same but different, the horizontal angle between these is known as *magnetic declination*. The magnetic north at a place may be either towards east or west of true north (Figure 1.16). If it is towards east, it is known as eastern or +ve declination, and if towards west, it is called as western declination or –ve declination. It means, eastern declination is to be added and western to be subtracted to the observed magnetic bearings to obtain true meridian. Magnetic declination varies from time to time and also from place to place

True bearing = magnetic bearing ± magnetic declination (E or W) (1.3)

To know the magnetic declination at a location, true meridian is established from astronomical observations (Sun or Stars) and magnetic meridian is determined by a compass. In other words, if we know the magnetic declination and magnetic bearing of laine, its true bearing can be computed. Engineering maps are made with respect to magnetic meridians, while maps used by army are made with respect to true meridian.

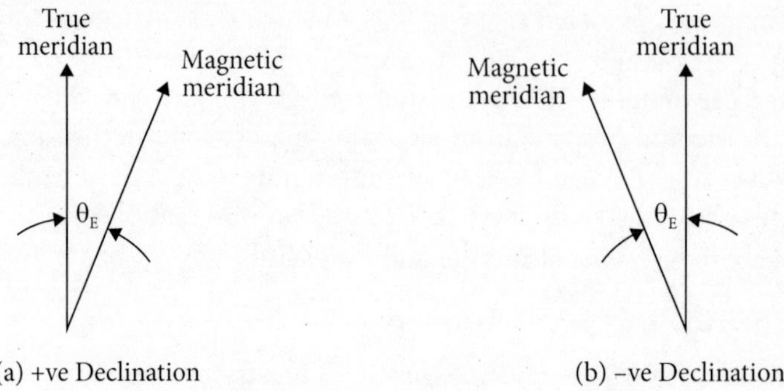

Fig.1.16: Determination of magnetic meridian

1.14.5 Local attraction

A properly balanced and freely suspended magnetic needle is expected to show the direction of magnetic meridian. However, many-times the local objects, like electric wires, magnetic ore, iron pipe line, and metallic buttons can attract magnetic needle towards themselves, and can give erroneous results. This deviation in reading is called *local attraction*. The materials/objects which can cause error due to local attraction could be magnetic rock or iron ore, steel structures, iron poles, rails, pipe-lines, electric poles and wires, key bunch, ring, knife, iron buttons, steel framed spectacles, and survey chain, arrows, hammer, clearing axe, etc. Special care is required to be taken to avoid error due to local attraction. However, in many circumstances, it is neither possible to remove the metallic objects from the survey field, nor changing the station of magnetic compass observations. In such cases, the fore bearing and back bearing of the line is taken, and those stations are considered unaffected by local attraction if the difference of fore bearing and back bearing of the lines is exactly 180°. The difference in value is the error. Using this approach, the stations affected or unaffected by local attraction. The error due to local attraction is distributed at the affected stations, such that now all the lines will satisfy the check, i.e.,

Difference in fore bearing and back bearing of a line = 1800 (1.4)

1.14.6 Computation of included angles from bearings

In a closed traverse, bearings of lines may be calculated if bearing of one of the line and the included horizontal angles between various lines are known, using the relationship:

Bearing of a line= given bearing + included horizontal angle (1.5)

Figure 1.17 illustrates the computation of horizontal angles from bearings. If at any point, bearings of any two lines are known, the included angle between these two lines can easily be found by drawing a sketch, and then taking the difference of angles. The examples are given below.

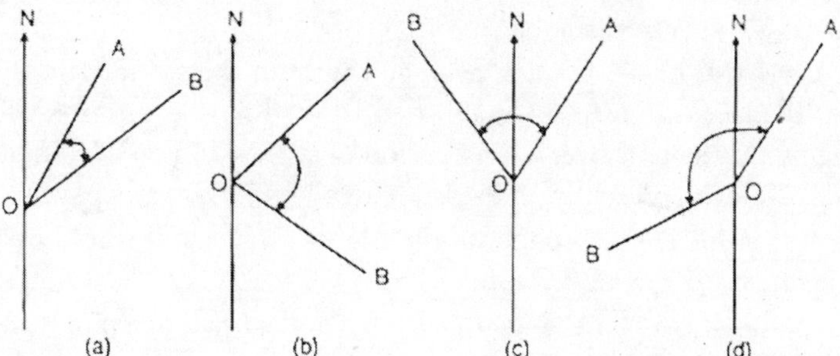

Fig.1.17: Relationship between horizontal angles and bearings

If the lines are on the same side of the meridian and in same quadrant (Figure 1.17a) the included angle ∠AOB= the difference of the reduced bearings of OA and OB. If the lines are on the same side of the meridian but in different quadrants (Figure 1.17b), the included angle ∠AOB= 180°– sum of the reduced bearings of OA and OB. If the lines are not on the same side of the meridian but they are in the adjacent quadrants (Figure 1.17c), the included angle ∠AOB= sum of the reduced bearings of OA and OB. If the lines are not on the same side of the meridian and also not in the opposite quadrants (Figure 1.17d) the included angle ∠AOB= 360°– difference of the whole circle bearings of OA and OB.

The above relationship can also be used to compute the bearings of lines, if all included angles and one bearing of a line is known. Due to local attraction problem, sometimes the bearings of all the lines are not measured for all the traverse lines, but angles can be measured with greater accuracy, in such cases, bearing of reaming lines in a traverse can be determined, and subsequent check applied for bearings.

1.15 MEASUREMENT OF LEVELS

Levelling is a branch of surveying which deals with measurements in vertical planes. It is the operation performed to determine and establish the elevations of points on the surface or beneath the Earth. It is also done to determine differences in elevation between points to draw the contours, find out the slope, and control the grades during construction. Levelling is of prime importance to the engineers for both design as well as execution of a project on the ground. The success of many engineering projects would depend upon accurate determination of elevations. The pre-requisite to compute the elevations of ground points is that the elevation of at least one point should be known with respect to which the elevations of other points can be computed. The point whose elevation is known to start the levelling work is known as the Bench Mark (BM). The instrument used to measure the elevations is known as Level, which is used along with a Levelling staff or Levelling rod.

The object is of levelling, therefore, is to; (*i*) determine the elevations of given points with respect to a datum, and (*ii*) establish the points of required height above or below the datum line. The levelling is used for many applications, such as; (*i*) to determine or to set the plinth level of a building, (*ii*) to decide or set the road, railway, canal or sewage line alignment, (*iii*) to determine or to set various levels of dams, towers, etc., and (*iv*) to determine the capacity of a reservoir.

1.15.1 Technical terms used in levelling

It is necessary to understand few technical terms often used in levelling. These are explained below (Refer to Figure 1.18).

a. **Level surface**: It is a surface perpendicular to the direction of gravity at all points. The surface of a still lake is a good example of level surface.

b. **Mean Sea Level (MSL)**: MSL is the average height of the sea for all stages of the tides. At any particular place, the MSL is established by finding the mean sea level (free of tides) after averaging tide heights over a long period of at least 19 years. In India, MSL used is that established at Karachi, presently, in Pakistan. In all important surveys this is used as datum.

c. **Level line**: It is a line lying throughout on a level surface and would be a part of the circle representing the mean sea level (MSL).

d. **Horizontal plane**: It is a plane tangential to the level surface at a point, and is also normal to the direction of gravity at that point.

e. **Horizontal line**: It is a line lying in a horizontal plane. The horizontal line is therefore, perpendicular to the vertical line at that point. A horizontal line is always tangential to the level line at that point.

f. **Datum surface**: It is a level surface used as reference surface, with respect to which elevations of other points are computed.

g. **Vertical plane:** It is a plane, which contain the vertical line at that point. Infinite number of vertical planes can contain a vertical line.

h. **Altitude:** The vertical distance of a point above the datum is called the altitude of that point. If the datum surface considered is the MSL, the altitude and the elevation will be the same.

i. **Elevation or Reduced level (RL)**: It is the level or altitude of a point with reference to the datum surface, measured along the direction of gravity.

j. **Bench Mark (BM)**: It is a point of known elevation above MSL. Usually the BM is marked on the ground after erecting a small concrete pillar and marking the centre of the top surface with a fixed brass plate is fixed.

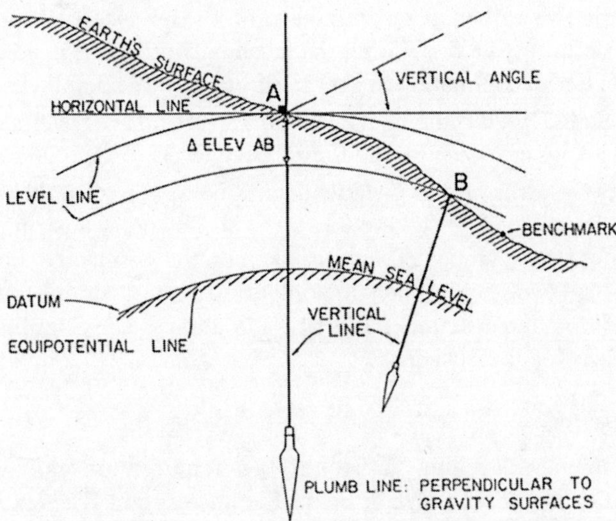

Fig.1.18: Representation of various terms (Schultz, 1987)

k. **Line of collimation:** It is the imaginary line passing through the centre of the eye piece and intersection of diaphragm (cross-hairs) and its further extension in the telescopic tube (Figure 1.19). The line of collimation may not pass through the centre of objective lens of telescopic tube, if the diaphragm is out of adjustment, and hence may give erroneous results.

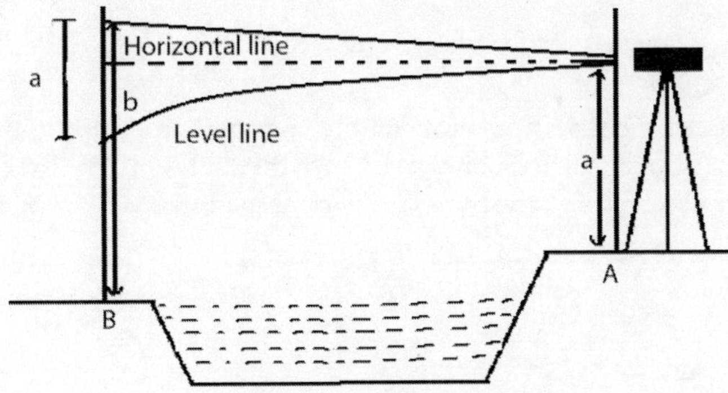

Fig.1.19: Line of collimation

l. **Line of sight**: In an adjusted telescopic tube, the line of sight is the imaginary line passing through the centre of the eye piece, centre of cross-hairs and centre of objective lens. All these points must lie in one straight line.

m. **Height of instrument:** It is the height of instrument above the ground, generally measured with a tape from horizontal axis of telescopic tube. It should not be confused with the elevation of height of telescope as the latter will be the height above a given datum where the instrument is setup.

n. **Back Sight (BS):** It is the sight taken on a level staff held at the point of known elevation (BM or datum). It is the first reading after the instrument is set in an area for levelling work. This reading is normally added to RL of the point to get the RL of line of collimation.

o. **Fore Sight (FS):** This is the last reading taken on the level staff from the instrument station before shifting it or just before ending the levelling work. This is normally subtracted with the elevation of line of collimation to get the elevation of the point where BS reading is taken.

p. **Intermediate Sight (IS):** It is the sight taken on a level staff after back sight and before the fore sight. These readings are taken to find the reduced levels of the points where staff was held.

q. **Change Point (CP):** It is also known as turning point (TP). On this point, both back sight and fore sight are taken. After taking fore sight on this point, instrument is shifted to some other convenient station, and back sight is taken on the staff held at the same point from the new station. These two readings help in computing the elevations of unknown points on the ground. In addition, the instrument line of collimation error is eliminated is the BS distance and FS distance is kept nearly equal.

1.15.2 Levelling staff

Along with a level instrument, a levelling staff (or rod) is required for taking the measurements to determine amount by which a point is above or below the line of sight. The levelling staff is a straight rectangular wooden or metallic (aluminium) rod having graduations (Figure 1.20). It is protected with a metal shoes at its bottom from wear and tear in the field. The foot of the shoe on levelling staff represents zero reading. These staffs are available in 3 m, 4 m, 5 m length, and may be divided into three groups;

a. solid staff -which is made up of one single solid piece, and hence sometimes difficult to transport and move in the field during long hours of observations,

b. folding staff- which is made up of two pieces hinged together, and can be folded when not in use, and

c. telescopic staff- which is made up of several pieces connected together; one piece slides into another and so on).

The folding and telescopic staffs are convenient to transport and work in the field. The vertical distance between the line of sight and the point over which the staff is held vertically is the staff reading which is recorded by the observer. The least count of normal solid staff is 0.005 m or 5 mm.

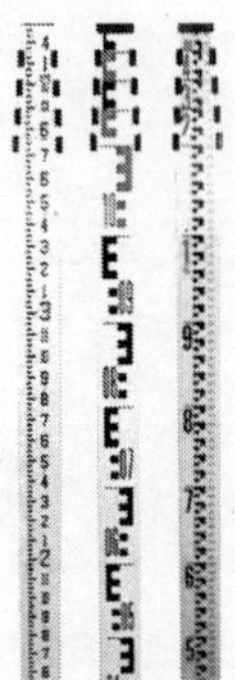

Fig.1.20: Various levelling staff

1.15.3 Levels

The instrument used for collecting the elevation measurements from the field are called levels. The purpose of a level instrument is to establish a horizontal line of sight in order to take the measurement to compute the elevations. The level consists of the following essential parts:

1. A telescopic tube which provides a line of sight.
2. A diaphragm (or cross-hairs) which holds the cross hairs (fitted near the eyepiece within telescope).
3. Eyepiece which is fitted at one end of telescope, and used to magnify the image formed in the plane of the diaphragm. The observer keeps one eye closer to it to read the staff reading during leveling. The eyepiece is fitted with a screw which is moved clockwise or anticlockwise to focus the diaphragm.
4. Objective lens, which is fitted at the other end of the telescope, is used to collect the incoming rays from the levelling staff (objects) that form the image inside the telescopic view.
5. A focusing screw which is used to focus the image at the cross hairs.
6. A level bubble tube which is attached to telescopic tube to make the line of sight horizontal.
7. A levelling head which supports the level instrument, and is fitted with a circular bubble tube, used for making the base of instrument approximately levelled.
8. A tripod for fixing the level instrument.

Various types of leveling instruments are:

a. Dumpy level
b. Tilting level
c. Engineers level
d. Automatic level
e. Digital level
f. Laser level

a. Dumpy level

The Dumpy level consists of a telescope, generally external focusing type, rigidly fixed to the vertical spindle. The telescope can be rotated in the horizontal plane about its vertical axis. A level tube is attached with the telescope. The level tube is fixed with its axis parallel to telescope tube, so that when bubble is centred, the telescope becomes horizontal. The level tube is graduated on either side of its centre to estimate how much the bubble is out of its centre. The instrument has a levelling head consisting of two parallel plates held apart by three (or four) levelling screws. The upper plate is called tribrach and the lower one is called trivet. A circular bubble tube (or bulls's eye) is fitted with the tribrach. The instrument base is approximately levelled using these foot screws which are rotated in a particular sequence. Exact levelling of line of sight is done by using bubble tube fitted with the telescope. With the availability of Auto levels, Laser levels, and Total Station equipment, the Dumpy level is no more used in levelling work.

b. Tilting level

The telescope of a tilting level is not rigidly fixed to the vertical spindle as in case of a Dumpy level (Figure 1.21). The telescopic tube, generally internal focusing type, can be tilted on a pivot about horizontal axis in the vertical plane upwards or downward through a small angle $\pm 3^{\circ}$ by means of a tilting screw. The line of collimation and the vertical axis of a tilting level need not be at right angles as required in the case of a Dumpy level. The bull's eye or circular bubble tube is fixed to the upper plate of levelling head for approximate levelling the base of instrument by using the three foot screws. For temporary adjustment of a Level, please see section 1.15.4.

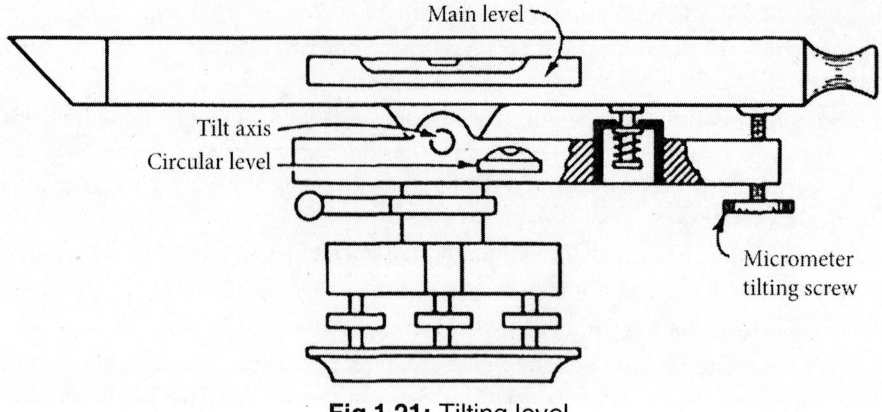

Fig.1.21: Tilting level

After the temporary adjustments are over, the levelling staff is now kept a point whose elevation is to be determined. The cross-hairs are focused and then leveling staff is bisected and focused. The exact levelling of the instrument is then done using the tilting screw before taking every reading. The reading where the horizontal hair intersects the staff is noted. An inclined mirror attached to the level tube enables the observer to view the centering of the bubble from the eyepiece end of the telescope without moving the eye close to bubble tube.

Tilting levels have the advantages over Dumpy level that the line of sight can be tilted upward or downward by ±3° so that observations are possible to take from the levelling staff, in case of undulating ground. These levels are more robust, lighter, compact and accurate than the Dumpy levels. The tilting arrangement saves a lot of time for making the instrument ready to take the readings. Tilting levels are most useful when only few readings to be taken from one setting of the instrument.

c. Engineer's level

An engineer's level primarily consists of a telescope which is internal focusing type. Their construction is like other level but very compact and light weight. Engineers level and auto-levels provides the advantage of taking faster observation, as the line of sight each time is not required to be levelled with the bubble tube.

d. Auto level

Auto level or Automatic level is a professional levelling instrument used by contractors, builders, land survey professionals, and engineers who require very accurate levelling work. This level is used to verify the elevation of foundations, footings, and walls; design proper drainage systems for homes and other structures; determine proper elevation for a floor; determining the height of doors and windows, building suspended ceilings, design of swimming pool, roadwork, and excavations.

The tripod of an auto level is kept on a stable ground, and instrument clamped on it with the clamping screw. The auto level needs to be adjusted temporarily, as explained in temporary adjustment of a level. If repeating this process does not solve the problem, the auto level might be out of adjustment. Once the base is levelled, Auto level will be automatically levelled along the line of sight due to incorporation of a self-levelling feature. Once the circular bubble at the base is centered manually, an *automatic compensator* would level the line of sight, and continue to keep it level. The automatic compensator consists of prisms suspended from wires to create a pendulum. The wire lengths, support locations, and nature of the prisms are arranged such that only horizontal rays would reach the intersection of cross hairs. Thus, a horizontal line of sight is achieved even though the telescope itself may be slightly tilted away from horizontal. Damping devices would bring the pendulum to come to rest quickly, saving the operator's time. Various components of an Auto level are shown in Fig. 1.22.

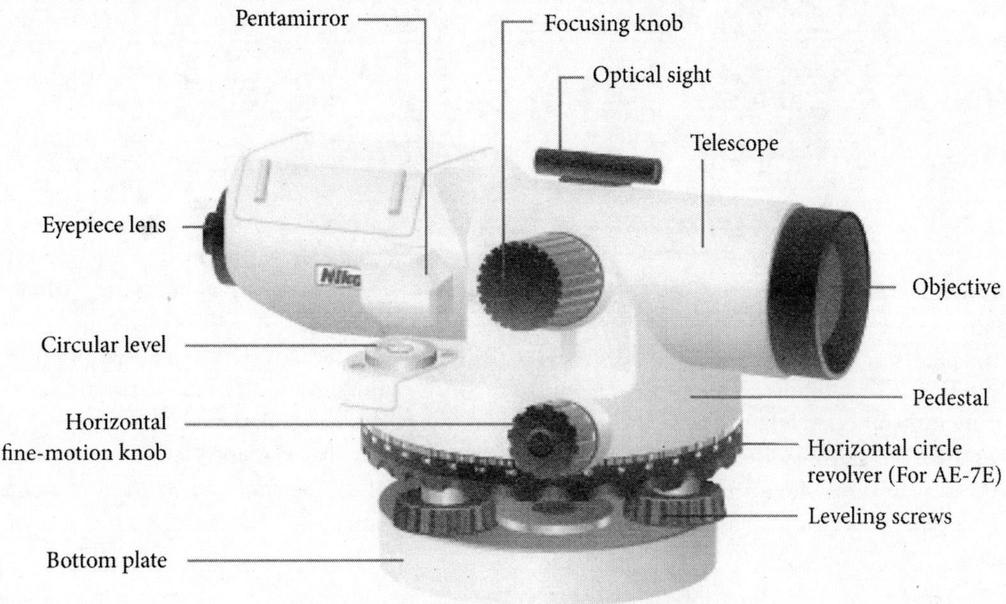

Fig.1.22: Various components of an Auto level

Auto levels are quick to level with greater accuracy which have made it more demanding amongst the surveyors. For general use, these are popular because of their ease and rapid operation, however the reading is manually recorded. They are fast and easy to use, and therefore can save a great deal of time and money in the field. An auto level has additional benefits, as it gives the upright reading that can be seen from the eyepiece, as compared to Dumpy/Tilting level where reading is inverse which is sometimes difficult and confusing to read, and becomes an important source of error. In addition, the line of sight is automatically adjusted with an internal compensator. The drawback is that some automatic compensators are affected by magnetic fields, which can result in systematic errors in staff readings.

e. Digital level

The electronic digital level is further modification of automatic level. It also uses a pendulum compensator to level itself, after an operator accomplishes rough levelling with a circular bubble (same as in case of auto levels). The instrument could be used to take readings manually, just like an automatic level, however, it is designed to operate by employing digital image processing approach. After levelling the instrument (Figure 1.23a), its telescope is focused toward a special bar-coded rod (Figure 1.23b). At the press of a functional button, the image of bar code is captured in the telescope's field of view, and processed. The processing is done by an onboard computer by comparing the captured image with the levelling staff's entire pattern, which is already stored in memory. When a match is found, the staff reading is displayed digitally. The reading can be recorded manually or automatically stored in a controller of the instrument.

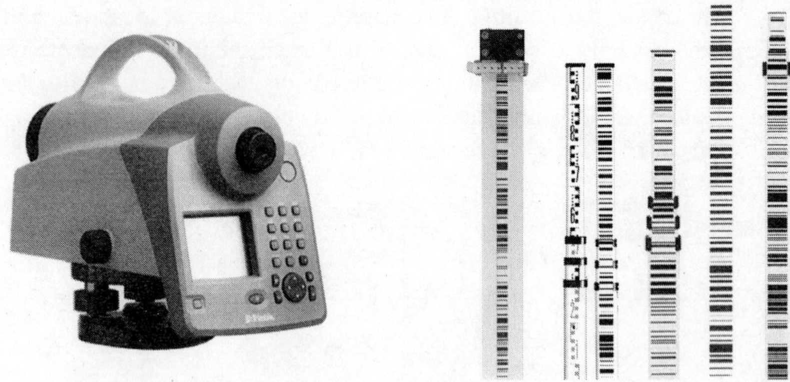

Fig.1.23: (a) Digital level, and (b) Bar code leveling staff

Digital levels have the provision of transferring the digital data to a computer where the measurements are processed to get the elevations, profiles or DEM. The length of the staff appearing within the telescope's field of view is a function of the distance from the staff. Thus, the instrument is also able to automatically compute the distance from staff; a feature quite useful for balancing the backsight and foresight distances. Normally, the instrument's maximum distance range is approximately 100 m, and its accuracy in staff readings is ±0.5 mm. The bar-coded staff is also available with English or metric graduations on reverse side of the bar code. This graduated side of the staff can be used to manually read it in situations where instrument can't read from the bar codes, such as when the staff is used under heavy bush or long grasses.

f. Laser level

Laser level has an electronic device that emits a laser beam, and a sensor fitted with the levelling staff can sense the laser beam and digitally display the measurements. The laser beam concept was introduced in levelling instruments for automisation. In surveying, the laser has long been recognized as a carrier wave, in electro-optical distance measurement. However, in the last ten years or so, laser levels have been frequently used for levelling operations required in measurement of land surface, and determination of the height and depth of large or small object on the terrain.

The laser levels enhance the working with speed of data collection and automisation. They provide more accurate reduced levels, and are frequently used for levelling, plumbing, machine control, excavation work, landscaping, construction, measuring elevation, alignment, grading of sites, construction stake out, concrete levelling, drainage design, and many more jobs (Ali and Al-garni, 1996). The architects have benefited extensively from the use of laser-based instruments in monitoring interior height control of buildings, decorations, setting out of individual walls and suspended ceilings and control of elevator guide rails.

A laser level generates a laser line across the point where it's pointed. It can be visible as a bright red or green colour, or completely invisible for survey work at places with high movement of people, such as roads. These laser levels are available in manual-levelling, auto levelling and self-levelling modes. Further advancements in laser level have led to the development of Rotary laser levels which emit laser in 360° rotation, covering vertical or horizontal planes, instead of just emitting a single line in one direction. Since the continuous rotation of the laser beam about its vertical axis determines the projection of the beam, it is important to keep laser level accurately levelled. Rotary levels (Figure 1.24a) have a greater range than line levels, and are more ideal for larger and exterior work sites. An integral part of using a rotor laser level is a laser detector which is typically mounted to level rod (Figure 1.24b) and detects the laser beam during outdoor survey. It can detect signals that make easier to find laser lines. In outdoor woks, they are to be used along with a laser detector and a graduated rod. The best laser levels for surveying are the ones that can generate several laser lines at the same time in both horizontal and vertical directions. The use of laser levels in construction industry has dramatically reduced the time to complete the project. It helps engineer check the perfect angles of a building during construction.

Fig.1.24: (a) Self-levelling rotary laser, (b) Laser detector and levelling staff

1.15.4 Temporary adjustment of level

Temporary adjustment of a level is required to be carried out prior to taking any observation on staff from an instrument station. It involves some operations which are required to be carried out in a proper sequence. It consists of three steps; Setting, Leveling and Focusing.

a. **Setting**: Firstly, the tripod stand is set up at a convenient height so that its head is horizontal (through eye judgement only). If the ground is flat, the three legs of tripod are kept at equal spacing to form equilateral triangle so that the levelling of the base is quick (Figure 1.25). The Level is then fixed on the tripod head by screws. The Level instrument is set up over a ground station by suspending the Plumb bob with the instrument base, and ensuring that the pointed tip of the Plumb bob touches the instrument station.

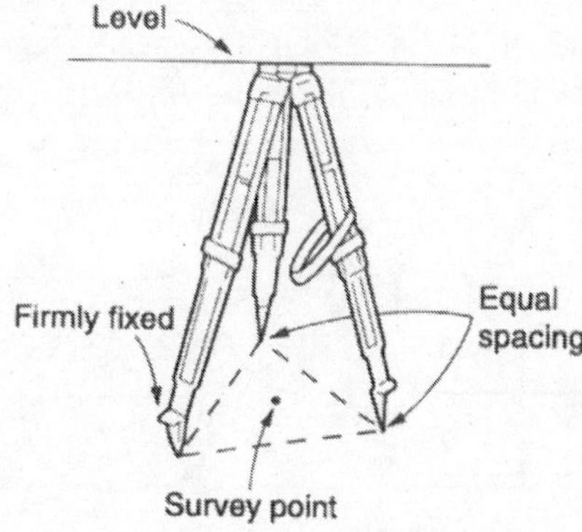

Fig.1.25: Setting up a tripod

b. **Leveling**: Secondly, the instrument is leveled approximately with the help of leveling screws till the circular bubble comes in the centre. Then, the bubble of the level tube is brought to the centre by using the designated screw. For approximate levelling of base, the telescope of the instrument is kept parallel to any two of the three levelling screws, and using these two screws the bubble should be brought in the centre of the levelling bubble tube, either by uniform inward or outward rotation of screws (as shown in Figure 1.26a). And once the bubble is centered, the telescope is brought perpendicular to its current position. Now, using the third screw, the bubble is brought exactly inside the circle, by either clockwise or anticlockwise rotation of levelling screw (Figure 1.26b). The telescope is now rotated 90° (original position as at Figure 1.26a), and the bubble is checked. If the bubble doesn't stay in the center of the circle, the previous steps are repeated to bring it in the centre. Normally, two-three repeat processes will bring the bubble perfectly in the centre. Turn the telescope by 360°, and check the bubble. It should ideally remain in the centre in all positions of telescope.

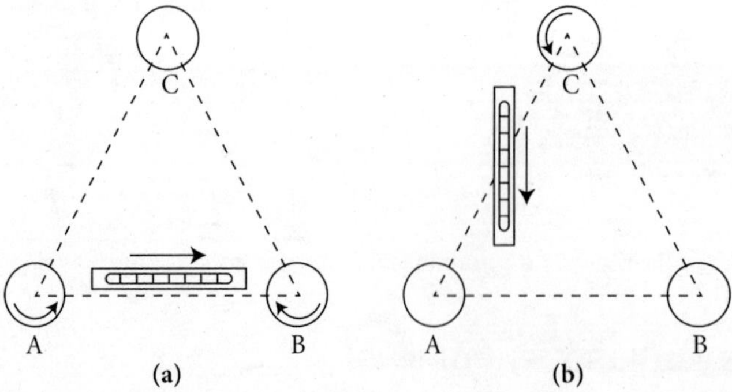

Fig.1.26: Levelling the base of the instrument using three foot screws

c. **Focusing:** Thirdly, focusing is required to be done in order to form the clear image through objective lens at the plane of the diaphragm and to view the clear image of through the eye-piece. The diaphragm is a cross-hairs at which the object is focused. These are available in different shapes, as shown in Figure 1.27. The focusing is carried out in two steps by removing parallax with proper focusing of objective lens and eye-piece. For focusing the eye-piece, screw attached to it is rotated till the diaphragm or cross-hairs are seen sharp and distinct. Focusing of cross-hairs might change from observer to observer, as it depends on the vision of the observer. For focusing the objective lens, the telescope is first pointed towards the Levelling staff (object), and the focusing screw is turned until the image of the Staff (object) appears clear and sharp, and also there is no relative movement between the image and the cross-hairs. Focusing of objective lens is required to be done before taking each and every observation.

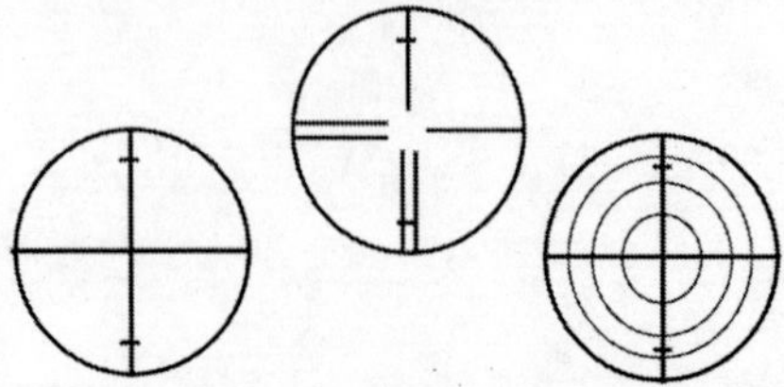

Fig.1.27: Various diaphragms

1.15.5 Reduction of levels

The reduction of level is the operation where the observed staff readings are used to find out the elevation (RLs) of points. There are two methods for reduction of levels:

a. Rise and Fall method, and
b. Height of instrument method

a. Rise and Fall Method

Since the ground is undulating, the Staff reading will be more if kept at a lower point, and less if kept at a higher point. Thus, the staff readings provide information regarding relative rise and fall of ground points. The difference in the staff readings indicates a rise or fall according to if the staff reading is smaller or greater than that at the preceding point, respectively. The difference between consecutive points is calculated by comparing each point after the first with that immediately preceding it. The RL of each point is then determined by adding rise or subtracting fall to/from the RL of the preceding point. This is the basic concept behind rise and fall method for finding out the elevations of unknown points.

If BS–FS is positive. then it is the rise, and if BS–FS is negative, it is fall

RL of the point = RL of previous point ± (rise or fall)

At the end of all computations, check is applied as:

$$\Sigma \text{ BS} - \Sigma \text{ FS} = \Sigma \text{ Rise} - \Sigma \text{ fall} = \text{First RL} - \text{Last RL} \tag{1.6}$$

This method is a laborious method as staff reading of each point on the ground after the first is compared with that preceding it, and the difference of level is entered as a rise or fall. It is slow and simple method. But, it is more precise because intermediate sights are also considered in calculations and checking the results. There are three checks for arithmetical calculations. The method is popularly used for earthwork calculations and other precise levelling operations.

b. Height of Instrument method

Height of Instrument (HI) method deals with obtaining the RL of the line of collimation by adding BS reading of a point whose RL is known, as shown in Figure 1.28. The RL of line of collimation is called the Height of Instrument. From this, the staff readings of all intermediate stations are subtracted to get the RL at those points. It is always measured from the benchmark. The HI method involves less computation in reducing the levels, so when there are large numbers of intermediate sights, it is used. It is a faster than the rise and fall method, but it has a disadvantage of not having check on intermediate sights.

HI = RL of BM + BS

RL of point = HI–FS of that point

At the end of all computations, check is applied as:

$$\Sigma \text{ BS} - \Sigma \text{ FS} = \text{First RL} - \text{Last RL} \tag{1.7}$$

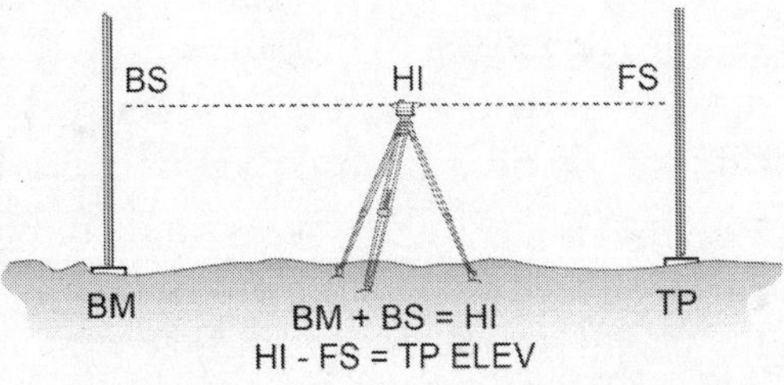

Fig.1.28: Height of instrument method

1.15.6 Types of direct levelling

The following are the different types of direct levelling used in the field:

a. Simple levelling
b. Differential levelling
c. Fly levelling
d. Profile and Cross section levelling, and
e. Reciprocal levelling.

a. Simple levelling

This method is used for finding out the difference between the levels of two nearby points. Figure 1.29 shows one such case in which level of BM is given as 100.000 m. RL of unknown point is computed as-

RL = BM + BS–FS

= 100.000 + 0.973) – 4.987

= 95.986 m

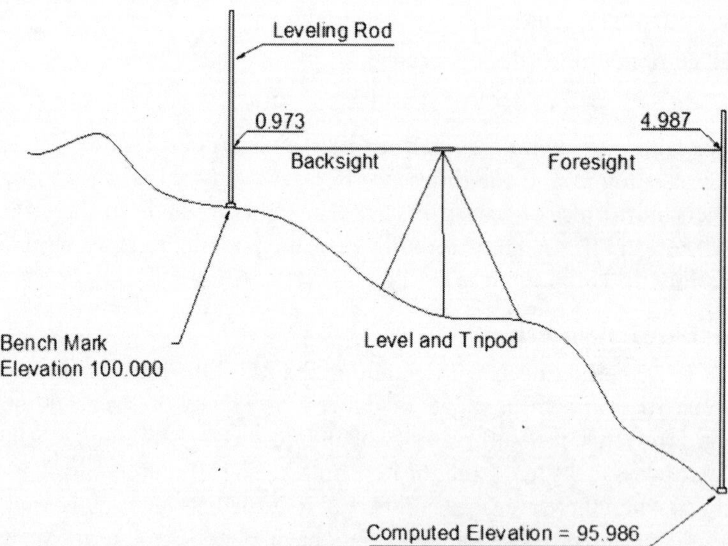

Fig.1.29: Simple levelling observations

b. Differential levelling

Differential levelling is used for finding the difference between the levels of two far-off points. If case of a large distance between two points A and D, it may not be possible to take the readings on A and D from a single setting of instrument, so differential levelling is used (Figure 1.30). In this method, the level instrument is set at more than one stations along the survey line, and at each shifting station both BS and FS readings are taken to determine the elevation difference between A and D. The following example in Figure 1.30 shows how the RLs of various points on the ground are calculated from BS and FS readings. The last point is another BM which could be used to check the levelling work.

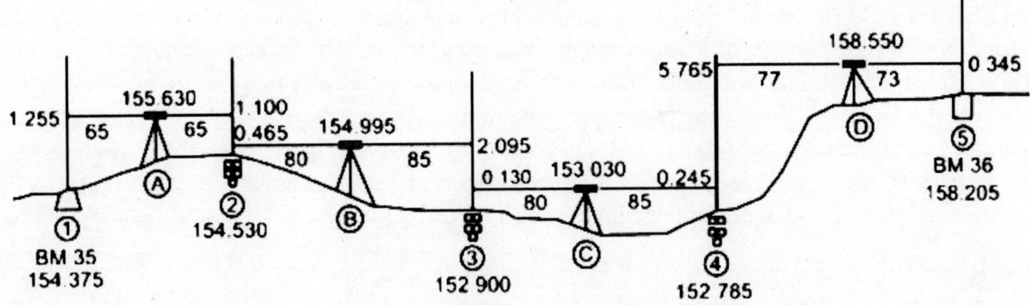

Fig.1.30: levelling observations

c. Fly levelling

If the survey site is away from the bench mark, the level instrument is set up near the BM and a back sight reading is taken on the BM. The survey work proceeds towards the site by taking fore sights and back sights on a number of change points till a temporary BM is established at the survey site. Now the levelling work is carried out at the site as per differential levelling method. At the end of the work, last reading is taken again on temporary BM. It will help determining the closing error in levelling work. The purpose of fly levelling is to connect the survey site to a BM away from the site, and then carry out the levelling work using differential levelling approach.

d. Profile and cross-section levelling

This type of levelling, also known as longitudinal levelling, is required to be carried out in highway, railway, canal, pipeline, transmission line, or sewage line projects to know the profile of the ground along selected alignments. It is most popular levelling method to estimate the amount of earth work (cutting or filling) required in large number of engineering projects dealing with the land surface. At regular interval, level readings are taken and RLs of various points along and across the alignment are determined. For drawing the profile of the route, distance is plotted against x-axis and RLs are plotted along y-axis. The vertical scale is usually larger as compared to scale for horizontal distances. It gives clear picture of the profile of the route. If the graph is plotted taking the values along the route, it is known as *longitudinal profile*, whereas if the elevation values across the route are plotted, it is known as *cross-section profile*. The cross-section profiles are usually drawn from the centre of the route on either side. Figure 1.31 shows the plan view of the scheme of profile levelling and cross-section levelling of the route.

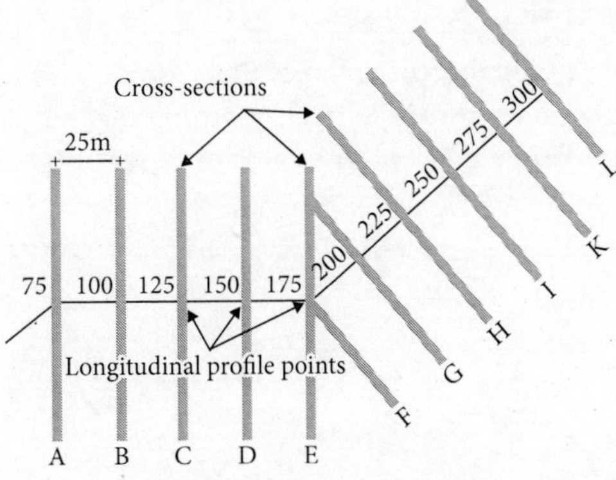

Fig.1.31: Longitudinal and cross-section profiles observations

In many engineering projects, not only longitudinal profile but also cross-section profiles are required at regular intervals. Both these profiles are used in calculating the earth works involved in the execution of projects. Figure 1.31 shows the scheme of such work in which cross-section profiles are taken at 25 m interval along the alignment line, and readings are taken in the perpendicular direction at an interval of 3 m on either side of the alignment line (from A to L). The distances on the cross-sections will be defined as left or right of the lines while doing the survey in forward direction. The length of the cross-section would depend upon the type of project occupying the land width.

e. Reciprocal levelling

In levelling, the distance of back sight and distance of fore sight is kept equal, as far as possible. It is done to eliminate the following errors:

a. Error due to non-parallelism of line of collimation and axis of bubble tube, and
b. Errors due to curvature and refraction.

While doing the levelling across features, like river, big ponds and reservoirs, it is not possible to maintain equal distance for fore sight and back sight distances. In such situations, the reciprocal levelling is used and the height difference between two points is precisely determined by two sets of correlation observations. In this method, the instrument is set up at a point A on one side of the valley, and readings are taken on the staff held near A (Figure 1.32a) and on the staff held at B on the other side of the valley. Let these readings be h^a and h_b, respectively. Reading h_a is considered to be accurate, but reading h_b may have an error 'e' due to collimation. The instrument is then moved at the other side of valley at point B, and again the reading is taken on the staff held near B and on the staff held at A on other side of valley. Let these readings be $h_{b'}$ and $h_{a'}$, respectively (Figure 1.32b). Reading $h_{b'}$ is considered to be accurate while reading $h_{a'}$ may have an error 'e' due to collimation. Let h be the true difference in height between A and B.

In Figure 1.32a

$h = (h_b - e) - h_a$

In Figure 1.32b,

$h = h_{b'} - (h_{a'} - e)$

Adding the above two relationahips, we get-

$2\,h = (h_b - h_a) + (h_{b'} - h_{a'})$

$h = 1/2\,[(h_b - h_a) + (h_{b'} - h_{a'})]$

$$\text{or } e = 1/2\,[(h_b - h_a) - (h_{b'} - h_{a'})] \quad (1.8)$$

In the above relations, it is assumed that the collimation error is the same when making observations from both the sides. However, in reality if we take reading from one end, there will a time difference to shift time to the instrument to the other side, and the value of refraction may change over time.

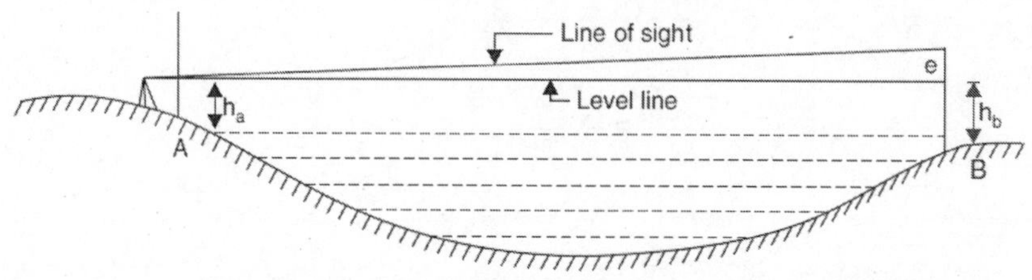

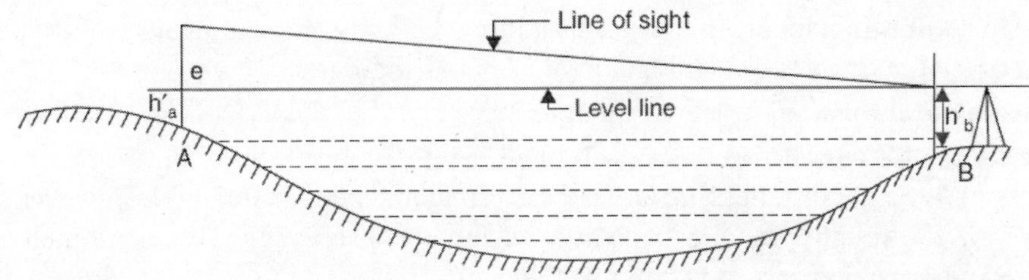

Fig.1.32: Reciprocal levelling observations (a) from side A and (b) from side B

1.15.7 Different types of errors

While taking observations, several types of errors might occur, such as instrumental errors, personal errors, natural errors. These are briefly discussed below;

a. **Instrumental errors:** These are of several types:.

   i. **Error in permanent adjustment of level:** For any major surveying work, the instrument needs to be tested and if required, gets to be adjusted. For small works, bubble of the level tube should be brought to the centre before each reading and balancing of sights are to be maintained.

   ii. **Staff defective and/or of non-standard quality**: If the graduation in staff are not standardized that may cause error in reading. In an ordinary leveling, the error may be negligible but in case of precise leveling, the defective graduations in staff may cause significant error.

   iii. **Error due to defective level tube:** The bubble of the level tube may be in the center, otherwise it may cause error. Sometimes improper curvature of the bubble tube can also cause error.

   iv. **Error due to defective tripod:** The tripod should be set up on a stable and firm ground. The tripod stand should be strong and stable otherwise it takes considerable time to make the instrument level. The nuts provided at the joints of the legs to the tripod head should be well-tightened before mounting the instrument.

b. **Personal errors:** These include;

   i. **Due to imperfect temporary adjustment of the instrument:** These errors are caused due to careless setting up of the level, improper leveling of the instrument, lack in focus of eyepiece or/and objective and error in sighting of the staff.

   ii. **Error in sighting**: This occurs when the horizontal cross-hair does not exactly coincide with the staff graduation or it is difficult to see the exact coincidence of the cross hairs and the staff graduations. The error can be minimised by keeping the small sight distance.

   iii. **Error due to staff held incorrectly:** If the staff is not held vertical, the staff reading obtained is greater than the correct reading. To reduce the error, the staff should be held exactly vertical taking help of a bubble level tube.

   iv. **Error in reading the staff:** These errors occur if staff is read upward, instead of downward, read against the top or bottom hair instead of the central hair, mistakes in reading the further graduations wrongly.

   v. **(v) Error in recording:** The common errors are entering a wrong reading (with digits interchanged or mistaking the numerical value of a reading), recording in wrong column, e.g., BS as IS, omitting an entry, entering the inverted staff reading without a minus sign etc.

vi. **Error in computing:** Errors may cause by adding the fore sight reading instead of subtracting it and or subtracting a back sight reading instead of adding.

c. **Error due to natural causes:** These may include;

d. **Error due to Earth's curvature**: In case of small sight distance, error due to the curvature is negligible, but if the sight distances are large, the error should be accounted for. However, the error can be minimized through balancing of sight or reciprocal observations. The combined error due to curvature and refraction (e_{comb}) is thus given by:

$$e_{comb}\ (m) = 0.0675\ D^2 \qquad (1.9)$$

where D is the distance in km

It is finally subtractive in nature as the combined effect provides increase in staff reading.

i. **Error due to wind:** Strong wind disturbs the leveling of an instrument and verticality of staff. Thus, the levelling work should not be performed in strong winds.

ii. **Error due to sun:** Due to bright sunshine on the objective, sometimes the staff reading cannot be read properly. To avoid such error, it is recommended to cover objective lens with a shed.

iii. **Error due to temperature:** Temperature of the atmosphere disturbs setting of parts of instrument as well as causes fluctuation in the refraction of the intervening medium. These lead to error in staff reading. The instrument therefore is placed under the shade or survey umbrella is used.

1.16 Contouring

A map represents 3D Earth surface onto 2D paper. While, the planimetric position of any object/feature is depicted by their latitude and longitude, the third dimension (i.e., elevations) on the maps are represented by the contours. Contour line is an imaginary line on a map, drawn through the points of equal elevation (Figure 1.33). In addition, there might be few BMs present on the maps whose elevations are known. The elevations of various objects/features can be obtained by levelling, and the spot levels of well distributed points to provide a fair representation of minimum and maximum elevations in the area. Their utility in the present form is, however, limited as they don't convey a meaningful information about the variation of topography. The color; hachures, shading, and tinting have also been used to show the relief, but these methods are not quantitative enough, and thus are generally not suitable for surveying and engineering work.

Contour lines, on the other hand, represent much more valuable information about third dimension on the map. They also give a fairly good idea about the topographic slope and the characteristics of terrain present. If a level surface at an elevation is intersected by the ground surface, the outer shape of the ground depicts the contour line at that elevation. The best example of a contour line is the outer line of a lake. Various contour lines at different heights can be drawn on a map by selecting a suitable interval, called *contour interval.*

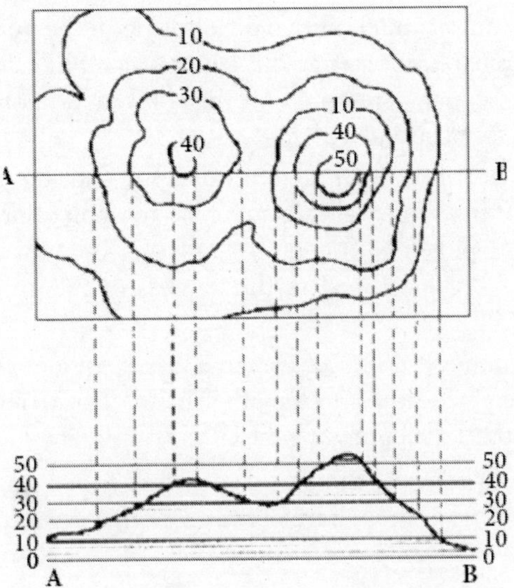

Fig.1.33: Representation of contours

1.16.1 Contour interval

The constant vertical distance between the successive contours is called the contour interval. Smaller the contour interval, greater will be the fieldwork to collect the elevation data. A contour interval sufficient to show features in a hilly region would be insufficient in flat regions. A contour interval for an area/project depends upon several factors, such as the nature of ground, the purpose and extent of survey, the scale of mapping, and the time & funds available.

In a fairly flat area, a large map scale may be appropriate with small contour interval to depict the small ground variation, while in hilly areas, a small scale map with large contour interval will be adequate for representation of details. The contour interval therefore is inversely related to the scale of the map. Small scale maps require large contour interval otherwise the amount of work will tremendously increase. As an example, the recommended values of the contour intervals are as follows:

For building sites: 0.5 m

For reservoirs, town planning schemes: 1 to 2 m

For location surveys: 2 to 5 m

For topographical surveys: 5 m and above.

1.16.2 Factors deciding the contour interval

Contour interval on a map is decided by the following factors:

a. **Scale of the map:** The contour interval is kept inversely proportional to the scale of the map. Smaller the scale of map, larger the contour interval. On the other hand, if the scale of the map is large, the contour interval should be small. If, on a small scale map a small contour interval is taken, the horizontal distance between two consecutive contours, known as *Horizontal equivalent,* would also be small and these contours might unite together. It therefore necessitates selecting the large contour interval on small-scale maps.

b. **Purpose of the map:** The contour interval also depends upon the purpose for which the map is to be utilised. If the map is prepared for setting out a highway on hill slopes, a large contour interval might be required. But, if the map is required for the construction of a university campus, a small contour interval is needed for accurate work.

c. **Nature of the ground:** The contour interval depends upon the general topography of the terrain. In flat ground, contours at small intervals are surveyed to depict the general slope of the ground, whereas high hills can be depicted with contours at larger contour interval. In other words, the contour interval is inversely proportional to the flatness of the ground i.e., steeper the terrain, larger the contour interval.

d. **Availability of time and funds:** If the time available is less, greater contour interval is selected to complete the project in the given time. On the other hand, if sufficient time is available, a smaller contour interval might be taken, keeping in view all the other factors as already described.

It may be noted that contour interval should be such that depending upon the scale of the map, purpose of the map, availability of time and nature of the ground, correct topography of the terrain may be depicted clearly.

1.16.3 Characteristics of contour lines

1. Concentric closed contours with decreasing values towards centre indicate a pond or depression.
2. Concentric closed contours with increasing values towards centre indicate a hill.
3. The same contour would appear on either side of a ridge or a valley.
4. Contour lines with the concavity (inverted V-shape) towards the higher ground indicate a ridge.
5. Contour lines cross with convexity (V-shape) towards the higher ground indicates a valley.
6. Closed contours indicate a steeper slope of the ground.
7. Irregular contours indicate uneven surface.
8. Equally spaced contours indicate uniform slope of the ground.
9. If the contours are straight, parallel and widely spaced, the ground is fairly level with gentle slope.
10. The shortest distance between successive contours indicates the direction of steepest slope, i.e., the steepest slope at a point on a contour is therefore at right angles to the contour.
11. The slope between two adjacent contour lines is assumed to be uniform.
12. A contour cannot branch into two contours at the same elevation.
13. Contours can't have sharp turns.
14. Contour lines of the same elevation can't merge together or intersect, except only in case of a vertical cliff, an overhanging cliff or a cave.
15. A contour must close, not necessarily in the limits of a map.
16. The contour lines don't cross the water surface, (e.g., river, pond).

Contour lines of different terrain shapes are presented in Figure 1.34

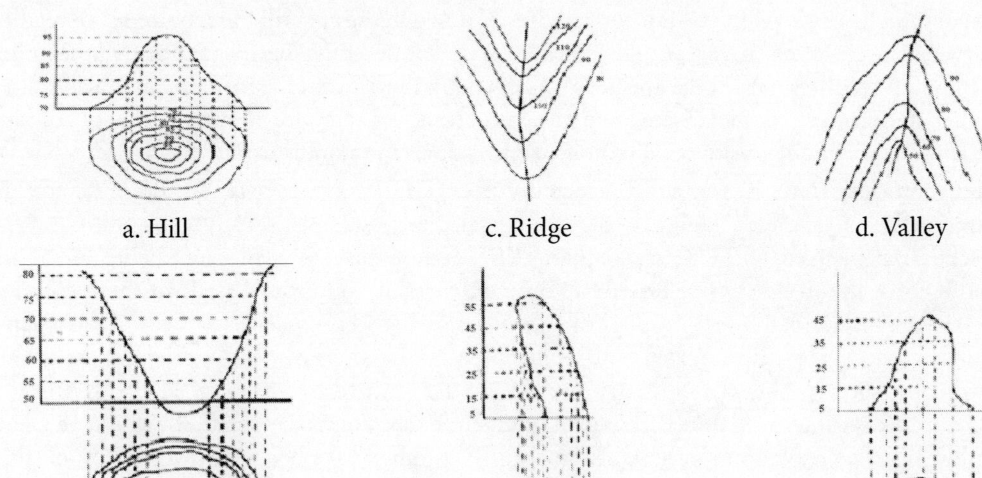

Fig.1.34: Contours depicting various shapes of the ground (e-Krishi Shiksha, 2019)

1.16.4 Uses of a contour map

Maps with contours on them are extremely useful for various engineering projects, as listed below:

1. Contours are most often used to depict the relief. Contours depict the nature and characteristics of the ground, hence useful to identify the suitable site for the project.
2. Profiles of the ground along a line and its cross-section can be drawn using contours. These help in computing the volume of earthwork (i.e., cutting and filling), if the formation level of project is known.
3. The optimum route of the railway, road, canal, pipe line or sewer line can be decided which can minimize the earthwork or balance the earthwork.
4. Intervisibility between any two points can be ascertained from the longitudinal profile of the ground, which is very important in many projects, such as airport planning.
5. Catchment boundary of a river can be drawn and area determined. It helps in determining the quantity of water available at any point along a river. Such study is very important in deciding the location of the bunds, dams, etc.
6. From the contours, the capacity of a reservoir is determined.

1.16.5 Methods of contouring

There are two main methods of surveying:

a. **Direct method-** This method finds the elevation of points in the area which lie on the selected contour line. The level instrument is set at a commanding position in the area and the elevation of line of collimation of instrument is determined. The required staff reading for a given contour line is then computed. The levelling staff is moved up and down in the area till the required staff reading is observed. The elevation of that point is marked on the plotted map. In this manner, all the points with same staff readings in the area are established, and in the process the instrument is also shifted to another station to cover more area. The same process is repeated for the next value of contour. This method is accurate but slow and tedious, hence suitable for small areas.

b. **Indirect methods**- Here, RLs of some selected points (also known as *spot levels*) are taken. The contour lines are interpolated using the RLs of these selected points. For selecting the points, several approaches may be used: (*i*) Method of squares, (*ii*) Method of cross-section, and (*iii*) Radial line method. Out of the three approaches, first approach is most commonly used. In the method of square, the area is divided into a number of squares and all grid points are marked on the ground at regular interval. The grid interval is to be decided by the surveyor, depending on the terrain and scale of mapping. The RLs of all grid points are determined by levelling. The square grid is plotted on the drawing sheet, and RLs of grid points marked. Points of specific elevation value are drawn by linear interpolation method. These points are joined together with smooth contour line and its elevation value is written along the contour line. French curves may be used for drawing the smooth lines. To provide, more readability to contour values, every fifth contour line is made slightly thicker.

Contours are drawn on maps by interpolating between points whose positions and elevations have been measured and plotted. Although, computerized mapping and contouring systems are replacing the manual plotting methods, but the principles of interpolating contours are still basically the same.

1.16.6 Digital elevation models (DEMs)

Elevation data form one of the important sources of ancillary data, which is most widely used in Earth-related investigations. The most common form for representation of elevation data in digital format is the grid type or raster format. The elevation matrix consists of a well- known regular grid of elevations representing real topography. Several names are in current use, including Digital Terrain Model (DTM), Digital Elevation Model (DEM), Digital Elevation Data (DED) and Digital Ground Model (DGM).

A DEM is an ordered array of numbers that represents the spatial distribution of elevations above some arbitrary datum in the landscape. It is a numerical representation of the spatial variation in land-surface elevation, which represents the land-surface as a matrix of elevation values (Z) implicitly located by their geographic coordinates (X, Y). Any point in a DEM can be related to its neighboring cells if the data storage is regular. While DEM consists of an array of values that represent topographic elevations measured at equal intervals on the Earth surface, the Digital Terrain Model (DTM) comprises of any terrain data. A DTM is a topographic map in digital format, consisting not only of a DEM, but also the types of land use, settlements, types of roads and drainage lines and so on.

Applications of DEM

Contours from topographic maps, stereo-air photos or stereo-satellite images can be used to generate DEM. The requirements of elevation and positional accuracy at different scales are given in Table 1.5. Once the DEM is in digital form, large number of other products may be derived from it useful for road planning and layout purposes.

Table 1.4: Requirements of contour interval for standard maps (Garg, 2021)

Map scale	Positional accuracy (m)	Elevation accuracy (m)	Typical C.I. (m)
1:1,000,000	300	30	100
1:5,00,000	150	15	50
1:250,000	75	8	25
1:1,00,000	30	6	20
1:50,000	15	3	10
1:25,000	8	2	5

DEMs have several applications, which include–

i. Site selection for engineering projects
ii. Analysing and comparing different types of terrain
iii. Selecting the alignment of roads, railways, canals, pipelines, etc.
iv. Cut-and-fill analysis in road design and other engineering projects.
v. Computing terrain parameters (e.g., slope, aspect, profiles, catchment area, etc.) to assist runoff and erosion studies.
vi. Displaying landforms in three dimensions, for design and planning of landscape
vii. Generating drainage network for hydrological studies.
viii. Ascertaining the intervisibility between two points
ix. Improving classifications when combined with satellite data.
x. Storing elevation data for digital topographic maps in national databases.

DEM can be considered analogous to digital remote sensing images, except that each pixel represents an elevation measurement, rather than a brightness value. Various derived products can be obtained from DEM such as -

1. Block diagrams, profiles and horizons
2. Volume estimation by numerical integration
3. Contour maps
4. Line of sight maps
5. Maps of slope convexity, concavity, and aspect
6. Shaded relief maps
7. Drainage network and drainage basin delineation

1.16.7 Area and volumes

The land is always dealt with area. For any engineering project, land is acquired first on the basis of area requirement. In some projects, land may be required along a corridor of defined width (e.g., rail, road, pipeline etc.), while in other projects a large catchment area may be involved, such as flood protection devices, reservoir, etc.). Thus, determining the requirement of area in a project is an essential part of surveying. Similarly, the volume of earthwork involved in projects, like road, rail, canal etc., or capacity of a reservoir, are to be determined by surveying methods.

a. Computation of areas

It is very easy to determine the area of a regular figure (plot of land), such as a triangle, a square, a rectangle, etc. However, most of the time, the project land is irregular in shape with no defined shape, so computation of area is not straight forward in such cases. For this purpose, from a survey line offsets are taken at regular intervals, and area is calculated from any of the following methods:

i. Trapezoidal rule
ii. Simpson's rule
iii. Using planimeter
iv. From coordinates

i. Trapezoidal rule

In trapezoidal rule, the area is divided into a number of trapezoids, (as shown in Figure 1.35) and boundaries are assumed to be straight between pairs of offsets. The area of each trapezoid is determined and added together to compute the entire area. If there are 'n' offsets at equal interval of 'd' then the total area is computed as-

$$A = d\left(\frac{O_1 = O_n}{2} + O_2 + O_3 + ... + O_{n-1}\right) \tag{1.10}$$

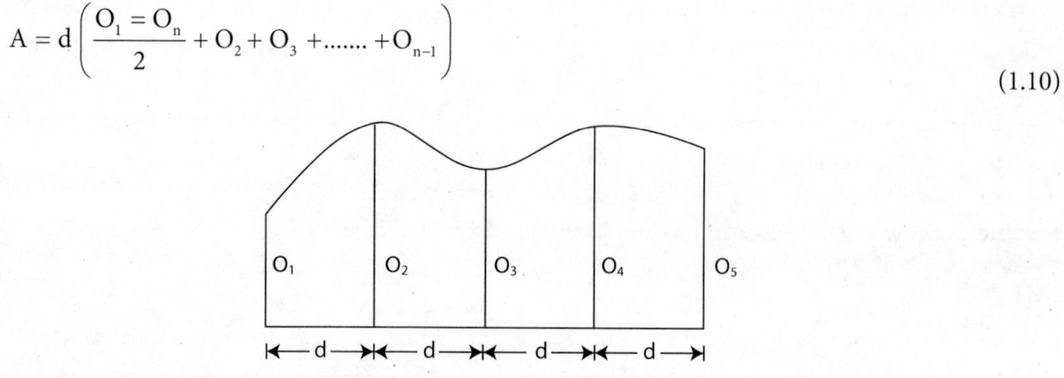

Fig.1.35: Area of trapezoids

ii. Simpson rule

In Simpson's rule, it is assumed that the irregular boundary is made up of parabolic arcs. The areas of the successive pairs of intercepts are added together to get the total area, as below:

$$A = \frac{d}{3}\left[\left(O_1 + O_n\right) + 4\,(O_2 + O_4 + ... + O_{n-2}) + 2\,(O_3 + O_5 + ... + O_{n-1})\right] \tag{1.11}$$

iii. Using Planimeter

Planimeter is a mechanical instrument used for measuring the area of a plan. It consists of a hard steel roller and a disc. The axis of roller coincides with the axis of tracer arm, hence it rolls only at right angles to the tracer arm. The roller carries a concentric drum which has 100 divisions, and is provided with a vernier to read tenth of roller division. A suitable gear system moves a pointer on disc by one division for every one revolution of the roller. Since the disc is provided with 10 such equal divisions, the reading on the integrating unit has four digits: (*i*) Unit read on the disc, (*ii*) Tenth and hundredth of a unit read on the roller, (*iii*) Thousandth read on the vernier. If the reading on disc is 2, reading on roller is 42 and vernier reads 6, then the total reading F = 2.426.

To find the area from a map, anchor pin of the planimeter may be placed either outside the map if the area on the map is small or inside the map if the area on the map is large. It is placed outside the plan. On the boundary of the area, a point is marked and tracer arm is set on it. The planimeter reading is taken. Tracer arm is carefully moved over the outline of the area in clockwise direction till the starting point is reached. The reading of planimeter is noted. Now the area of the plan may be determined as-

$$\text{Area} = M\,(F - I + 10\,N + C) \tag{1.12}$$

where M is the multiplying constant is equal to the area of the plan (map) per revolution of the roller i.e., area corresponding to one division of disc

F is the final reading

I is the initial reading

N is the number of completed revolutions of disc. Plus sign to be used if the zero mark of the dial passes index mark in clockwise direction and minus sign if it passes in anticlockwise direction.

C is the constant of the instrument, which when multiplied with M, gives the area of zero circle. It is added only when the anchor point is inside the area.

Multiplying constants, M and C, are normally written on the planimeter. The user can verify these values by (*i*) Measuring a known area (like that of a rectangle) keeping anchor point outside the area, and (*ii*) Again measuring a known area by keeping anchor point inside a known area.

iv. From coordinates

The area of an irregular closed polygon is usually computed by the coordinate method. In this procedure, coordinates of each turning point in the figure must be known. They are normally obtained by traversing, but any other method (e.g., GPS or LiDAR) that provides the coordinates can also be used. The coordinate method is easily visualized; it reduces to one simple equation that applies to all geometric configurations of closed polygons, and can be easily programmed for obtaining solution through computer-based system. The procedure for computing areas by coordinates can be developed with reference to Figure 1.36 as shown below.

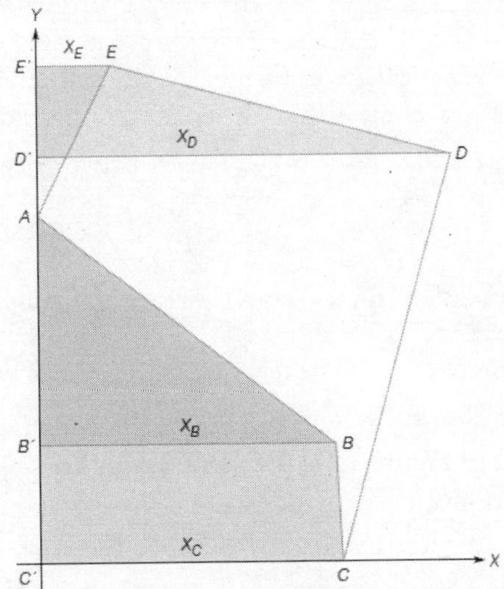

Fig.1.36: Computation of area of closed irregular polygon

$$2(\text{area}) = +X_AY_B + X_BY_C + X_CY_D + X_DY_E + X_EY_A$$
$$-X_BY_A - X_CY_B - X_DY_C + X_EY_D + X_AY_E \qquad (1.13)$$

The above equation can easily be remembered by listing the x and y coordinates of each point in in two columns, as shown below, with coordinates of the starting point repeated at the end. The products noted by diagonal arrows are ascertained with dashed arrows considered plus and solid ones with minus. The algebraic summation of all products is computed, and its absolute value is divided by 2 to get the area.

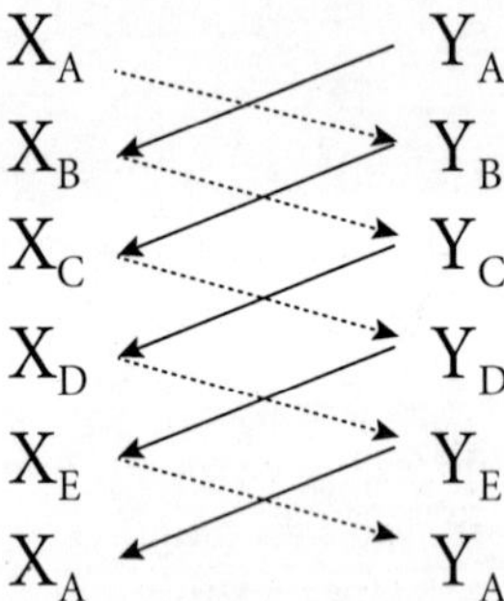

b. Computation of volumes

The computation of volumes from the measurements taken in the field is required in design and planning of several engineering works. For example, volume of earthwork is required for feasible alignment of road, canal and sewer lines, etc. The computation of volume of various materials, such as coal, sand, gravel is required to check the stockpiles. For estimation of volume of earthwork, normally the cross sections are taken at right angles to a fixed alignment, which runs continuously through the earthwork. The spacing of the cross sections will depend upon the accuracy required. The volume of earthwork is computed from these cross-sections, using either the Trapezoidal rule or Prismoidal rule, as given below.

i. Trapezoidal Rule

It is also known as average or mean sectional area formula. This method is based on the assumption that the mid-area of a pyramid is half the average area of the ends, and the end sections are in parallel planes.

If A_1 and A_2 are areas of the ends and d is the length between two sections, the volume of the prismoid is given by-

$$V = d/2 \,[\, A_1 + A_n + 2\,(A_2 + A_3 + A_4 + A_5 + \ldots\ldots\ldots\ldots + A_{n-1}) \qquad (1.14)$$

where '*n*' are number of segments at interval of '*d*', Area at L = *nd*, being A*n*.

ii. Prismoidal Rule

The prismoidal formula can be used to compute volume of all geometric solids that are considered as prismoids. A prismoid, as shown in Figure 1.37, is a solid having ends that are parallel but not similar and trapezoidal sides that are also not congruent. Most earthwork solids obtained from cross-section data fit into this classification.

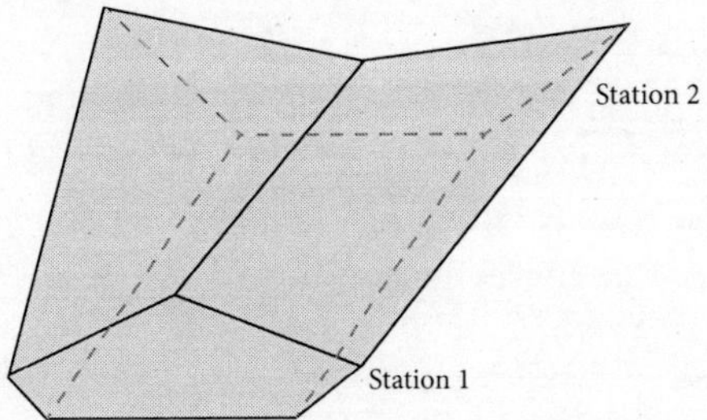

Fig.1.37: Prismoidal section

$$V = d/3\,[(A_1 + A_n) + 4\,(A_3 + A_5 + ... + A_{n-1}) + 2\,(A_2 + A_4 + ... + A_{n-2})] \quad (1.15)$$

where n is number of even segments.

1.17 PLANE TABLING

Plane Table is called the father of "Mapping". It is used to prepare various maps in the field. It is a graphical method, where observations, mapping and contouring, proceed simultaneously. It is a time-consuming and laborious process to create a map, but is a more accurate approach. The plane table consists of a wooden drawing board with arrangement of fixing it on a tripod. With the availability of Total Station, GPS, satellite images and LiDAR data, Plane tabling methods of surveying are almost obsolete now-a-days.

1.17.1 Advantages and disadvantages of plane table surveying

Advantages

a. The map is prepared in the field; therefore, the office work is minimal.

b. The observations and plotting are done simultaneously, and the surveyor can see the terrain before him/her, so chances of losing the details are rare.

c. Errors in plotting the details on plane table can be checked by drawing the check lines using several methods.

d. The angles and distance measurements can be obtained graphically, and hence there is no need to carry out measurements in the field.

e. Many far-off objects/details can be plotted accurately using intersection method to prepare the map in shortest time.

f. It is advantageous in magnetic areas where compass survey is not reliable.

Disadvantages

i. Plane tabling work is dependent on weather conditions (e.g., rains, high winds).

ii. The plane table is not very much suitable in a dense forest area or urban area.

iii. If survey map is to be replotted at some different scale, entire work has to be re-done.

iv. Its accessories are heavy to carry in the field from one location to another.

v. The paper maps are subject to shrinkage or expansion while working in the field for long time.

vi. It is laborious and time-consuming to prepare map of a large area

vii. It requires large manpower for doing various activities from observations to plotting.

1.17.2 Methods of plane tabling

There are three steps to work with plane table; (i) Levelling the plane table, (ii) Centering the plane table, and (iii) Orienting the plane table. Levelling is done using the sprit bubble tube. Centering is done with the help of plumb bob. Orientation is done either with a magnetic compass or with a back ray method. The equipment required for plane table surveying are shown in Figure 1.38:

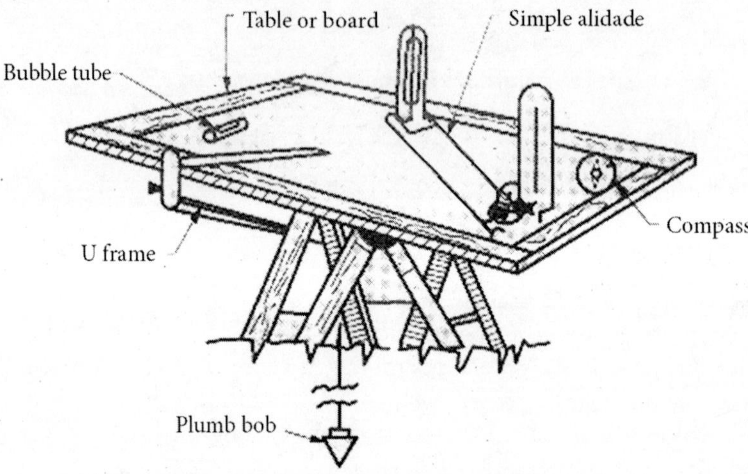

Fig.1.38: Accessories of plane table survey

There are four methods of plane table surveying:

i. Radiation,
ii. Intersection,
iii. Traversing and
iv. Resection.

The radiation method uses of a distance and a direction to locate the object on the map. The plane table is set up at a station and various points in the vicinity are located on the map by measuring the distances to the different stations using a tape (or tacheometry), and subsequently radiating (drawing) the rays from the instrument station to those points which are to be located on the map. The intersection method is used extensively to locate an object by two directions from two stations at the end of a base line of known length. The method is widely used for plotting the far-off details on the maps. From the first station of a base line, directions are drawn to the various stations, and the same is done by shifting the plane table at other end of base line. The intersections of these two directions of the same object will give the location on the map. The traversing method is used in areas where visibility is obstructed by high-rise building/forest or topography. Plane table occupies each traverse station and details are plotted either by radiation method or intersection method. The resection method is used to accurately locate the position of ground station (unknown point) on the map, occupied by plane table. It is done by drawing the rays from two or three already known and plotted point on the map. Once the position of ground station is plotted on the plane table, any of the method for plotting the details could be used. For resection methods, two-point problem or three-point problem may be used.

Now-a-days, plane tabling is considered as obsolete, and therefore is not carried out in the field. The field data is directly collected and stored using either Digital Levels, GPS or Total Station, and mapping is carried out using the capabilities of the software in a computer. Alternatively, photogrammetry and remote sensing images are used to create various types of maps required in various civil engineering projects.

1.18 THEODOLITE SURVEYS

Theodolite is the most versatile instrument used in survey work. It is a multi-purpose equipment which can be used for measuring the angles, prolongation of a straight line, determination of levels and determination of distances. In general, it is used for measuring the vertical and horizontal angles of given objects. Inception of a theodolite was done basically for providing the horizontal and vertical controls in survey work, i.e., the coordinates. For mapping a larger area, locations of ground controls are essentially required which can be determined using angles and other observation.

1.18.1 Various types of theodolites

There are various models of theodolites, such as:

a. Vernier theodolite

b. Optical theodolite

c. Electronic theodolite (or Total Station)

Vernier theodolite is a commonly used instrument for measuring the horizontal and vertical angles. It can also be used for prolonging a line, levelling work, determining the indirect distances (through Techeometry), determining the elevations of distance objects (Trigonometrical levelling). In a vernier theodolite, readings for vertical and horizontal angles can be read through main scale and vernier scale. It has two verniers for horizontal readings and another two verniers for vertical readings, and using these verniers, both the angles can be read accurately up to 20".

Precise theodolites use optical principle for more accurate results. In optical theodolites, vertical and horizontal circles are made of glass, and the lowest readings are read through a micrometer. Least count ranges from 0.2 to 10 seconds. Now-a-days electronic theodolites (or Total Stations) are used which read and display the angles and other measurements. The electronic theodolite has an opto-electronic system. The encoders count the pulse of the movement and displays the measurements digitally, with the least count from 0.1" to 5". The electronic device of digital display provides less fatigue with less chances of reading errors. Details of Total Stations are given in Module 3.

1.18.2 Various parts of a vernier theodolite

Figure 1.39 shows a sectional view of a typical vernier theodolite. The main parts of a vernier theodolite are:

a. **Telescope:** A telescope, which is mounted on a horizontal axis (trunnion axis), can rotate in the vertical plane. Its function is to provide a line of sight for measuring angles.

b. **Upper plate and Lower plate:** There are two horizontal circular plates in vernier theodolite (Figure 1.39). On lower side, the upper plate is attached to an inner spindle which rotates in the outer spindle of lower plate. Using upper clamping screw, upper plate can be clamped to lower plate. Using slow motion tangent screw, slight relative motion between the two plates can be given, even after clamping the plate. Two diametrically opposite verniers, A and B, are fixed to upper plate. These are provided with magnifying glasses, and used in reading horizontal circle readings with an accuracy of up to 20". The lower plate, attached to the outer spindle carries a graduated circle from 0 to 360°. Graduations are up to an accuracy of 20′. It can be clamped at any desired position using lower clamp screw. If upper clamp is locked and the lower one is loosened, the two plates will rotate together. But, if the upper clamp is loosened and lower clamp is locked, upper plate alone will rotate. This arrangement is utilised in measuring the horizontal angle as well as setting up the initial reading as 00° 00' 00".

c. **Vertical circle:** A vertical circle, graduated up to an accuracy of 20', is rigidly connected to the telescope, and hence it moves along with the rotation of the telescope in vertical plane. It is used to read vertical angles. The graduations on vertical circle are in quadrantal system, 0° line being horizontal.

d. **Plate bubble tube:** The plate bubble tube is mounted on the upper plate. It helps in making the vertical axis of the instrument truly vertical.

e. **Vernier frame:** It is a T-shaped frame, and is also known as T-frame or Index frame. It consists of a vertical arm of T-shape and a horizontal arm, supporting telescope. With the help of the clamping screw the vertical frame and hence the telescope can be clamped at an angle. The vertical arm it carries verniers C and D to read the graduations on vertical circle. Verniers C and D are provided with glass magnifiers to read angles with magnified view. The A-shape frame supports the telescope, and it allows telescope to rotate on its trunnion axis in vertical plane. The T-frame and the clamps are also fixed to this frame.

f. **Altitude bubble**: An altitude bubble tube is fitted over the horizontal arm. The bubble is brought to center with the help of a screw, before taking each reading of vertical angle.

g. **Levelling head:** It consists of two parallel triangular plates known as *tribrach plates*. The entire instrument rests on these plates. The upper tribrach plate is provided with three levelling screws. By operating these screws, levelling of the theodolite can be done. The lower tribrach can be fitted into a tripod head.

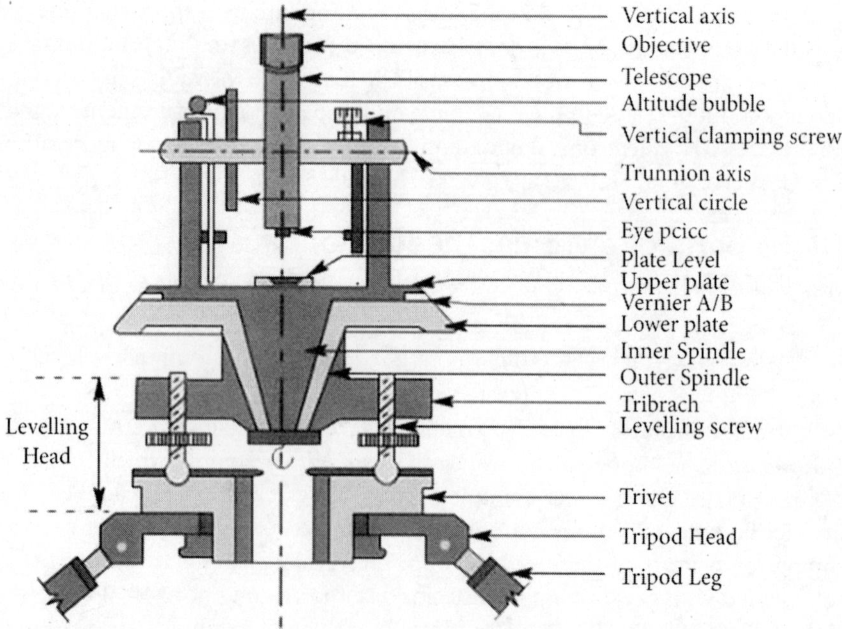

Fig.1.39: Various part of a vernier theodolite (Dangi, 2020)

h. **Shifting head:** It is provided below the lower plate, and used to slide one plate over another over a small area of about 10 mm radius. The two plates can be tightened in the desired position. It facilitates the exact centering of the instrument.

i. **Tripod:** It is just like any other tripod used for fixing the instrument. Theodolite is always mounted on a tripod before its use. At the lower end, the tripod legs are provided with steel shoes to get a good grip with the ground. The top of tripod is provided with external screw to which the lower tribrach plate can be clamped. When not in use, the tripod head is protected with a steel cap provided for this purpose.

j. **Plumb bob:** A plumb bob is a triangular metallic weight which is suspended with a thread and hook with the lower base of the equipment. The pointed tip of the plumb bob indicates facilitates exact centering of the theodolite at a station.

1.18.3 Technical terms

a. **Centering:** It involves setting up the theodolite exactly over an instrument station so that its vertical axis lies immediately above the station mark. It can be done by means of plumb bob suspended from a small hook attached to the vertical axis of the theodolite. Some instruments having shifting arrangement of centre facility can provide easy and rapid centering of equipment.

b. **Line of collimation:** It is also known as the *line of sight.* It is the imaginary line joining the intersection of the cross-hairs of the diaphragm to the optical centre of the object-glass and in its continuation.

c. **Axis of the telescope:** It is also an imaginary line joining the optical centre of object-glass to centre of the eye-piece.

d. **Vertical axis:** It is the axis about which the telescope can be rotated in the horizontal plane (Figure 1.39).

e. **Horizontal axis.** It is also called the *trunnion axis* or the *transverse axis.* It is the axis about which the telescope can be revolved in the vertical plane (Figure 1.39).

f. **Transiting:** It is also known as *plunging* or *reversing.* It is the process of turning the telescope about its horizontal axis through 180° in the vertical plane, thus bringing it upside down and making it point exactly in opposite direction.

g. **Swinging:** It involves turning the telescope about its vertical axis in the horizontal plane. A swing is called *right swing* if the telescope is rotated clockwise, whereas the *left swing* of the telescope involves the rotation in anticlockwise direction.

h. **Face left:** If the vertical circle of the instrument is on the left of the observer while taking a reading, the position is called *face left* and the observations taken on horizontal or vertical circle are known as *face left observations.*

i. **Face right:** If the vertical circle of the instrument is on the right of the observer while talking a reading, the position is called *face right,* and the observations taken on horizontal or vertical circle are known as *face right observations.*

j. **Changing face:** It is the operation of changing the vertical circle from left to the right of observer, and vice-versa. It is done in two steps: transit the telescope and then swing it. Firstly, the telescope is revolved through 180° in a vertical plane and then rotated through 180° in the horizontal plane.

k. **One set of observations:** The combined observations taken on face left as well as face right make one set of observations.

l. **Fundamental axes**: There are three fundamental axes of a theodolite; the vertical axis, the axis of telescope or line of collimation, and the horizontal axis or trunnion axis. In an adjusted theodolite, all the above three axes are mutually perpendicular to each other.

1.18.4 Measurement of horizontal angles

Theodolite is generally used for accurately measuring the horizontal and vertical angles. For this, the theodolite is centered on the desired station point, levelled and telescope is focused. This process of centering, levelling and focusing is called *temporary adjustment of the instrument*, and it is similar to any optical telescope, such as levels. Now the object is bisected at the intersection of cross-hairs so that both the horizontal and vertical angle readings are taken. The reading is read from the main scale as well as vernier scale, and both are added together to get final reading. The vernier scale is a small scale used for determining the fractional parts of the smallest division of the main scale more accurately than it can be done by simply estimating with eye. The vernier scale carries an index mark (arrow) which represents the zero of the vernier divisions.

There are two methods of measuring the horizontal angles:

a. Reiteration method, and
b. Repetition method

a. Reiteration method

Reiteration method is generally preferred for measuring several horizontal angles around the instrument station. It measures several angles in continuation, and finally closes the horizon at the starting point. These angles are measured on both left face and right face. The final readings of the verniers should be almost same as their initial readings. If error is observed, it is equally distributed among all the measured angles.

Suppose it is required to measure the angles AOB, BOC and COD by reiteration method, as shown in Figure 1.40a. The steps involved are:

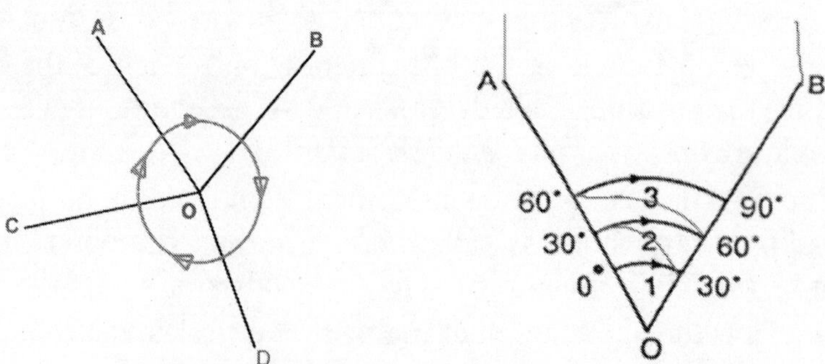

Fig.1.40: (a) Reiteration method (b) Repetition method

i. Set up the instrument over station O and centre it and level it accurately.

ii. Make sure that the instrument face is left face.

iii. Set the vernier A to 0° by using the upper clamp and its tangent screw.

iv. Direct the telescope to object A, and bisect it accurately by using the lower clamp and its tangent screw. Ensure that the reading at vernier A is still 0°. Note down the reading in vernier B also.

v. Loosen the upper clamp and turn the telescope clockwise until the point B is exactly bisected by using the upper tangent screw. Read both the verniers. The mean of the two vernier readings (after deducting 180° from the reading at vernier B) will give the value of the horizontal angle AOB.

vi. Similarly, bisect objects C and D successively, read both the verniers at each bisection, and find the value of horizontal angles BOC and COD.

vii. Finally, close the horizon by sighting towards point A.

viii. The vernier A should have the same reading as the starting reading, but generally it is observed that there is a small error. This error is either equally distributed in all the angles or distributed in individual angle in proportion to its magnitude. If the error is very large, all the readings are discarded and a new set of readings is taken.

ix. Change the face of the instrument from left to right.

x. Again measure the horizontal angles in the same manner (as with face left) by turning the telescope in anticlockwise direction.

xi. Find the closing error, and apply the necessary correction to all the horizontal angles in face right.

xii. The mean of the two results for each angle (face left and face right) is taken as its true value.

The observation table is shown in Table 1.6.

Table 1.5: Reiteration method of recording observations

Inst. at	Sighted to	Face left								Swing Right			Face right								Swing left			Average horizontal angle		
		A			B		Mean			Horizontal angles			A			B		Mean			Horizontal angles					
		°	'	"	'	"	°	'	"	°	'	"	°	'	"	'	"	°	'	"	°	'	"	°	'	"
0	P	00	00	00	00	00	00	00	00	43	53	10	00	00	00	00	00	00	00	00	43	53	10	43	53	10
	Q	43	53	10	53	10	43	53	10	54	30	50	43	53	00	53	00	43	53	10	54	30	50	54	30	50
	R	98	24	00	24	00	98	24	00	126	47	20	98	24	00	24	00	98	24	00	126	47	20	26	47	20
	S	225	11	20	11	20	225	11	20	134	46	40	225	11	20	11	20	225	11	20	134	48	40	134	48	40
	P	360	00	00	00	00	360	00	00				360	00	00	00	00	360	00	00						

b. Repetition method

This method is used to determine very accurate horizontal angle. It is preferred when only one or two angles are to be measured around the instrument station. Here, the same angle is measured several times and the correct value of the angle is obtained by dividing the accumulated reading with the number of repetitions. The minimum number of repetitions made is usually three, individually on face left and face right. This way, horizontal angle can be measured to a finer degree of accuracy than the least count of the vernier.

Suppose a horizontal angle AOB is to be measured by repetition method (Figure 1.40b):

i. Set up the theodolite at station-point O and centre it and level it accurately.
ii. Make sure that the face of the instrument is left face.
iii. Set the vernier A to 0° by using the upper clamp and its tangent screw. Then loosen the lower clamp, direct the telescope to bisect object A by using the lower clamp and its tangent screw. Ensure that the reading of vernier A has not changed. Note down the reading in vernier B also.
iv. Loosen the upper clamp, bisect the object B by clockwise movement of telescope and using the upper clamp and its tangent screw.
v. Read both the verniers, and take the mean value.
vi. Loosen the lower clamp and turn the telescope anitclockwise until the object A is bisected again. Check the vernier readings which must be the same as at point B.
vii. Loosen the upper clamp, turn the telescope clockwise and again bisect B using the upper tangent screw. The verniers will now read nearly twice the value of previous angle.
viii. Repeat the processes (vi and vii) until the angle is measured with 3 repetitions. Read both verniers, and determine the average value of angle AOB by dividing final value with the number of repetitions (in this case 3).
ix. Change the face of the instrument to right face. Repeat steps (iii to viii) and determine one more value of angle AOB.
x. The average of the two values of the angle (one on face left and another on face right) will give the precise value of angle AOB.

The observations are recorded in the tabular from as given in Table 1.7.

Table 1.6: Repetition method of recording observations

Inst. at	Sighted to	Face left										Swing Right			Face right										Swing left			Average horizontal angle		
		A			B		Mean			No. of repetition	Horizontal angles			A			B		Mean			No. of repetition	Horizontal angles					
		°	'	"	'	"	°	'	"		°	'	"	°	'	"	'	"	°	'	"		°	'	"	°	'	"
O	P	00	00	00	00	00	00	00	00					00	00	00	00	00	00	00	00							
	Q	56	43	20	43	20	58	43	20	1	58	43	20	58	43	40	43	40	58	43	40	1	58	43	40			
	Q	168	19	40	19	40	168	19	40	3	58	43	33	168	19	20	19	20	168	19	20	3	58	43	27	58	43	30

1.18.5 Measurement of vertical angles

A vertical angle is an angle between the inclined line of sight and the horizontal. It may be an angle of elevation or depression, according as the object is above or below the horizontal plane. In both the methods of measuring horizontal angles, if the object is bisected at the intersection of cross hairs (centre), vertical angles can also be read at the same time on verniers C and D. The average of both the verniers readings are taken. It is important that the altitude bubble is in the centre for each reading of vertical angle. After taking vertical angles at face left and face right, mean value is adopted. The observations of vertical angles are also recorded in the same manner as the horizontal angles readings.

To measure the vertical angle of an object A at a station O (Figure 1.41a), the following steps are followed:

i. Set up the theodolite at ground point O and centre it and level it accurately. The instrument should be on face left.
ii. Set the zero reading at verniers of vertical circle by using the vertical clamp and tangent screw. The line of sight is thus made horizontal.
iii. Loosen the vertical circle clamp screw and bisect object A exactly by using the vertical circle tangent screw. Bring the bubble in the altitude level in centre position by using the screw.
iv. Read both the verniers C and D on the vertical circle. The mean of the two vernier readings gives the value of the required vertical angle AOA' directly.
v. Change the face of the instrument from left to right and repeat the process at (iii) and (iv) above. Thus, one more value of vertical angle AOA' on face right.
vi. The average of the two values of the angle is taken which is the required value of vertical angle. The vertical angle could be angle of elevation (+ive) of angle of depression (-ive), as shown in Figure 1.41b.

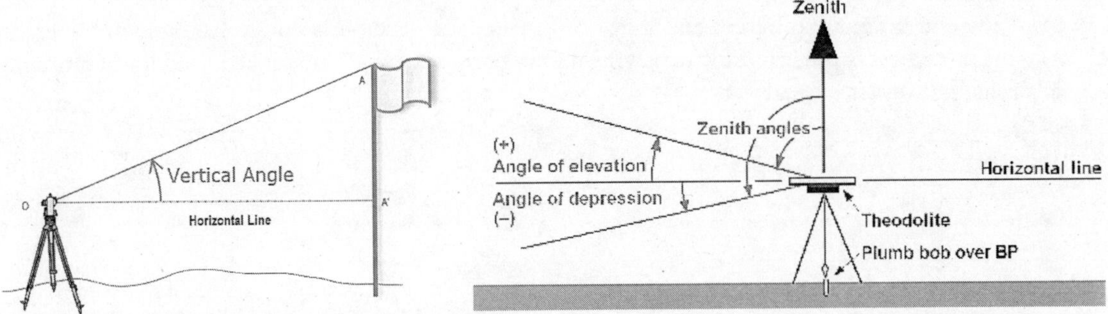

Fig.1.41: (a) Measurement of vertical angle AOA', and (b) Angle of elevation and angle of depression

1.18.6 Prolonging a straight line

Theodolite can be used to prolong a given line, such as in alignment surveys. The procedure to prolong a line AB up to Z (say) is given below (Figure 1.42).

i. Set up the theodolite at A and centre it and level it accurately.
ii. Bisect point B accurately by keeping a survey flag rod.
iii. In the same line of sight, establish a point C at some convenient distance, away from B, as shown in Figure 1.42. Use theodolite to bisect so that vertical hair of the diaphragm overlaps with point B and point C.
iv. Shift the theodolite at B, take a fore sight on C and establish another point D at some convenient distance in line beyond point C.
v. Repeat the above processes until the last point (Z) is established.

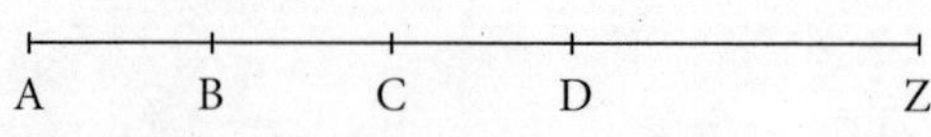

Fig.1.42: Prolonging a line with a theodolite

1.18.7 Levelling with a theodolite

The theodolite can be used as a level, if the motion of the telescope in vertical plane is clamped such that the vertical circle verniers, C and D, read zero value. If the instrument is in perfect adjustment, the line of sight will be horizontal when the bubble is in its central position and vertical angle reading is zero. The instrument is now ready to be used as level, the level difference between two points can be determined by taking the levelling staff reading at both these points.

1.18.8 Traversing with a theodolite

Traversing is that type of survey in which a number of connected survey lines form the framework and the directions and lengths of the survey lines are measured with the help of an angle measuring and distance measuring instrument. There are two types of traverse surveying (Figure 1.43):

a. **Closed traverse**: when the interconnected lines form a closed figure and the last line ends at the beginning of first line, it is known as a closed traverse. The closed traverse is generally used for locating the boundaries of area, lakes, forest, and survey of large areas.

b. **Open traverse**: when the interconnected lines cover a large area along a corridor, and start point form and end point are different and they don't meet each other, it is said to be an open traverse. The open traverse is preferred for surveying a long narrow strip of land as required for alignment of a road, railway line, canal, etc.

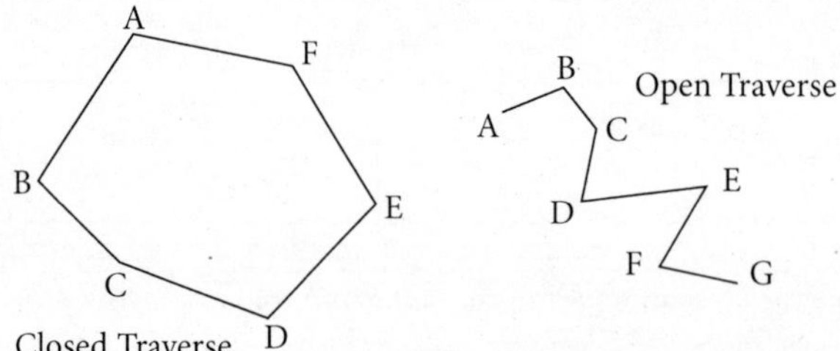

Fig.1.43: Closed traverse (left) and open traverse (right)

In theodolite traversing, the field work consists of (i) reconnaissance survey of area, (ii) selection of traverse stations, (iii) marking and locating the traverse stations, (iv) taking observations for traverse lines and angles, (v) adjusting the errors, (vi) computation of coordinates of traverse stations, (vii) plotting the coordinates of traverse stations, and (viii) take more observations of surrounding features to plot them on a map. A theodolite is normally used for determining the angles of the traverse lines, and an EDM or tacheometry is commonly used for taking the linear measurements of traverse lines.

1.18.9 Methods of theodolite traversing

The measurement of angles between two successive lines is generally carried out for large traverses where the high degree of accuracy is required. In this method, the angles between the successive lines are measured, and the bearing of a line is observed. The bearings of the remaining lines can be computed from the observed bearing and angles. In a closed traverse, the angles measured are either exterior or interior, accordingly the traverse is run in a clockwise direction or in anti-clockwise direction (Figure 1.44). The sum of the two angles (exterior and interior) at a traverse station is 360°. The angles can be measured by the method of repetition for greater precision. The distance between traverse sides is determined either by tacheometry (if it is short) or EDM/Total Station (if it is long). The closed traverse is preferably run in anti-clockwise direction.

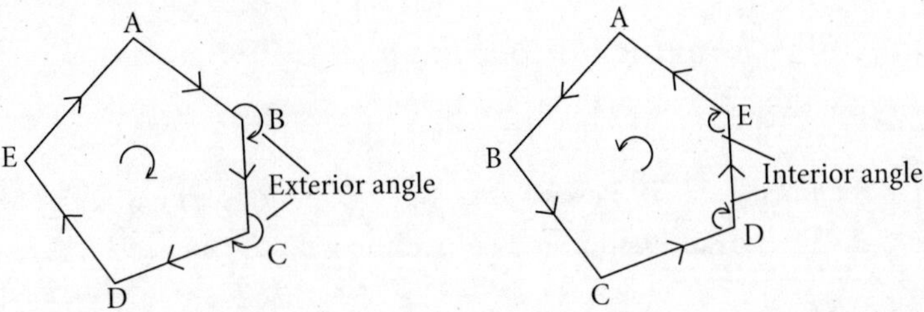

Fig.1.44: Exterior angles (clockwise traverse) and interior angles (anti-clockwise traverse)

In running the closed traverse ABCDEA, set up the theodolite at station A and measure the angle BAE. Distance AB is also computed by tacheometry or EDM. Also observe the bearing of line AB using magnetic compass. Shift the theodolite at B and measure the angle ABC and side BC. This way, visit all the traverse stations to measure the angles and sides.

In case of an open traverses, deflection angle is measured. Deflection angle is the angle made by a traverse line with the extended preceding traverse line. It can be deflection angle to the right or deflection angle to the left, depending on the direction (anticlockwise or clockwise) of angle measurement, respectively, as shown in Figure 1.45. Angle α_1 is the deflection angle to the right and α_2 is the deflection angle to the left. This is much suitable when the survey lines make small deflection angles with each other as in the case of surveys for roads, railways, pipe lines etc. In running a traverse as in the Figure, set up the theodolite at the starting station A and observe the bearing of line AB. Shift the instrument to station B, set the vernier A reading to zero and take a back sight reading on A. Then transit the telescope, loosen the upper clamp, turn the telescope clockwise and take a fore sight on C. Read both the verniers, the mean of these readings is the required deflection angle α_1 of BC from extended line AB. Also note down the direction of angle measurement (right in this case). Now shift the theodoilte, and set up at C station, and take the observations in the same way as before to measure angle α_2 (direction will be right here). Likewise, the deflection angle along with its direction (left or right) at each station is measured till the last station is covered.

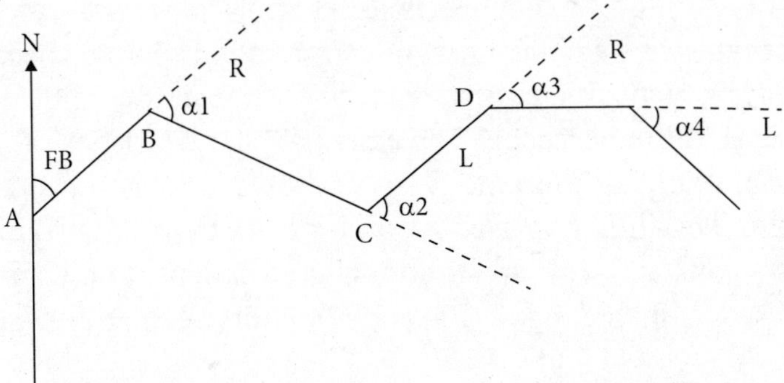

Fig.1.45: Deflection angle measurement in open traverse

1.18.10 Errors in theodolite observations

No observation is free from error. There are errors in theodolite observations also, which are to be minimized and removed. Some errors can be minimized by taking certain precautions while using the instrument and method of observation, while the errors, once they are present, can be adjusted in the observations before using them for any computational work.

I. *Errors eliminated by changing the face the theodolite:*

a. Error due to the line of collimation not being perpendicular to the horizontal axis of the telescope.

b. Error due to the horizontal axis of the telescope not being perpendicular to the vertical axis.

c. Error due to the line of collimation not coinciding with the axis of the telescope.

II. *Errors eliminated by reading both verniers and averaging the readings:*

a. Error due to the axis of the vernier-plate not coinciding with the axis of the main scale plate.

b. Error due to the unequal graduations.

III. *Error eliminated by measuring the angle on different parts of the horizontal circle:*

a. Error due to the unequal graduations

1.19 TACHEOMETRY

In tacheometry, horizontal and vertical distances are determined by angular observations with a techeometer. Tacheometry is more accurate than the chaining/taping, and more rapid in rough and difficult terrain where levelling is tedious and measuring distance by chaining/taping is not only inaccurate but slow and laborious. It is a best suited method when taking observations for steep and undulating/broken ground, river, water or swampy areas. Tacheometry is preferably used for traversing, but it is also used for contouring.

1.19.1 Instruments used

The main instruments used in tacheometry are a tacheometer, and a levelling rod. A tacheometer is a transit theodolite where telescope is fitted with a special diaphragm, called stadia hairs, i.e., a diaphragm fitted with three horizontal hairs; one at the top, another in the middle and third at the bottom of diaphragm. These horizontal hairs are equidistant from the central one. The types of stadia diaphragm commonly used in tacheometers are shown in Figure 1.46. The term tacheometer is restricted to a transit theodolite which is provided with an anallactic lens in the telescope. The essential characteristics of a tacheometer are that the value of the multiplying constant ($K = f / I$) should be 100 and additive constant ($C = f + d$) should be zero. Levelling rod used is similar to as used in levelling work. To make the value of additive constant zero, an additional convex lens, known as anallatic lens, is provided in the telescope. By having K=100 and C=0, the calculation work is considerably reduced.

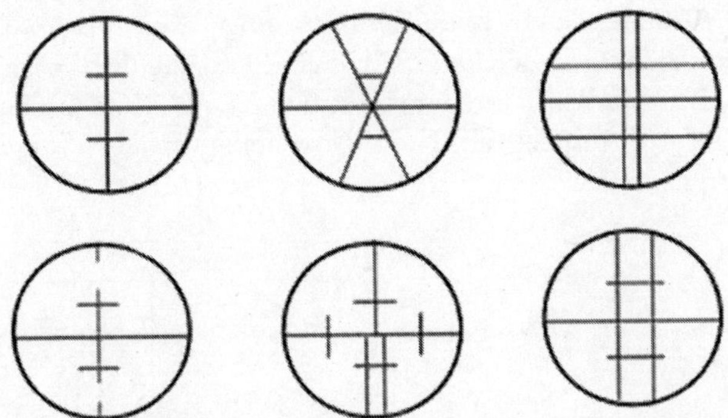

Fig.1.46: Stadia diaphragm commonly used in tacheometers

1.19.2 Methods of tacheometry

The principle used in tacheometry is that the horizontal distance between an instrument-station and a point where levelling rod is kept can be determined with the staff intercept (difference of top reading and bottom reading). For determining the horizontal distance between two points, tacheometer is kept at one point and levelling staff is kept at another, and stadia readings on the levelling staff are read.

Case I - When the line of sight is horizontal

The staff intercept reading is multiplied by the instrument constant (K) and added with an additive constant (C) to get the horizontal distance between tacheometer and levelling staff. The horizontal distance (D) is computed as;

$$D = K\,S + C \tag{1.16}$$

Where K is the multiplying constant (usually 100), S is the staff intercept and C is the additive constant (usually zero). In a simplified from, the above equation can be written as-

$$K = 100S \tag{1.17}$$

There could be situations when the ground is undulating and the levelling staff is either above or below the line of sight. In such cases, a vertical angle subtended by the point at instrument station is also measured, in addition to stadia readings, to determine the horizontal distance using trigonometrical relationship. This is a fast method to determine the horizontal distance.

The vertical distance (elevation difference) between instrument and levelling staff can also be determined, if we take one more levelling staff observation at the BM.

RL of levelling staff point = RL of instrument axis– Central hair staff reading (1.18)

Case II - When the line of sight is inclined

In case the ground is undulating and horizontal sights are not possible, inclined sights are taken. In this case, the staff may be held either vertical or normal to the line of sight. In general, most commonly adopted method is when the staff is held vertical as it is simpler in calculation.

When the staff is held vertically.

In Figures 1.47 and 1.48, the staff is held vertical; in one case the point is at higher elevation and in other case, the staff is at lower elevation than the tacheometer. From tacheometer, all the three stadia readings and vertical angle θ subtended by middle wire reading at C is observed. From C, if we draw a perpendicular line to line of sight O'C will cut O'A line at A' and extended O'B line at B'. Let A'O'C be angle α, then by geometry B'O'C will also be angle α. In two Δs, AA'C and CB'B, angles AA'C and CB'B are equal to (90° + α) and (90° – α), respectively. The angle α being very small, angles AA'C and CB'B may be considered practically equal to 90°.

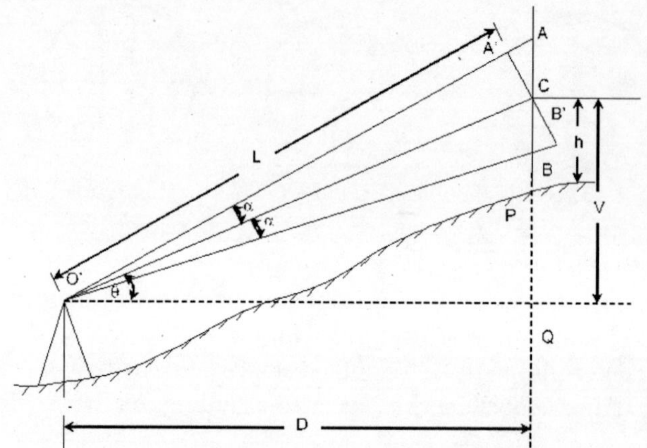

Fig.1.47: Staff held vertical at higher elevation

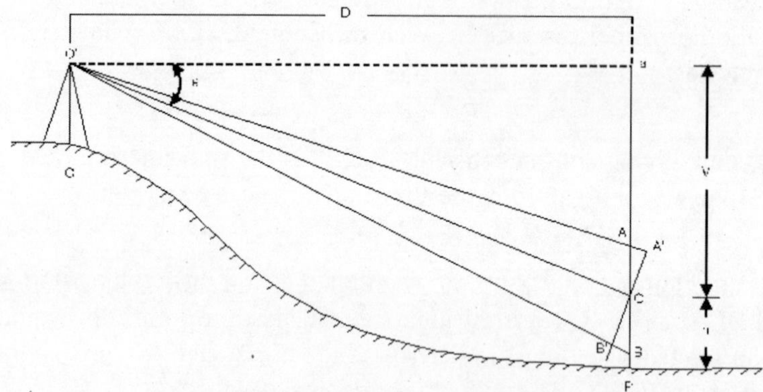

Fig.1.48: Staff held vertical at lower elevation

$A'B' = AB \cos \theta = S \cos \theta$

$D = L \cos \theta \qquad (1.18)$

$$= \frac{f}{i}(A'B')\cos\theta + (f+d)\cos\theta$$

$$= \frac{f}{i} \times S\cos^2\theta + (f+d)\cos\theta$$

And $V = L \sin 0 \qquad (1.20)$

$$= \frac{f}{i} \times S\sin\theta\cos\theta + (f+d)\sin\theta$$

$$= \frac{f}{i} S \frac{\sin 2\theta}{2} + (f + d)\sin\theta$$

Also V = D tan θ (1.21)

Knowing the value of V, the RL of the staff point is calculated as-

When θ is an angle of elevation (Figure 1.48)

RL of staff station P = RL of instrument axis + V – h (1.22)

When θ is an angle of description (Figure 1.49).

RL of staff station P = RL of instrument axis – V – h (1.23)

1.20 TRIGONOMETRICAL LEVELLING

It is an indirect method of levelling in which the elevation of the point is determined from the observed vertical angles and the measured distances. It is commonly used in topographical work to find out the elevations of the top of buildings, chimneys, churches etc., from a distance. This is a faster method to get the elevations of top of structures and objects. Elevation of a BM in the area must be known.

1.20.1 Finding height of an object which is accessible

Let PP' is a tower whose elevation of the top is to be determined (Figure 1.49). Set up the theodolite at a convenient ground point A so that the top of tower and a staff kept on the BM can be bisected. Measure vertical angle α of the top of tower as well as take the staff reading at the BM. Measure D, the horizontal distance between theodolite station and tower.

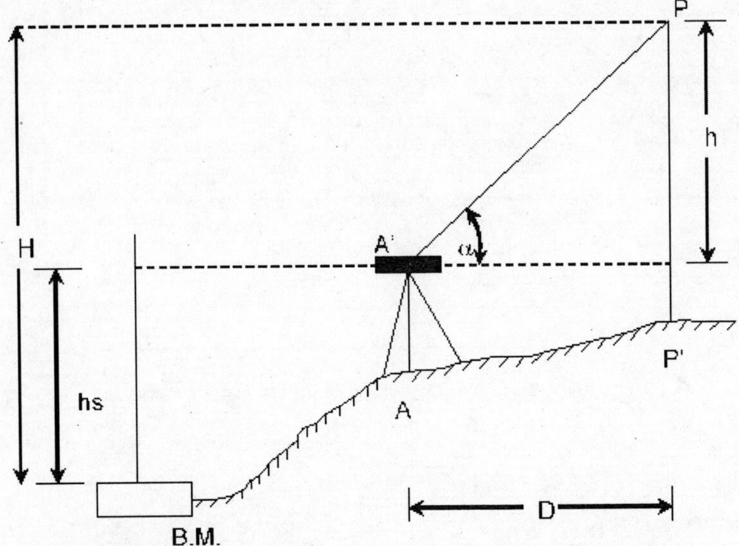

Fig.1.49: Measurement when the object is accessible

To find the height of the object above a BM:

Let H = height of the object above the BM

h = height of the tower above the instrument axis

h_s = height of the instrument axis above the BM

α = vertical angle of the top of tower at the instrument station

D = horizontal distance from the instrument station to the base of the tower.

Then, $h = D \tan \alpha$

$$H = h + h_s = D.\tan \alpha + h_s \tag{1.24}$$

If the distance D is large, correction for curvature and refraction, i.e., $\left\{0.00673\left(\frac{D}{1000}\right)^2\right\}$ is to be applied.

1.20.2 Finding height of an object which is inaccessible

To find the height of the tower PP' above a BM, select two stations A and B suitably on a fairly level ground so that these points lie in a vertical plane with the tower, and measure the distance AB with a tape (Figure 1.50). Set up the theodolite at station A to take a staff reading kept on the BM. Read the vertical angle α. Shift the theodolite at B point and take similar observations as taken at A point. Read the vertical angle β.

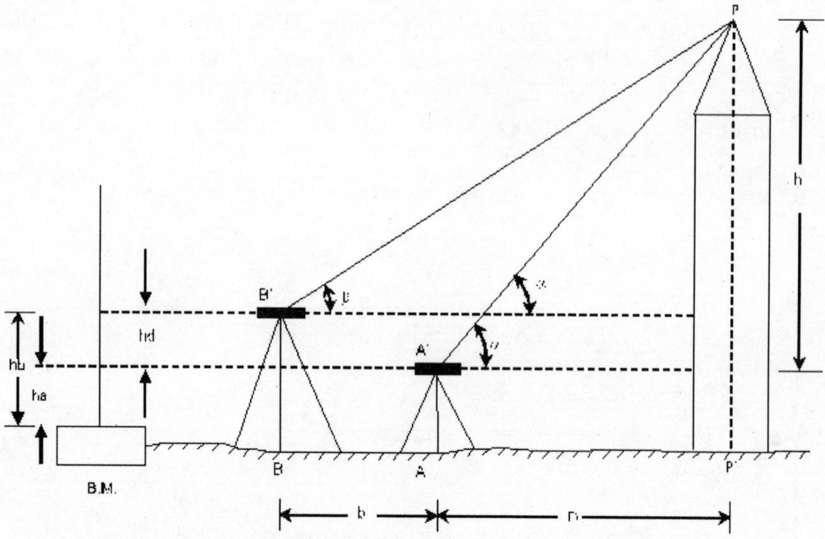

Fig.1.50: Measurement when object is inaccessible

Let

b = horizontal distance between A and B.

D = distance of the object from A point.

h = height of the tower P above instrument axis at A'.

h_a = staff reading at the BM when the instrument is at A.

h_b = staff reading at the BM when the instrument is at B.

h_d = the level difference between A and B of the instrument axes = $h_a \sim h_b$.

When the instrument at farther station B is higher than that at the near station A (Figure 1.51).

$h = D \tan \alpha$

$h - hd = (D + b) \tan \beta$

Putting the value of h from (i) and (ii),

$D \tan \alpha - h_d = (D + b) \tan \beta$

or $D (\tan \alpha - \tan \beta) = h_d + b \tan \beta$

$$\text{or } D = \frac{b \tan \beta + h_d}{\tan \alpha - \tan \beta} \tag{1.25}$$

Put this value of D in $h = D \tan \alpha$:

$$h = \frac{b \tan \beta + h_d}{\tan \alpha - \tan \beta} . \tan \alpha \tag{1.26}$$

Height of the tower above the BM,

$H = h + h_d$

When the base of tower is inaccessible and instrument can't be kept in same vertical plane

Let A and B be the two instrument stations not in the same vertical plane as that of a tower P (Figure 1.51). Select two stations A and B on a level ground and measure the horizontal distance b between them. Set the instrument at A and level it. Use it as a level and take a backsight h_s on the staff and kept at BM. Now, measure the angle of elevation α_1 to P, and horizontal angle BAP (θ_1).

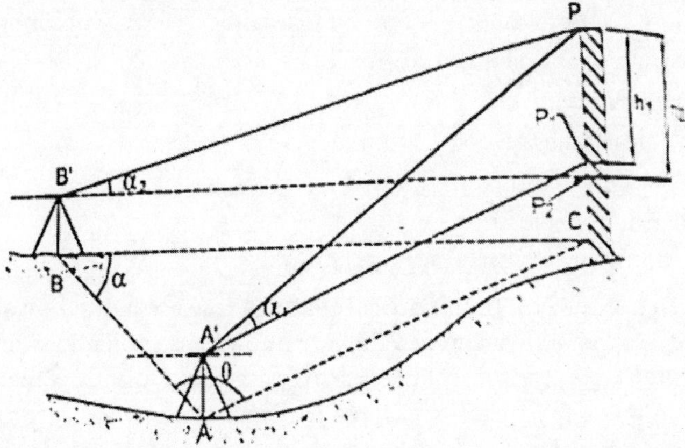

Fig.1.51: The base of tower is inaccessible and instrument is kept in different vertical planes

Now, shift the instrument to point B, and measure the angle of elevation α_2 to P. Also measure the horizontal angle ABP at B as θ_2.

Let

α_1 = angle of elevation from A to P

α_2 = angle of elevation from B to P

θ_1 = Horizontal angle BAC (or BAP) at station A

θ_2 = Horizontal angle ABC (or ABP) at station B

$h_1 = PP_1$ = height of the object P from instrument axis of A

h = PP = height of the object P from instrument h_2 PP_2 axis of B

In triangle ABC

Angle ACB = $180° - (\theta_1 + \theta_2)$

AB = b

$BC = b \sin \theta_1 / [\sin (180° - (\theta_1 + \theta_2))]$

$AC = b \sin \theta_2 / [\sin (180° - (\theta_1 + \theta_2))]$

By knowing the AC and BC from equation, we get-

$h_1 = AC \tan \alpha_1$

$h_2 = BC \tan \alpha_2$

RL of P = height of the instrument axis at A + h_1 (1.27)

or

R.L of P = height of the instrument axis at B + h_2 (1.28)

Height of the instrument axis at A = RL of BM + BS

Height of the instrument axis at B = RL of BM + BS

1.21 TRAVERSE COMPUTATIONS

One the field observations are completed for a traverse; the next task is to compute the coordinates of traverse stations. These coordinates are required to be plotted to carry out the detailed mapping of the area. For the computation of coordinates (x, y and z), following observations are to be taken in the field:

i. Magnetic bearing of at least one traverse line
ii. Length of at least one traverse line
iii. Elevation of at least one traverse station
iv. Included angles between traverse lines
v. Vertical angles of traverse lines
vi. Real-world coordinates of one traverse station

Once the coordinates of traverse points are computed, these are plotted on a plan with reference to x-axis and y-axis. If the length and bearing of a line are known, its projections on the y-axis and x-axis may be done, called *latitude* and *departure* of the line, respectively. Latitude is measured northward, and is also known as *northing*, and departure, if measured eastward is known as *easting*. The latitude of a line is determined by multiplying the length of the line with the cosine of its reduced bearing; and departure is computed by multiplying the length with the sine of its reduced bearing (Figure 1.52). If *l* is the length of the line and θ is its reduced bearing, then latitude and departure are calculated as-

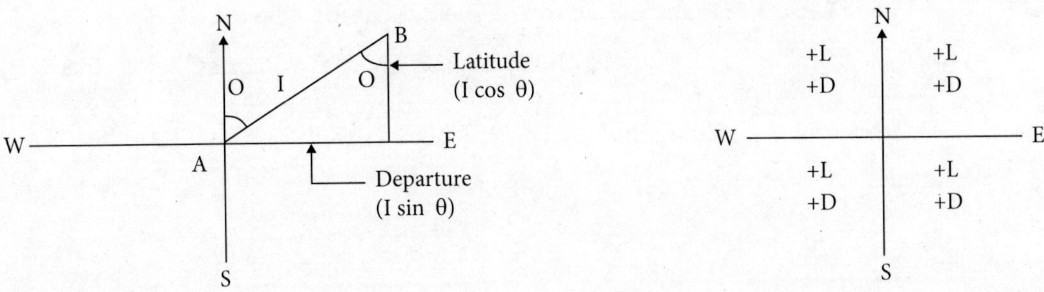

The latitude and departure of lines arc also expressed in the following ways:
Northing = latitude towards north = + L
Southing = latitude towards south = – L
Easting = departure towards cast = + D
Westing = departure towards west = – D

Fig.1.52: Comutation of latitude and departure of a line

The reduced bearing of a line will determine the sign of its latitude and departure, the first letter N or S of bearing defines the sign of the latitude and the last letter of bearing E or W defines the sign of the departure. If the WCB of a line is known, it can be converted into RB to determine the sign of latitude and departure. By knowing the bearing of one line and the included horizontal angle, the bearings of remaining lines can be computed. The latitude and departure of any point with reference to the proceeding traverse station are called *consecutive co-ordinates of the station.* The coordinates of a traverse station with reference to a common origin are called *independent coordinates.*

1.21.1 Adjustment of a closed traverse

Due to errors present in the observations, the coordinates of a closed traverse stations when plotted may not close itself, but will have a small difference. The errors in the linear and angular observations therefore are to be adjusted before using them for computational purpose. It is also called *Balancing a Traverse.* These errors include:

a. Adjustment of angular errors
b. Adjustment of bearings.
c. Adjustment of closing error of traverse

a. Adjustment of angular error

In a closed traverse, the sum of all interior angles should be equal to $(2n-4) \times 90^\circ$, and that of the exterior angles should equal $(2n + 4) \times 90^\circ$, where 'n' is the number of sides in a closed traverse. The difference between this sum and the sum of the measured angles in a closed traverse is called the *angular error of closure.* The angular error of closure should not exceed the least count of theodolite (x) used, i.e., x $\sqrt{n}$. If it exceeds, observations are to be repeated. These permissible errors are shown in Table 1.6. To distribute the error, one of the approaches is to distribute it equally among all the angles, if all the angles are measured with equal precision and under similar conditions, this error. The other approach is to distribute the error in each angle according to its magnitude. This approach is considered to be more accurate and requires computation, however, the first approach is simple and fast and may not be as accurate.

Table 1.7: Permissible errors in Theodolite traversing :

Traversing for	Permissible Angular Error	Permissible Linear error
Land surveys and location of roads, railways, etc	$1' \sqrt{N}$	1 in 3000
Survey work for cities and important boundaries	$30'' \sqrt{N}$	1 in 5000
Important Surveys	$15'' \sqrt{N}$	1 in 10, 000

Where N = Number of angles

b. Adjustment of bearings

Many times, bearings of a traverse are measured, instead of angles. In such cases, the closing error in bearings may be determined by comparing the fore bearing of a line and back bearing of that line of a closed traverse, as they should differ by 180°. The difference is the error which has to be adjusted in the bearings. Alternatively, we compare the known bearing of the traverse line with the measured bearing, and difference, if found, is adjusted in the bearings. At the end, we must ensure that the back bearing and fore bearing of all the lines differ by 180°.

If we know the RB of a traverse line, the RBs of remaining lines are computed using the adjusted (corrected) included angles.

c. Adjustment of closing error

For all the sides of traverse, latitude and departure are computed using the adjusted RB of lines, and proper sign is used as per the quadrant of traverse line. Ideally, the sum of all latitudes and sum of all departures must be zero in a closed traverse. But due to errors in the field measurements (e.g., bearings, distances, etc., the sum of all latitudes and sum of all departures, individually, may not come out to be zero). That means, the traverse will not close at the starting point. The distance by which the end point of a survey fails to meet with the starting point is called the *closing error* or *error of closure.* Figure 1.53 shows the plotting of an anticlockwise closed traverse ABCDEA, where A and A_1 are the starting and end points, respectively, and AA_1 represents the closing error.

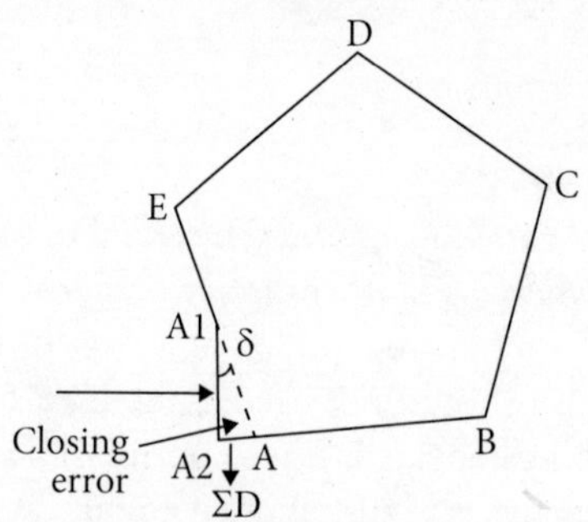

Fig.1.53: Representation of closing error

The magnitudes of two components of this error (A_1A_2 and AA_2) perpendicular to each other may be determined by finding the algebraic sum of the latitude (ΔL), as well as departures (ΔD). Since the triangle A_1A_2A is right-angled at A_2, the linear closing error (AA_1) is computed as:

$$\text{Closing error} = \boldsymbol{A}\ _1 = \sqrt{(A_1A_2)^2 + (A\ _2)^2} = \sqrt{(\Sigma L)^2 + (\Sigma D)^2} \tag{1.29}$$

The direction of the closing error is given by the relation,

$$\tan\theta = \frac{\Sigma D}{\Sigma L}, \tag{1.30}$$

where θ is the reduced bearing. The signs of ΔL and ΔD will define the quadrant of the closing error.

The latitudes and departures are now adjusted by applying the correction to them in such a way that the algebraic sum of the latitudes and departures should be equal to zero. Any one of the two rules (Bowditch Rules and Transit Rules) may be used for finding the corrections to balance the survey:

(1) *Bowditch Rule:* It is also known as the *Compass rule*. It is used to adjust the traverse when the angular and linear measurements are equally precise. By this rule, the correction in each latitude or departure of line is computed as:

Correction to latitude or departure of any side

= Total error in latitude or departure

$$\times\left(\frac{length\ \not{p}\ \ that\ side}{perimeter\ \not{p}\ \ traverse}\right) \tag{1.31}$$

(2) *Transit Rule:* The Transit rule is used to adjust the traverse when the angular measurements are more precise than the linear measurements.

(i) Correction to latitude of any side

= total error in latitude

$$\times\left(\frac{latitude\ \not{p}\ \ that\ side}{arithmeticl\ \ sum\ \not{p}\ \ the\ all\ latitudes}\right) \tag{1.32}$$

(ii) Correction to departure of any side

= total error in departure

$$\times\left(\frac{departure\ \not{p}\ \ that\ side}{arithmeticl\ \ sum\ \not{p}\ \ all\ departures}\right) \tag{1.33}$$

1.21.2 Computation of coordinates

Very popular method of showing the adjustment of closing error as well as computation of coordinates is systematically done using Gale's Traverse Table.

Gale's Traverse Table:

The computations for a closed traverse may be made in the following steps and entered in a tabular form known as Gale's Traverse, as shown in Table 1.8.

i. Adjust the included angles, if the sum is not equal to (2n-4) x 90°, as explained above
ii. Calculate the adjusted RB of all traverse liens, as explained above.

iii. Compute the latitudes and departures (consecutive coordinates) of traverse lines, as explained above.

iv. Apply the necessary correction to the latitude to the latitudes, and departures so that the sum of the corrected latitudes is zero and sum of corrected departures is equal to zero.

v. By knowing the exact coordinates of a traverse point, find the independent coordinates of the remaining traverse points from the consecutive coordinates.

vi. Plot all the traverse stations on a plan for map preparation.

Table 1.8: Sample Gales traverse table for computation of corrected coordinates

Inst. Stn	Angles	Correction	Corrected Angles	W.C. bearings	Reduced Bearings	Quadrant	Line	Length	Point	Consecutive Co-ordinates				Corrections				Corrected Consecutive Co-ordinates				Independent Co-ordinates		Remarks
										Lat.		Dep.												
										+Northing	-Southing	+easting	-Westing	Northing	Southing	Easting	Westing	Northing	Southing	Easting	Westing	Northing	Easting	

1.22 TRIANGULATION SURVEYS

Triangulation is one of the methods of fixing accurate controls. It is based on the trigonometric proposition that if one side and two angles of a triangle are known, the remaining sides can be computed. A triangulation system consists of a series of triangles in which long side is normally called as *base line*. The base line is measured and remaining sides are calculated from the angles measured at the vertices of the triangles; vertices being the control points are called as triangulation stations (Figure 1.54). Applications of triangulation surveys includes; (i) establishment of accurate control for plane and geodetic surveys as well as photogrammetric surveys covering large areas, (ii) determination of the size and shape of the Earth, and (iii) determination of accurate locations for setting out of civil engineering works, such as piers and abutments of long span bridges, fixing centre line, terminal points and shafts for long tunnels, measurement of the deformation of dams, etc.

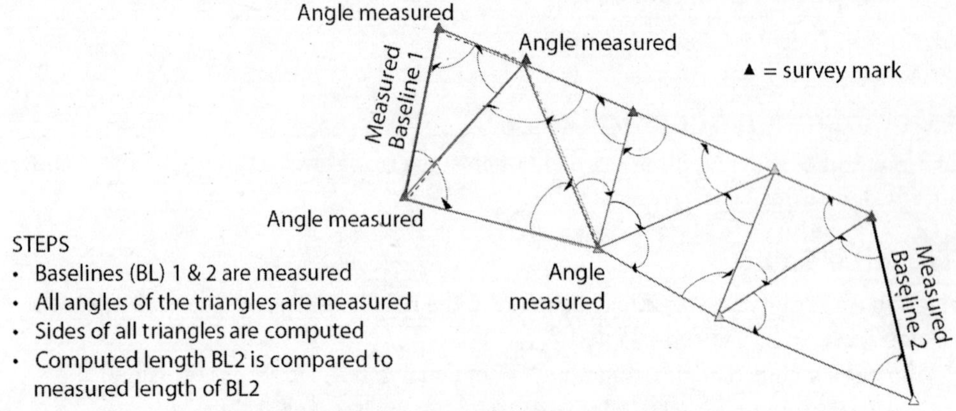

Fig.1.54: Triangulation survey scheme

The area to be covered by triangulation scheme must be carefully studied to select the most suitable positions for the control stations. Existing maps, especially if contoured can be of great value since the size and shape of triangles formed by the stations can be difficult to visualize in the field.

The following factors are considered in selecting the stations-

1. The best shape of a triangle is an isosceles triangle whose base angles are 56° 14' each.
2. Simple triangles should be preferably equilateral triangles which are may treated as well conditioned triangles.
3. The angles of simple triangles should not be less than 30° or more than 120°
4. Braced quadrilaterals are preferred figures but no angle should be less than 30°
5. Centered polygons should be regular, and no angle should be less than 40°
6. No angle of the figure, opposite a known side should be small
7. The sides of the figures should be of comparable length
8. Stations are placed on the highest points of the elevated places such as hilltops, house rooftops, etc., to ensure intervisibility.
9. Easy access to the stations with instruments and equipment.
10. Stations should be useful for providing intersected points and also for a subsequent detailed survey.
11. Very far-off stations should be avoided for plane surveys.
12. Grazing line of sight should be avoided and line of sight should not pass over the industrial area to avoid atmospheric refraction.
13. Cost of clearing and cutting of vegetation along the line of sight should be minimum.

1.22.1 Trilateration

Trilateration is a surveying procedure in which the lengths of the sides of a triangle of triangulation scheme are measured, generally electronically, and included angles are determined trigonometrically. The field techniques for trilateration are similar to those for triangulation, with the exception that only lines are measured while large number of angles are calculated. Trilateration in surveying is used to determine the horizontal positions, in addition to other methods, like triangulation, intersection, resection and GPS positioning.

The trilateration consists of series of joined or overlapped triangles that forms triangles, quadrilaterals and polygons. It employs measurement of sides of the triangle, using the EDMs. With the development of EDMs, trilateration has become highly accurate and precise for expanding and establishing the horizontal controls. The areas that is subjected to seismic activity, trilateration is employed to study the gradual and secular movement in Earth's crust. It is used for determining the position of objects on the surface of Earth, involving the distance and angles. Trilateration is also used in control expansion or densification for future metropolitan growth; coastline control; inland waterways; control extension; densification for land subdivisions and construction; and deformation surveys of dams, geothermal areas, structures, regional/local tectonics, and landslides. It is used to achieve rapid control expansion with utmost accuracy, and is less expensive as compared to triangulation.

1.22.2 Principle of triangulation

In triangulation, entire area to be surveyed is divided into framework of triangles. Vertices of the individual triangles are known as triangulation stations. Triangulation is more accurate than the theodolite transverse, as there is less accumulation of error than that in theodolite traverse. A series of triangulation may consist of (i) a chain of simple triangles, (ii) braced quadrilaterals, and (iii) centered polygons, as shown in Figures 1.55a, b and c. Simple triangles of a triangulation system do not provide many checks on the accuracy of observations as there is only one route through which distance can be computed. In this layout, checking baselines and azimuth at frequent intervals are necessary to avoid the errors.

Although triangle is the smallest figure in triangulations scheme, but braced quadrilateral containing four corner stations is preferred. Braced quadrilaterals system is treated to be the best arrangement, as it provides a means of computing the length of the sides using different combination of the sides, diagonals and angles, and thereby offering more checks on measured/computed data. The braced quadrilateral system is considered as the best arrangement of triangles as it provides several ways to compute the length of sides using various combinations of sides and angles. A series of triangulation scheme may also consist of centered polygons, where one station is within the polygon. Such figure provides many more checks than a simple quadrilateral, so it offers more accurate solution, but takes more computational time. The figures containing centered polygons (such as quadrilaterals, pentagons, or hexagons) and centered triangles is known as *centered figures*. This layout is generally preferred for covering a vast area. This system provides a proper check on the accuracy of work, but the progress of work is slow as more settings of instrument are required.

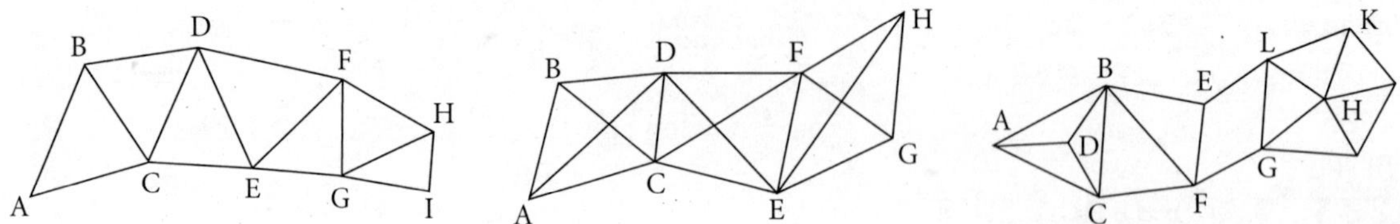

Fig.1.55: (a) Chain of triangles, (b) quadrilaterals, and (c) centred polygons.

1.22.3 Types of triangulations schemes

The triangulation schemes are classified into three categories: (i) First order or Primary triangulation, (ii) Second order or Secondary triangulation, and (iii) Third order or Tertiary triangulation. The basis of the classification of triangulation schemes is the accuracy with which the length and azimuth of a line of the triangulation are determined. Triangulation systems of different accuracies depend on the extent and the purpose of the survey. The first order triangulation is of the highest order, and is employed either to determine the Earth's figure or to furnish the most precise control points to which secondary triangulation may be connected. The specifications of each category of triangulation are given in Table 1.9.

Table 1.9: Specifications of triangulation schemes

S.No.	Specifications	Primary triangulation	Secondary triangulation	Tertiary triangulation
1	Length of the baseline	5 km to 15 km	1.5 km to 5 km	0.5 km to 3 km
2	Length of the sides of the triangle	30 km to 150 km	8 km to 65 km	1.5 km to 10 km

3	Average triangular error or closure	< 1 sec	3 sec	6 sec
4	Maximum station closure	> 3 sec	8 sec	12 sec
5	Actual error of base	1: 300000	1:150,000	1:75,000
6	Probable error of base	1 in 10,00,000	1 in 5,00,000	1 in 250,000
7	Discrepancy between two measures	10mm/km	20 mm/km	25 mm/km
8	Probable error of computed distance	1 in 60000 to 1 in 250000	1 in 20,000 to 1 in 50,000	1 in 5,000 to 1 in 20,000
9	Probable error in astronomical observation	0.5 sec	2 sec	5 sec

In primary triangulation, the length of base line is between 5 to 15 km, and length of the sides of triangles is between 30 to 150 km. The secondary triangulation consists of a number of points fixed within the framework of primary triangulation. The stations are fixed at close intervals so that the triangles formed are smaller than the primary triangulation. The third-order triangulation consists of a number of points fixed within the framework of secondary triangulation, and forms the immediate control for detailed engineering and other surveys. The sizes of the triangles are small and instrument with moderate precision may be used.

The purpose of triangulation is to establish the accurate control points for plane and geodetic surveys of large areas, and locating engineering works accurately. The triangulation is based on the principle that if the length and bearing of one side and three angles of a triangle are measured precisely, the lengths and directions of other two sides can be computed. So, it is important to measure the base line precisely as it can be used at a starting point for computation. If the surveyed area is large, more than one base line is measured at suitable distances to minimize accumulation of errors in lengths. As a check, the length of one side of last triangle is also measured and compared with the computed length, known as the check line, as shown in Figure 1.56. If the coordinates of any vertex of the triangles and azimuth of any side are also known, the coordinates of the remaining vertices may be computed. This way, coordinates of all triangulation stations are computed.

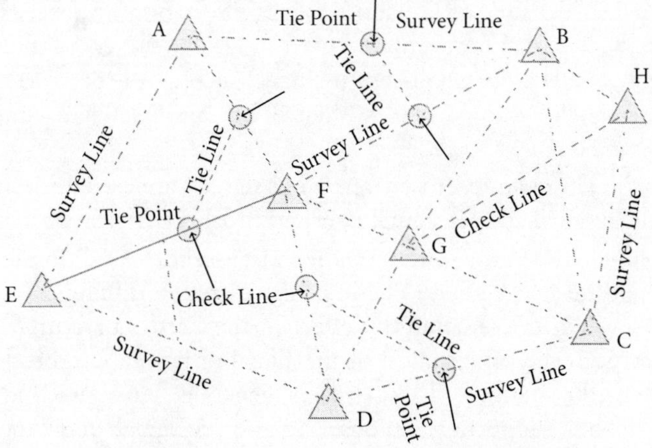

Fig.1.56: Various lines in a triangulations scheme

The following corrections for base line measurement are to be done

i. Correction for the absolute length
ii. Correction for temperature
iii. Correction for pull or tension
iv. Correction for sag
v. Correction for slope
vi. Correction for alignment
vii. Reduction to mean sea level
viii. Axis signal correction
ix. Correction for the unequal height
x. Reduction to chord to arc

1.22.4 Technical terms

a. **Triangulation station marks:** Stations are marked with a permanent mark buried below the surface on which a target/signal or instrument is to be centred for taking observations. This mark can be bronze or copper mark cemented into surface.

b. **Laplace station:** The triangulation station at which astronomical observation for azimuth is made is called Laplace station.

c. **Satellite stations:** In order to secure well condition triangle or have better intervisibility, objects such as church tops, flag poles or towers etc., are sometimes selected as triangulation stations, called satellite stations. Sometimes, it is not possible to set the instrument over the triangulation station, so a subsidiary station known as a satellite station or false station is selected as near as possible to the main station. Observations are made to the other triangulation stations with the same precision from the satellite station.

d. **Signals:** Signals are the devices erected on the ground to define the exact position of a triangulation station. It is placed at each station so that the line of sights may be established between the triangulation stations. The signals are classified in two broad classes: (i) luminous signals, and (ii) opaque signals.

   i. **Luminous signals**: These are further divided into two categories; sun signals, and night signals. Sun signals reflect the rays of the sun towards the station of observation, and are known as *heliotropes*. Such signals can only be used in clear weather. While making observations at night, night signals are used. These includes, such as various forms of oil lamps with a reflector that can be used for sights less than 80 km, and Acetylene lamps which are used for sights more than 80 km.

   ii. **Opaque signals**: The opaque, or non-luminous signals, are used during day. Most commonly used are the Pole signal, Target signal, Pole and Brush signals, Stone cairn, and Beacons.

e. **Towers:** Intervisibility between the stations is the most essential condition in triangulation. When the distance between the stations is too large or the elevation difference is less, both signal and station are to be elevated to overcome the effect of the Earth curvature. A tower is erected at the triangulation station when the station or the signal or both are to be elevated to make them intervisibile between the stations. These towers generally have two independent structures; Outer structure- which is supporting the observer and the signal, and Inner structure- which is supporting the instrument only. This arrangement does not disturb the instrument setting due to the movement of observer. These towers may be made of masonry, timber or steel, but timber scaffolds are most commonly used to heights over 50 m.

The computation of height of signal depends upon the (i) distance between the stations, (ii) relative elevation of the stations, and (iii) profile of the intervening ground. If the intervening ground gives the clear visibility, the distance of horizon from the station of known elevation is given by the following formula-

$h = 0.06735\ D^2$ (1.34)

or $D = \sqrt{(h / 0.06735)}$

Where h is the height of station above datum (in m), and D is the distance of visible horizon (in km).

Let, A and B be two stations having their ground portion elevated to A' and B', respectively (Figure 1.57).

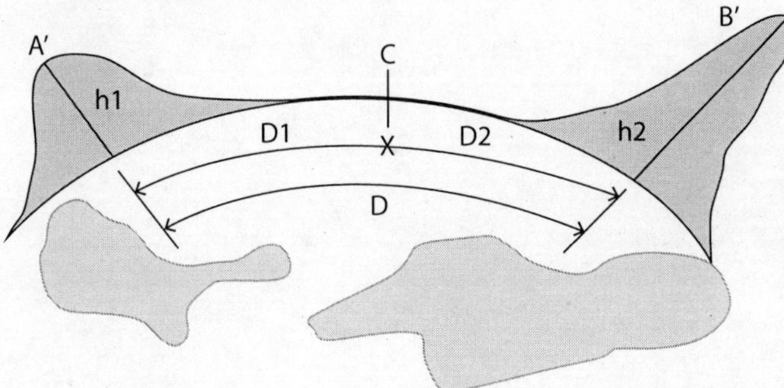

Fig.1.57: Intervisibility between two triangulation stations

h_1 = RL of station A'

h_2 = minimum required elevation of station B so that it is visible from A

D_1 = distance of visible horizon from A

D_2 = distance of visible horizon from B at an elevation h_2

D = known distance between A and B

Now, $D_1 = \sqrt{\dfrac{h_1}{0.06735}} = 3.858.\ \sqrt{h_1}$

and, $D_2 = 3.858.\ \sqrt{h_2}$

But, $D = D_1 + D_2$ or $D_2 = D - D_1$

Normally the line of sight is kept at 3 m above the ground to avoid grazing sights as the refraction is maximum near the ground.

f. **Axis signal correction:** Signals are installed at each station with different heights to ensure the intervisibility of stations and measuring the vertical angles. The height of instrument axis and the signal at the station sighted are also different, which may result in error in the observed vertical angles. The corrections applied to vertical angles is known as axis signal correction, or eye and object correction, or instrument and signal correction. Let A be the instrument station and B be the station at which signal is kept. The observed vertical angle from A to B will be the true vertical angle, if height of instrument at A (h1) is equal to the height of signal at B (S2). But, these two heights, i.e., h1 and S2 (Figure 1.58a), respectively, are never same, a correction has to be applied to the observed vertical angle α. Let α1 be the corrected vertical angle for the axis signal, and D be the horizontal distance AB (Figure 1.58b) equal to the spheroidal distance AA. The axis signal correction is α1 (BAE in Fig. 1.59b). In triangle ABO,

$\angle BAO = \angle A_1AO + \angle BAA_1 = 90° + \alpha$

$\angle AOB = \theta$

$\angle ABO = 180° - [(90° + \alpha) + \theta] = 90° - (\alpha + \theta)$

$\angle EBF = 90° - [90° - (\alpha + \theta)] = \alpha + \theta$

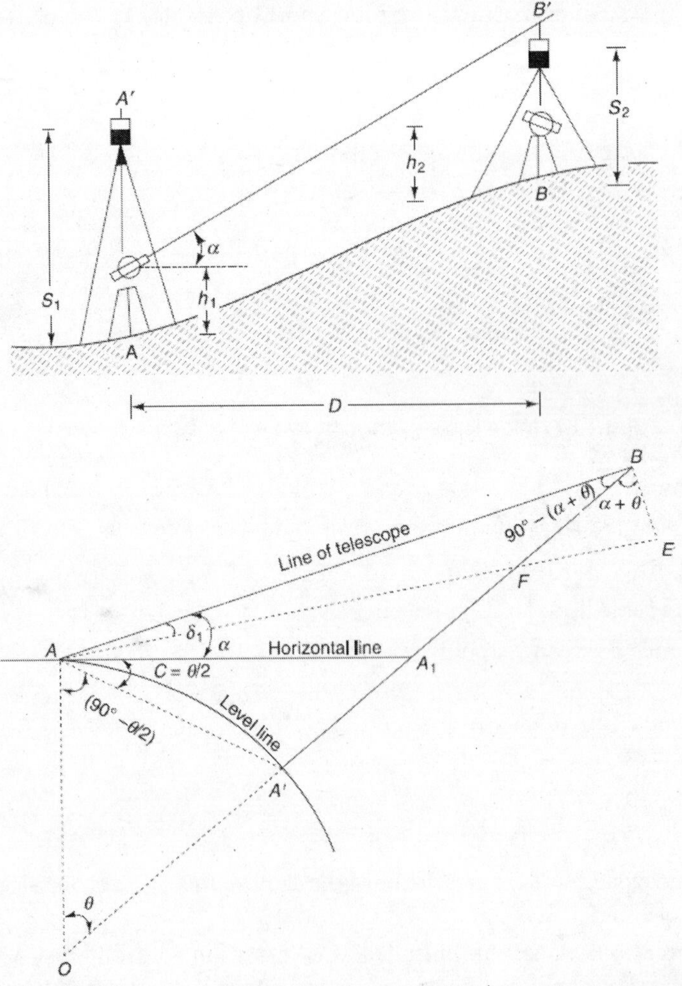

Fig.1.58: (a) Angle observations between triangulation stations, and (b) Axis signal corrections

The following relationship can be derived using Figures 1.59a and b. Usually θ is very small as compared to α, hence the equation can be derived as-

$$\tan \delta_1 = (S_2 - h_1) \cos^2\alpha / D \quad (1.35)$$

If the vertical angle is small, the equation reduces to-

$$\delta_1 = (S_2 - h_1) / (D \sin 1'') \qquad (1.36)$$

Similarly, if observations are made from B towards point A with β as the observed vertical angle and β_1 the corrected vertical angle for axis signal correction, it can be shown that

$$\tan \delta_2 = [(S_1 - h_2) \cos^2\beta] / D \qquad (1.37)$$

The axis signal correction is negative for an angle of elevation and positive for an angle of depression.

g. **Reduction to center:** The angles measured at satellite station are then corrected and reduced to what they would from the actual triangulation station. The operation of applying this correction due to the eccentricity of the station is generally known as reduction to centre. Distance between the actual triangulation station and satellite station is determined by using trigonometric levelling.

1.22.5 Triangulation survey work

It involves two major steps; (i) Field work, and (ii) Computational work. The field work involves (a) Reconnaissance survey, (b) Erection of signals, (c) Measurement of starting baseline, (d) Measurement of horizontal and vertical angles, (e) Observations to determine the azimuth of sides, and (f) Measurement of closing baseline. The computational work includes:

- Checking the observed angles
- Checking the triangulation errors
- Checking the closing horizon error at each station (i.e., sum =360°)
- Computation of corrected length of baseline
- Computation of sides of triangles
- Computation of the latitude and departure of each side of the triangulation network, and
- Computation of independent coordinates of the triangulation stations.

The reconnaissance of the area involves; (i) proper examination of the terrain, (ii) selection of suitable positions for baselines, (iii) selection of suitable positions of triangulation stations, and (iv) determination of intervisibility of triangulation stations.

1.22.6 Accuracy of triangulation

It can be computed using the relationship given below:

$$m = \sqrt{\frac{\Sigma E^2}{3n}} \qquad (1.38)$$

Where m is the root mean square error of unadjusted horizontal angle (in secs.) as obtained from the triangular error, ΣE^2 is the sum of the square of all the triangular errors in the triangulation series, and n is the total number of triangles in the series.

UNIT SUMMARY

This unit discusses 22 different topics of surveying, starting from setting–up the instruments, data collection, error minimisation to map preparation. Components of various surveying equipment, such as levels, theodolites, compass, etc., are explained. Distance, elevations, beraings and angles measurements play an important role for providing horizontal and vertical controls as well as map preparation in engineering projects. Observation methods have been explained and various parameters computed.

These observations need to be accurate enough to create maps useful for project planning. The utility of each equipment is highlighted.

SOLVED EXAMPLES

Example 1.1:

Compute the plotting accuracy of a map at 1:50,000 scale.

Solution:

Plotting accuracy = 0.25 mm x scale

= 0.25 x 50,000 = 12,500 mm

= 1250 cm = 12.5 m

So, any detail smaller than 12.5 m on the ground can't be plotted at 1: 50,000 scale.

Example 1.2:

How many Survey of India toposheets are there at 1:50,000 scale to cover equivalent area of 1: 250,000 scale?

Solution:

One Survey of India toposheet at 1:250,000 scale covers 1° latitude and 1° longitude, while on toposheet at 1:50,000 scale covers 15' x 15' area. So, 16 toposheets at 1:50,000 would be required to cover equivalent area at 1:250,000 scale.

Example 1.3:

How much length is covered by 1° latitude and 1° longitude at the equator?

Solution:

Each degree of latitude is approximately 69 miles (111 kilometers) apart. At the equator, the distance is 68.703 miles (110.567 kilometers). A degree of longitude is widest at the equator with a distance of 69.172 miles (111.321 kilometers), and it gradually reduces to zero as all longitude lines meet at the poles.

Example 1.4:

Convert the following whole circle bearing to reduced bearing (with quadrant)

a. 30° b. 160°

c. 200° d. 315°

Solution:

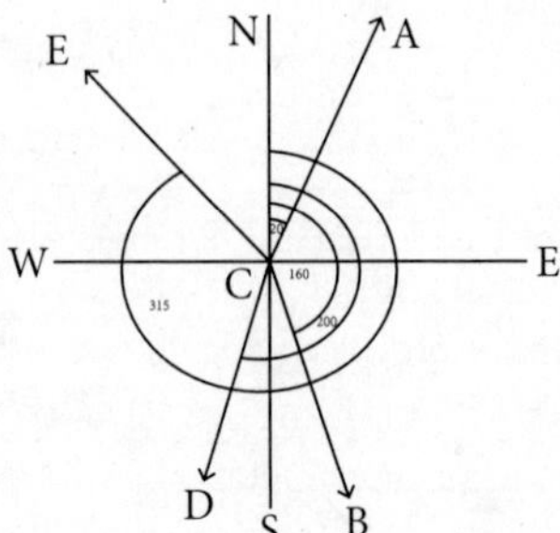

For 30°, Reduced Bearing (RB) = N 30° E

For 160°, RB = (180° – 160°) = S 20° E

For 200°, RB = (200° – 180°) = S 20° W

For 315°, RB = (360° – 315°) = N 45°W

Example 1.5:

Convert the following quadrantal bearing into whole circle bearing:

1. N 30°30' E
2. S 50° E
3. S 20°45' W
4. N 25° W

Solution:

Now, the whole circle bearing can be written directly as below:

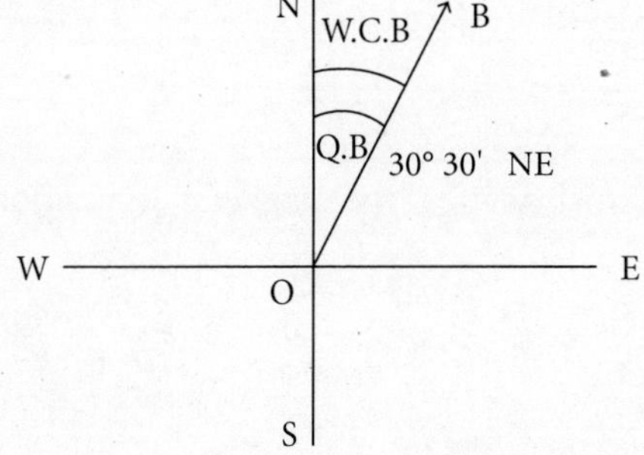

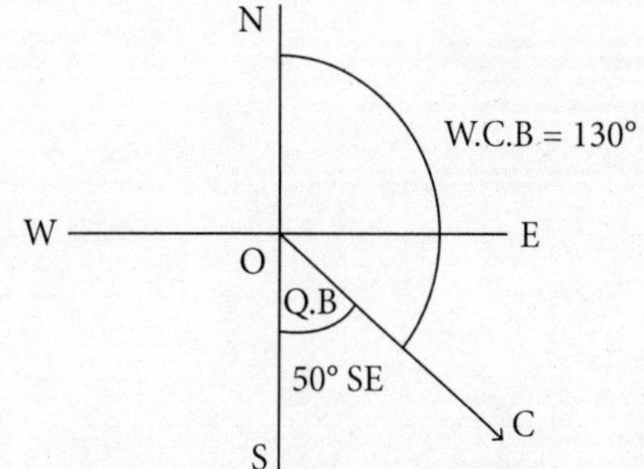

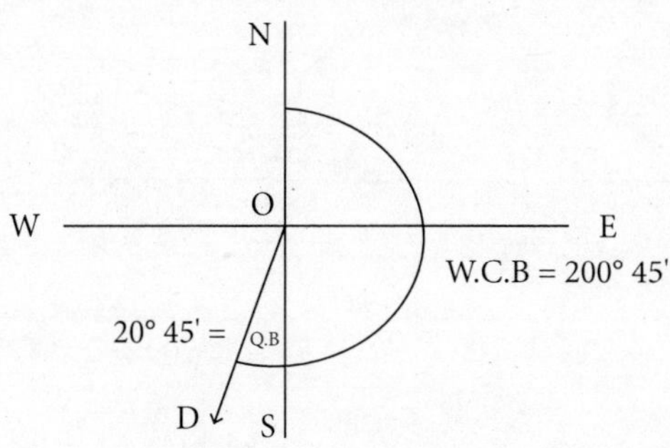

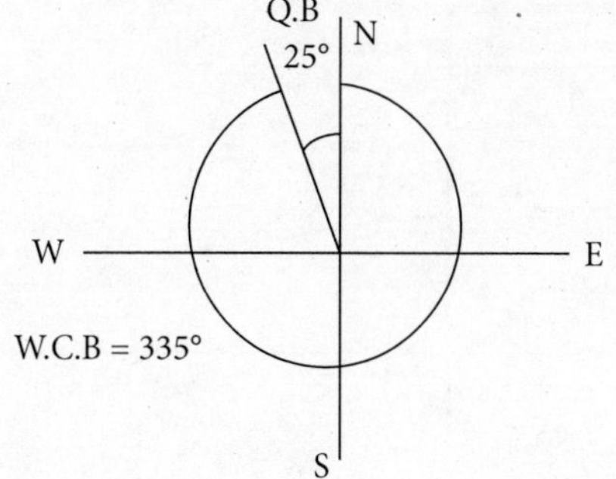

1. WCB = 30° 30'
2. WCB =180° – 50° = 130°
3. WCB = 180° + 20° 45' = 200° 45'
4. WCB = 360° – 25° = 335°

Example 1.6:

The fore bearings of the four lines AB, CD, EF and GH are, respectively, as under: (i) S 25° 30° W; (ii) N 30° W; (iii) S 40° 30' W; (iv) N 700 45'E. Determine their back bearings.

Solution:

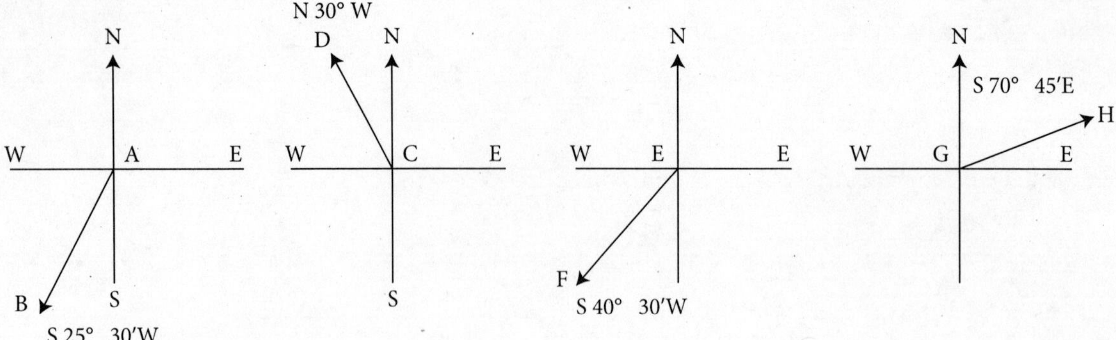

Line AB N 25° 30' E

Line CD N 30° 00' E

Line EF N 40° 30' E

Line GH N70° 45'W

Example 1.7:

The magnetic bearing of a line OP is 89° 45'. Determine the true bearing if the magnetic declination is (a) 5° 30' E, and (b) 4°15' W.

Solution:

a. In Figure (i)

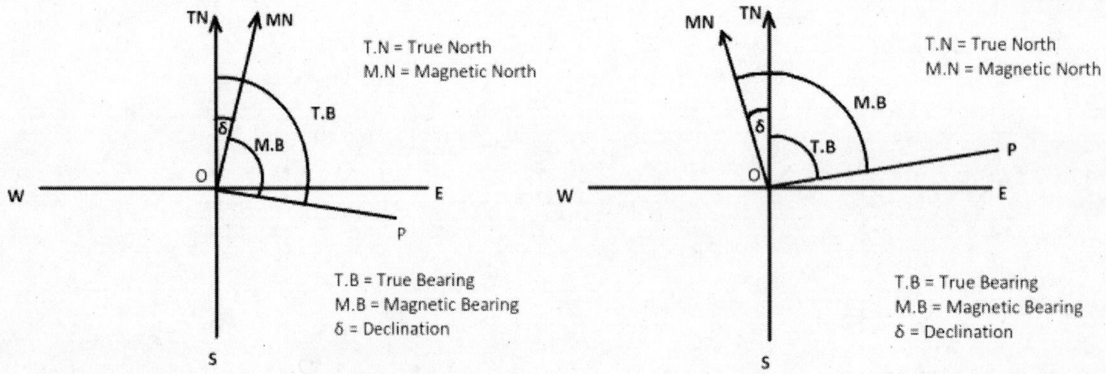

True Bearing = Magnetic Bearing + Declination

= (89°45' + 5°30') = 95°15'

b. In Figure (ii)

True Bearing = (Magnetic Bearing – Declination)

= (89°45' – 4°15') = 85°30'

Example 1.8:

The compass observations as follows were taken of a closed traverse ABCDEA, find the station affected by the local attraction and correct the observed bearing of the lines.

Line	Bearing
AE	319° 00'
AB	72° 45'
BA	252° 00'
BC	349° 00'
CB	167°15'
CD	298° 30'
DC	118° 30'
DE	229° 00'
ED	48° 00'
EA	135° 30'

Solution:

By inspecting the fore and back bearings of the lines, it is clear that the observed fore and back bearings of the line CD (i.e., bearings of CD and DC) exactly differ by 180°. Therefore, both the stations C and D are considered free from local attractions. Therefore, all the bearings observed from these stations should be correct bearings. The bearings of the line DE should be correct bearing.

Therefore, bearing of ED (back bearing of DE) = 229°–180° = 49°. But the observed bearing of the line ED = 48°, so station E is affected by local attraction by 10. To correct the bearing of the lines at the station E, a value of 1° should be added.

The error at the station E is said to be +1°. To correct the values of bearings observed at the station E, 1° should be added to the observed bearings, therefore, the correct bearings of lines ED and EA are 49° and 136° 30', respectively.

From the correct bearing of the line EA, the station A may be corrected: Bearing of AE (back bearing of EA) =136° 30' (bearing of EA) + 180° = 316° 30'. But the observed bearing of the line AE is 319°. Therefore, the station A is affected by local attraction. To correct the bearing of AE, observed bearing from station A, 2° 30' should be deducted. Therefore, the correction for local attraction–2° 30'.

The correct bearings of AE and AB are, therefore, 316° 30' and 70° 15', respectively. Therefore, the correct bearing of BA (back bearing of AB) = 70°15' (bearing of AB) + 180° = 250° 15'. But the observed bearing of BA is 252°. Therefore, the station B is affected by local attraction with the correction for local attraction as 1° 45'. Therefore, the correct bearings of BA and BC are 250° 15' and 347° 15' respectively.

Line	Bearing	Correction due to local attraction	Corrected bearings
AE	319° 00'	-2° 30'	316° 30'
AB	72° 45'	-2° 30'	70° 15'
BA	252° 00'	-1° 45'	250°15'

Line	Bearing	Correction due to local attraction	Corrected bearings
BC	349° 00'	-1° 45'	347°15'
CB	167°1 5'	0° 00'	167°15'
CD	298° 30'	0° 00'	298° 30'
DC	1180 30'	0° 00'	118° 30'
DE	229° 00'	0° 00'	229° 00'
ED	48° 00'	1° 00'	49° 00'
EA	135° 30'	1° 00'	136° 30'

Example 1.9:

Following data of a closed compass traverse PQRSP was taken in a clockwise direction:

i. Fore bearing and back bearing at station P= 55° and 135°, respectively
ii. Fore bearing and back bearing of line RS = 211° and 31°, respectively
iii. Included angles at Q = 100° and at R = 105°
iv. Local attraction at station R = 2° W

From the above data, calculate the local attraction at stations P and S, and calculate the corrected bearings of all the lines.

Solution:

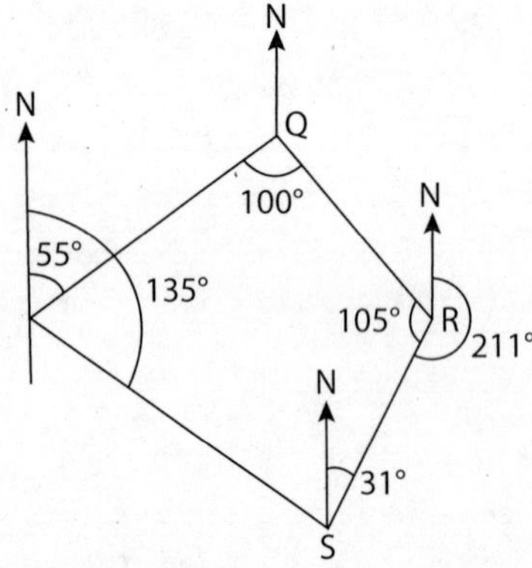

Figure shows the traverse, as FB and BB of the line RS differ exactly by180°, stations R and S are either free from local attraction or affected by it equally. As the station R is affected, the station S is also affected. The local attraction at S is given as 2° W. In other words, the bearing measured at R and S stations are added with 2° due to local attraction.

Therefore, corrected FB of RS = 211° – 2° = 209°

Interior ∠QPS = 135° – 55° = 80°

Interior ∠PSR = 360° – (80° + 100° + 105°) = 75°

The bearings of all the lines can be determined from the included angles and the corrected bearing of the line RS is 209°.

BB of RS = 209° – 180° = 29°

FB of SP = 29° + (360° – 75°) = 314°

BB of SP = 314° – 180° = 134°

FB of PQ = 134° – 80° = 54°

BB of PQ = 54° + 180° = 234°

FB of QR = 234° – 100° = 134°

BB of QR 134° + 180° = 314°

FB of RS 314° – 105° = 209° (*OK, checked*)

Example 1.10:

In a closed traverse ABCDE, the bearings of the line AB was measured as 150° 30'. The included angles were measured as under: ∠A = 130° 10', ∠B = 89° 45', ∠C = 125° 22', ∠ D = 135° 34', and ∠E = 59° 09'. Calculate the bearings of all other lines.

Solution:

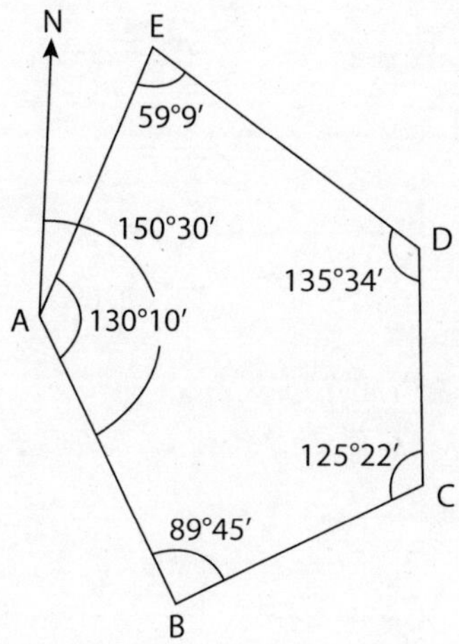

Bearing of BC = Bearing of A + ∠ABC

= (150° 30' + 180°) + 89° 45' = 420° 15' = 60° 15'

Bearing of CD = Bearing of CB + ∠BCD

= (60° 15' + 180°) + 125° 22' = 365° 37' = 5° 37'

Bearing of DE = Bearing of DC + ∠CDE

= (5° 37' + 180°) + 135° 34' = 321° 11'

Bearing of EA = Bearing of ED + ∠DEA = (321° 11' – 180°) + 59° 9' = 200° 20'

Check: For checking the calculations, it is advisable to calculate the bearing of the first line from bearings of the last line.

Bearing of AB = Bearing of AE + ∠EAB

= (200° 20' – 180°) + 130° 10' = 150° 30

Hence, checked.

Example 1.11:

The bearing of the sides of a traverse ABCDE are as follows:

Side	FB	BB
AB	97° 15'	277° 15'
BC	12° 00'	192° 00'
CD	271° 30'	91° 30'
DE	189° 15'	9° 15'
EA	124° 45'	304° 45'

Calculate the interior angles of the traverse.

Solution:

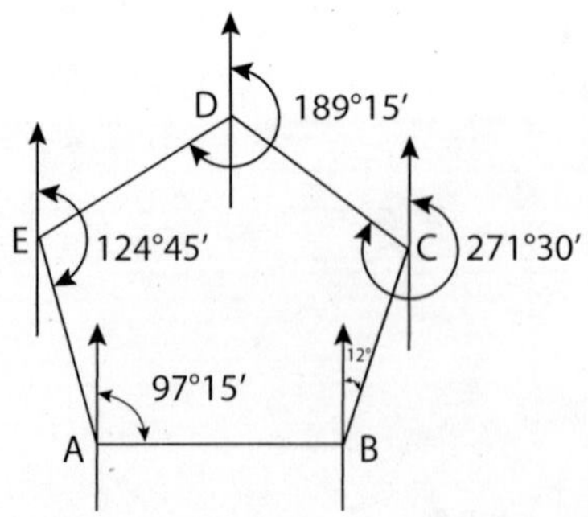

Bearing of AE = BB of EA – FB of AB

= 304°45' – 97°15' = 207°30' (exterior angle)

Interior angle at A = 360° – 207°30' = 152°30'

Bearing of BA = BB of AB – FB of BC

= 277°15' – 12°00' = 265°15' (exterior angle)

Interior angle at B = 360° – 265°15' = 94°45'

Bearing of CB = BB of BC – FB of CD

= 192°00' – 271°30' = -79°30'

Interior angle at C = 79°30'

Bearing of DC = BB of CD – FB of DE

= 91°30' – 189°15' = -97°45'

Interior angle at D = 97°45'

Bearing of ED = BB of DE – FB of EA

= 09°15' – 124°45' = 115°30'

Interior angle at E = 115°30'

Check = (2*5–4) * 90° = 540°

= 152°30' + 94°45' + 79°30' + 97°45' + 115°30'

= 540°

Example 1.12:

First four columns show the readings that are taken in the field. Compute the reduced levels of various points using Rise and Fall method.

Solution:

BS (m)	FS (m)	Rise (m)	Fall (m)	RL (m)	Station
1.575				100	10
0.355	1.355	0.22		100.22	1
0.605	2.685		2.33	97.89	TP1
0.485	2.355		1.75	96.14	TP2
1.37	1.68		1.195	94.945	2
1.025	1.57		0.2	94.745	3
1.15	1.99		0.965	93.78	TP1
1.71	1.84		0.69	93.09	TP2
1.38	1.955		0.245	92.845	4
1.31	1.55		0.17	92.675	5
1.31	1.61		0.3	92.375	6
1.635	1.13	0.18		92.555	7
2.865	0.66	0.975		93.53	8
2.95	0.055	2.81		96.34	TP
2.43	0.43	2.52		98.86	9
0.435	0.245	2.185		101.045	TP
	1.4		0.965	100.08	10

Checks:

Σ FS–Σ BS = Σ Rise – Σ Fall = Change in RL

Σ FS–Σ BS =0.08

Σ Rise – Σ fall =0.08

Change in RL =100.08-100= 0.08

Hence, checked.

Example 1.13:

The following readings were observed with a levelling instrument, the instrument was shifted after 5th and 11th reading, 0.585, 1.010, 1.735, 3.295, 3.775, 0.350, 1.300, 1.795, 2.575, 3.375 3.895, 1.735, 0.635, and 1.605 m. Determine the RLs of various points using height of collimation method if the RL of a point on which the first reading was taken is 136.440 m. Apply necessary checks.

Solution:

BS (m)	IS (m)	FS (m)	HI (m)	RLs (m)	Remarks
0.585			137.025	136.440	1, 2, 3, 4
	1.010			136.015	
	1.735			135.290	
	3.295			133.730	
0.350		3.775	133.600	133.250	5
	1.300			132.300	6, 7, 8, 9, 10
	1.795			131.805	
	2.575			131.025	
	3.375			130.225	
1.735		3.895	131.440	129.705	
	0.635			130.805	11, 12
		1.605		129.835	
ΣBS=2.670		ΣFS=9.275			

Check

ΣBS–ΣFS = Last RL – First RL

2.670 – 9.275 = 129.835 – 136.440

-6.605 = -6.605 m

Example 1.14:

The following consecutive reading were taken with a level and 4 m staff at a common interval of 30 m, as 0.725 on A, 0.935, 2.845, 3.745, 3.935, 0.965, 1.135, 1.785, 2.625, 3.845, 0.965, 1.575 and 2.015 on B. The elevation of point A is 320.50 m. Enter the levels in a table, calculate the reduced levels of points, and apply the checks. Also calculate the gradient of line AB.

Solution:

Interval	BS (m)	IS (m)	FS (m)	Rise (m)	Fall (m)	RL (m)	Remark
0	0.725					320.500	A
30		0.935			0.210	320.290	
60		2.845			1.910	318.380	
90		3.745			0.900	317.480	
120	0.965		3.935		0.190	317.290	CP-1
150		1.135			0.170	317.120	
180		1.785			0.650	316.470	
210		2.625			0.840	315.630	
240	0.965		3.845		1.220	314.410	CP-2

270		1.575			0.610	313.800	
300			2.015		0.440	313.360	B
Sum	**2.655**		**9.795**		**7.140**		

Checks:

Σ B.S. –Σ F.S. = 2.655 – 9.795 = -7.140

Σ Rise – Σ fall = 0.00 – 7.140 = -7.140

Last RL – First RL = 313.360 – 320.500 = -7.140

Distance AB = 30 * 10 = 300 m

Gradient of line AB = (First R.L. – Last RL) / Distance AB = (-7.140) / 300 = 0.0238

Example 1.15:

The following consecutive readings were taken with a level and 4 m levelling staff on a continuously sloping ground at 30 m intervals. 0.680, 1.455, 1.855, 2.330, 2.885, 3.380, 1.055, 1.860, 2.265, 3.540, 0.835, 0.945, 1.530 and 2.250 m. The RL of the starting point was 80.750 m. (i) Enter the above readings in a proper table, (ii) Determine the RL of various staff stations, and (iii) Compute the average gradient of measured ground.

Solution:

Interval	BS (m)	IS (m)	FS (m)	Rise (m)	Fall (m)	RL (m)	Remark
0	0.680					80.750	B.M.
30		1.455			0.775	79.975	
60		1.855			0.400	79.575	
90		2.330			0.475	79.100	
120		2.885			0.555	78.545	
150	1.055		3.380		0.495	78.050	C.P.1
180		1.860			0.805	77.245	
210		2.265			0.405	76.840	
240	0.835		3.540		1.275	75.565	C.P.2
270		0.945			0.110	75.455	
300		1.530			0.585	74.870	
330			2.250		0.720	74.150	
Sum	2.570		9.170		6.600		

Checks:

Σ BS –Σ FS = 2.570 – 9.170 = -6.600

Σ Rise – Σ fall = 0.000 – 6.600 = -6.600

Last RL – First RL = 74.150 – 80.750 = -6.600

Distance = 30 x 11 = 330

Gradient = (First RL – Last RL) / Distance = (80.750 – 74.150) / 330 = 0.020

Examples 1.16:

The following level readings are taken from a page of a level book. Some of the readings are missing. Fill up the missing readings and apply the arithmetic checks.

Station	BS	IS	FS	Rise	Fall	RL
A	3.125					
B	×		×	1.325		125.505
C		2.320			0.055	
D		×				123.850
E	×		2.655			
F	1.620		3.205		2.165	
G		3.625				
H			×			123.090

Solution:

The steps in the solution are as follows:

FS of station B= 3.125-1.325=1.800 m

BS of station B= 2.320-0.055=2.265 m

RL of BM= 125.505-1.325=124.180 m

Fall of station E= 125.850-125.115= 0.735 m

IS of station D = 2.655-0.735=1.920 m

BS of station E= 3.205-2.165=1.040 m

Rise of station H=123.090-120.945=2.145 m

FS of station H= 3.625-2.145=1.480 m

The missing entries are filled and presented in the following table:

Station	BS (m)	IS (m)	FS (m)	Rise (m)	Fall (m)	RL (m)	Remarks
A	3.125					124.180	BM
B	2.265		1.800	1.325		125.505	TP
C		2.320			0.055	125.450	
D		1.920		0.400		123.850	
E	1.040		2.655		0.735	125.115	TP
F	1.620		3.205		2.165	122.950	TP
G		3.625			2.005	120.945	
H			1.480	2.145		123.090	BM
Sum	8.050		9.140	3.870	4.960		

Checks:

ΣBS–ΣFS= 8.050–9.140= -1.09 m

ΣRise – Σfall =3.87–4.96= -1.09 m

LRL–FRL=123.09-124.180=-1.09 m

Since, ΣBS – ΣFS = Last RL – First RL= Σ Rise – Σ Fall

Therefore, the calculations are correct.

Example 1.17:

A level was set up at the mid-point between two pegs A and B, 50 m apart and the staff readings at A and B were 1.22 and 1.06 m, respectively. With the level set up near A, the readings at A and B were 1.55 and 1.37 m, respectively. Find, the collimation error per 100 m length of sight.

Solution:

Distance between A and B = 50 m.

When the level was set up at center, the difference in staff readings =1.22–1.06 = 0.160 m

When the level was set up near, the difference in staff readings A = 1.55 -1.37 = 0.180 m

Collimation error is the difference in above two values = 0.160–0.180 = -0.020 (- sign indicates downwards).

So, collimation error per 100 m length = (0.020 / 50)*100 = 0.04 m inclined downward.

Example 1.18:

Following observations are taken during a reciprocal levelling.

Instrument near	P	Q
Staff reading at P (in m)	1.824	0.928
Staff reading at Q (in m)	2.748	1.606

If the reduced level of P is 140.815 m, then compute the reduced level of Q.

Solution:

h_P =1.824 m, h_Q = 2.748 m

h'_P = 0.928 m, h'_Q = 1.606 m

Correct difference in elevation between P and Q is

$h = [(h'_Q - h'_P) + (h_Q - h_p)] / 2$

$= (1.606 - 0.928) + (2.748 - 1.824) / 2$

= 0.801 m > 0, and therefore point Q has lower elevation than point P.

The RL of Q = (140.815–0.801)

= 140.014 m

Example 1.19:

The following offsets were taken from a chain line to an irregular boundary line at an interval of 10 m: 0, 2.50, 3.50, 5.00, 4.60, 3.20, and 0 m. Compute the area between the chain line, the irregular boundary line and the end of offsets by: (a) the Trapezoidal rule, and (b) Simpson's rule.

Solution:

By Trapezoidal rule:

Here d=10 m

Required area=10 / 2{0 + 0 + 2(2.50 + 3.50 + 5.00 + 4.60 + 3.20)}

= 5*37.60 =188 m^2

By Simpson's rule:

d=10 m

Required area=10/ 3{0 + 0 + 4(2.50 + 5.00 + 3.20) + 2(3.50 + 4.60)}

= 10/3{42.80 +16.20}=10/3*59.00

= 196.66 m^2

Example 1.20:

An embankment of width 10 m and side slopes 1½:1 is required to be made on a ground which is level in a direction transverse to the centre line. The central heights at 40 m intervals are as: 0.90, 1.25, 2.15, 2.50, 1.85, 1.35, and 0.85 m, calculate the volume of earth work according to (i) Trapezoidal formula, and (ii) Prismoidal formula

Solution:

The cross-sections areas are calculated by

Δ= (b + sh)*h

Δ1= (10 + 1.5*0.90)*0.90 = 10.22 m^2

Δ2= (10 + 1.5*1.25)*0.90 = 14.84 m^2

Δ3= (10 +1.5*1.25)*2.15 = 28.43 m^2

Δ4= (10 +1.5*2.50)*2.50 = 34.38 m^2

Δ5= (10 +1.5*1.85)*1.85 = 23.63 m^2

Δ6= (10 +1.5*1.35)*1.35 = 16.23 m^2

Δ7= (10 +1.5*0.85)*0.85= 9.58 m^2

a. *Volume according to trapezoidal formula*

V= 40 / 2{10.22 + 9.58 + 2(14.84 + 28.43 + 34.38 + 23.63 + 16.23)}

= 20{19.80 + 235.02} = 5096.4 m^2

b. *Volume calculated in prismoidal formula:*

V = 40 / 3 {10.22 + 9.58 + 4(14.84 + 34.38 +16.23) + 2(28.43 + 23.63)}

= 40 / 3 (19.80 + 261.80 +104.12) = 5142.9 m^2

Example 1.21:

The areas enclosed by the contours in the lake are as follows:

Contour (m)	Area (m^2)
270	2050
275	8400

280	16300
285	24600
290	31500

Calculate the volume of water between the contours 270 m and 290 m by Trapezoidal formula.

Solution:

Volume according to trapezoidal formula:

$= 5 / 2\{2050 + 31500 + 2(8400 + 16300 + 24600)\}$

$= 330{,}250\ m^3$

Example 1.22:

The following readings were taken with a tacheometer on to a vertical staff, calculate the tacheometric constants.

Horizontal Distance (m)	Stadia Readings (m)
45.00	0.885 1.110 1.335
60.00	1.860 2.160 2.460

Solution:

$D_1 = KS_1 + C$ ....(1)

$D_2 = KS_2 + C$ ....(2)

So, 45 = K (1.335 – 0.885) + C

45 = K (0.45) + C ....(3)

60 = K (2.460 – 1.860) + C

60 = K (0.6) + C ....(4)

Equating C from equations 3 and 4, we get-

45 – K (0.45) = 60–K (0.6)

0.15K = 15

K =100

Now put the value of K in either equation 3 or 4, we get C =0

Example 1.23:

The tacheometer instrument was setup over a station P of RL 1850.95 m and the height of instrument above P was 1.475 m. The staff was held vertical at a station Q and the three readings were 1.050, 1.900 and 2.750 m with the line of sight horizontal. Calculate the horizontal distance of PQ and RL of Q point.

Solution:

D = KS + C

Now, S = 2.750 – 1.050 = 1.700 m

D = 100 (1.700) + 0 = 170 m

RL of Q = 1850.95 + 1.475 – 1.900

= 1850.525 m

Example 1.24:

A tachometer was setup at a station A and the following readings were obtained on a staff held vertical. Calculate the horizontal distance AB and RL of B, when the constants of instrument are 100 and 0.15.

Inst. Station	Staff Station	Vertical angle	Hair Reading (m)			Remark
A	BM	- 6° 40'	1.200	1.900	2.600	RL of BM = 850.500 m
A	B	+ 8° 20'	0.800	1.600	2.400	

Solution:

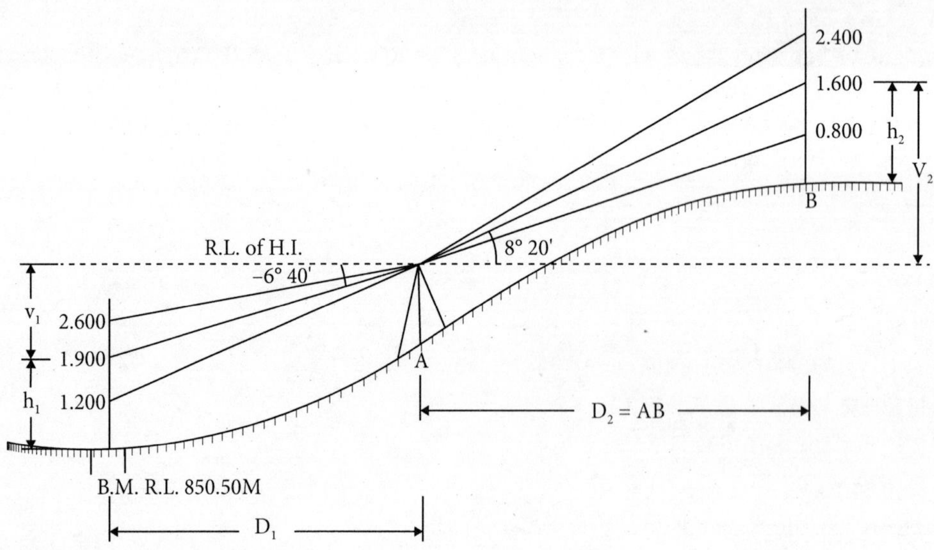

In first observation

$S_1 = 2.600 - 1.200 = 1.400$ m

$\theta_1 = -6^\circ 40'$ (Depression)

K = 100 and C = 0.15

Vertical Distance $V_1 = KS_1 \sin 2\theta / 2 + C \sin\theta$

$= 100 (1.400) \sin (2 \times 6^\circ 40') / 2 + 0.15 \sin 6^\circ 40'$

$= 16.143 + 0.0174$

$= 16.160$ m

In second observation

$S_2 = 2.400 - 0.800 = 1.600$ m

$\theta_2 = + 8^\circ 20'$ (Elevation)

Vertical Distance $V_2 = KS_2 \sin 2\theta / 2 + C \sin\theta$

$= 100 (1.600) \sin (2 \times 8^\circ 20') / 2 + 0.15 \sin 8^\circ 20'$

$= 22.944 + 0.022$

$= 22.966$ m

Horizontal distance $D_2 = KS_2 \cos^2\theta + C \sin\theta$

$= 100 (1.600) \cos^2 8°20' + 0.15 \sin 8°20'$

$= 156.639 + 0.148 = 156.787$ m

RL of instrument axis = RL of BM + h_1 + V_1

$= 850.500 + 1.900 + 16.160$

$= 868.560$ m

RL of B = RL of Inst. axis + $V_2 - h_2$

$= 868.560 + 22.966 - 1.600$

RL of B = 889.926m

Example 1.25:

To determine the gradient between two points P and Q, a tacheometer was set up at R station, and the following observations were taken keeping the staff vertical at P and Q. If the horizontal angle PRQ is 36°20' and RL of HI is 100 m, determine the average gradient between P and Q.

Staff station	Vertical angle	Stadia readings (m)
P	+4°40'	1.210, 1.510, 1.810
Q	-0°40'	1.000, 1.310, 1.620

Solution:

In the first observation (From R to P)

$S_1 = 1.810 - 1.210 = 0.6$ m

$\theta_1 = +4°40'$

Horizontal distance $D_1 = KS_1 \cos^2\theta + C \sin\theta$

$= 100 \times 0.6 \times \cos^2 4°40' + 0$

$= 59.60$ m

Vertical Distance $V_1 = KS_1 \sin 2\theta/2 + C \sin\theta$

$= 100 \times 0.6 \times \text{Sin} (2 \times 4°40') / 2 + 0$

$= 4.865$ m

In the second observation (From R to Q)

$S_2 = 1.620 - 1.000 = 0.62$ m

$\theta_2 = -0°40'$

Horizontal distance $D_2 = KS_2 \cos^2\theta + C \sin\theta$

$= 100 \times 0.62 \times \text{Cos}^2 0°40' + 0$

$= 61.99$ m

Vertical distance $V_2 = KS_2 \sin 2\theta/2 + C \sin\theta$

$= 100 \times 0.62 \times \sin (2 \times 0°40') / 2 + 0$

$= 0.721$ m

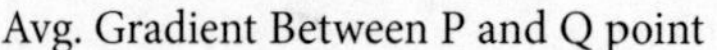

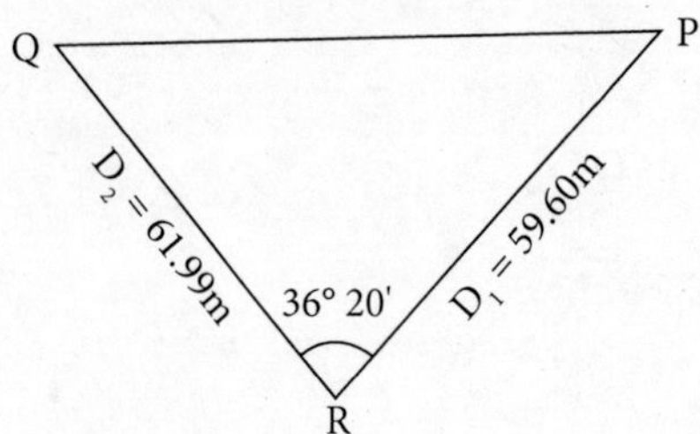

Distance D_1 = PR = 59.60 m

Distance D_2 = QR = 61.99 m

Angle PRQ = 36°20'

$PQ^2 = PR^2 + QR^2 - 2 \times PR \times QR \times \cos 36°20'$

$PQ^2 = (59.60)^2 + (61.99)^2 - 2 \times 59.60 \times 61.99 \times \cos 36°20'$

PQ = 37.978 m

Difference of elevation between P and Q

RL of P = RL of HI + $V_1 - h_1$

= 100 + 4.865 – 1.510

= 103.355 m

RL of Q = RL of HI – $V_2 - h_2$

= 100 – 0.721 – 1.310

= 97.969 m

Difference = 103.355 – 97.969 = 5.386 m

Average gradient between P and Q = Difference in RL between P & Q / Distance PQ

= 5.386 / 37.978

= 1 / 7.051

Exercise 1.26:

The vertical angles to vanes fixed at 1 m and 3 m above the foot of the staff held vertically at station Q were 3°20' and 6°40', respectively from instrument station P. If the elevation of the instrument axis at station P is 101.520 m, calculate (i) the horizontal distance between P and Q, and (ii) the elevation of the staff station Q.

Solution:

S = 3 – 2 = 1 m

θ_1 = 6°40'

θ_2 = 3°20'

h = 1 m

$D = S / [\tan \theta_1 - \tan \theta_2]$

= 1 / [tan 6°40'–tan 3°20']

= 34.13 m

$V = S \tan \theta_2 / [\tan \theta_1 - \tan \theta_2]$

= 2 x tan 3°20' / [tan 6°40'–tan 3°20']

= 1.99 m

Elevation of staff station Q = RL of HI + V – h

= 101.520 + 1.99 – 1.0

= 102.510 m

Example 1.27:

From the top of a light house, the angles of depression of two ships are 30° and 45°. The two ships, as it was observed from the top of the light house, were 100 m apart. Find the height of the light house.

Solution:

Height = Distance / [cot (original angle) – cot (final angle)]

Height of the light house = 100 / (cot 30° – cot 45°)

= 50 m

Example 1.28:

An instrument was set up at P and the angle of depression to a vane 2 m above the foot of the staff held at Q was 5° 36′. The horizontal distance between P and Q was known to be 3000 metres. Determine the R.L. of the staff station Q given that staff reading on a B.M. of elevation 436.050 was 2.865 metres.

Solution:

The difference in elevation between the vane and the instrument axis = D tan α

= 3000 tan 5° 36′ = 294.153 m

Combined correction due to cuvature and refraction C = 0.06735 D^2 metres

when D is in km = 0.606 m.

Since the observed angle is negative, the combined correction due to curvature and refraction is subtractive.

Difference in elevation between the vane and the instrument axis = 294.153 – 0.606 = 293.547 = h

RL of instrument axis = 436.050 + 2.865 = 438.915

RL of the vane = RL of instrument axis – h = 438.915 – 293.547 = 145.368

RL of Q = 145.368 – 2 = 143.368 m

Example 1.29:

In order to ascertain the elevation of the top (Q) of the signal on a hill, observations were made from two instrument stations P and R at a horizontal distance 100 metres apart, the station P and R being in the line with Q. The angles of elevation of Q at P and R were 28° 42′ and 18° 6′, respectively. The staff reading upon the bench mark of elevation 287.28 m were respectively 2.870 m and 3.750 m, when the instrument was at P and at R, the telescope being horizontal. Determine the elevation of the foot of the signal if the height of the signal above its base is 3 metres.

Solution:

Elevation of instrument axis at P = R.L. of B.M. + Staff reading = 287.28 + 2.870 = 290.15 m

Elevation of instrument axis at R = R.L. of B.M. + staff reading = 287.28 + 3.750 = 291.03 m

Difference in level of the instrument axes at the two stations S = 291.03 – 290.15 = 0.88 m

α_1 = 28° 42 and α_2 = 18° 6′

S cot α_2 = 0.88 cot 18° 6′ = 2.69 m

D = 152.1 m

h_1 = D tan α_1 = 152.1 tan 28° 42′ = 83.272 m

RL of foot of signal = RL of inst. axis at P + h_1- ht. of signal

= 290.15 + 83.272 – 3 = 370.422 m

Check :

(b + D) = 100 + 152.1 m = 252.1 m

h_2 = (b + D) tan α_2 = 252.1 x tan 18° 6′ = 82.399 m

RL of foot of signal = RL of inst. axis at R + h_2 + ht. of signal

= 291.03 + 82.399 – 3 = 370.429 m

Example 1.30:

There are two poles with different heights; one on each side of the road. The higher pole is 54 m high, and from its top, the angle of depression of top and bottom of the shorter pole is 30° and 60°, respectively. Find the height of the shorter pole.

Solution:

Let AB be the higher pole, and CD be the lower pole.

In triangle ABC,

tan 60 = AB / AC

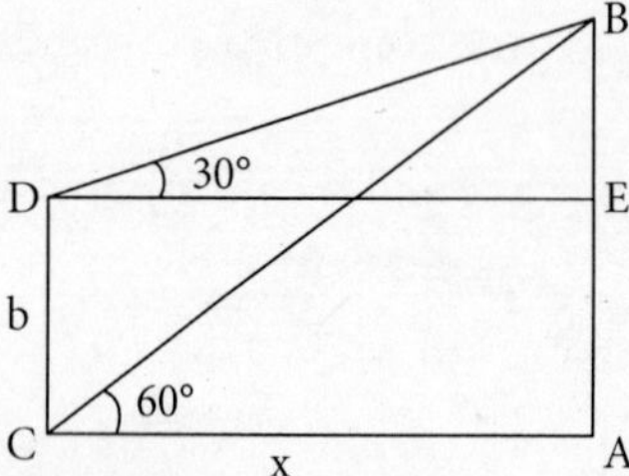

√3 = 54 / AC

AC=54/ √3 = DE

In triangle BED,

tan 30 = BE / DE

BE = tan 30 *DE

= (1/√3) * (54/ √3)

= 18 m

CD = AE = AB – BE

CD = 54 – 18 = 36 m

Therefore, height of the shorter pole = 36 m

Example 1.31:

In the following table, the WCB and the lengths of traverse lines of a closed traverse 1-2-3-4-5-6-7-8-9-10-1 are given. Compute the correct latitudes and departures of traverse lines, if the coordinates of point 1 are (1000, 2000) m:

Side	WCB (Φ)	Length (L)
1-2	327*56'7"	217.72
2-3	331*53'7"	113.33
3-4	82*39'2"	318.67
4-5	90*28'55"	137.77
5-6	109*53'9"	80.81
6-7	111*30'4"	71.13
7-8	169*58'59"	219.18
8-9	264*51'15"	162.36
9-10	256*55'51"	208.60
10-1	276*9'20"	101.260
Σ		1630.83

Solution:

First compute the latitude and departure of lines taking into account the proper sign of the quadrant.

Side	WCB (Φ)	Length (L)	L cos(Φ)	L sin(Φ)
1-2	327*56'7"	217.72	184.506	-115.582
2-3	331*53'7"	113.33	99.958	-53.405
3-4	82*39'2"	318.67	40.735	316.056
4-5	90*28'55"	137.77	-1.159	137.765
5-6	109*53'9"	80.81	-27.489	75.991
6-7	111*30'4"	71.13	-26.071	66.180
7-8	169*58'59"	219.18	215.839	38.124
8-9	264*51'15"	162.36	-14.562	-161.706
9-10	256*55'51"	208.60	-47.170	-203.197
10-1	276*9'20"	101.260	10.858	-100.676
Σ		1630.83	3.769	-0.456

Determine the closing error = $\sqrt{[(3.769)^2 + (-0.456)^2]}$ = 3.80 m

Apply corrections to each as per Latitude and Departure by Bowditch method

Correction in each Latitude CL =ΣL×(L/ΣL)

Similarly, correction in each Departure CD = ΣD * (L/ΣL)

The corrected Latitude and Departure as well as Coordinates of the points are computed as given below:

Station	Corrected Lcos Φ	Corrected Lsin Φ	Corrected latitude	Corrected departure
1	184.002	-115.522	1000	2000
2	99.696	-53.374	884.478	2184.002
3	39.998	316.144	831.104	2283.698
4	-1.477	137.803	1147.248	2323.696
5	-27.674	76.013	1285.051	2322.219
6	-26.235	66.2	1361.064	2294.545
7	-216.345	38.184	1427.264	2268.310
8	-14.937	-161.661	1465.448	2051.965
9	-47.652	-203.139	1303.787	2037.028
10	10.624	-100.648	1100.648	1989.376
1			1000	2000
	Σ Lcos Φ = 0	ΣLsin Φ = 0	Check	

Example 1.32:

The data for a closed traverse is given below. Balance the traverse by Bowditch method and Transit method.

Line	Length	WCB
AB	89.31	45°10′
BC	219.76	72°05′
CD	151.18	161°52′
DE	159.10	228°43′
EA	232.26	300°42′

Solution:

Perimeter = ΣL = Sum of lengths = 851.61 m

Magnitude of closing error = $\sqrt{((0.51)^2 + (0.224)^2)}$ = 0.557 m

Direction of closing error = $\tan^{-1}(0.224 / 0.51)$ = 23°42'42.6″

By Bowditch method

CL =ΣL×(L/ΣL)

Correction in latitude of line AB = 0.51×(89.31 / 851.61) =0.053

Correction in latitude of line BC = 0.51×(219.76 / 851.61) =0.132

Correction in latitude of line CD = 0.091

Correction in latitude of line DE = 0.095

Correction in latitude of line EA = 0.061

Similarly,

CD = ΣD * (L/ΣL)

Correction in departure of line AB = 0.224×(89.31/851.61) = 0.023

Correction in departure of line BC = 0.224×(219.76/851.61) = 0.058

Correction in departure of line CD = 0.224×(151.18/851.61) = 0.040

Correction in departure of line DE = 0.224×(159.10/851.61) = 0.042

Correction in departure of line EA = 0.224×(232.26/851.61) = 0.061

For corrected coordinates,

Corrected Latitude= Latitude -Correction (since (ΣL) is positive, correction will be negative)

Corrected departure= Departure-Correction

Line	Length	WCB	Consecutive coordinate		Correction		Corrected coordinate	
			Latitude (Lcos∅)	Departure (Lsin∅)	Latitude	Departure	Latitude	Departure
AB	89.31	45°10'	62.968	63.335	0.053	0.023	62.915	63.312
BC	219.76	72°05'	67.606	209.103	0.132	0.058	67.474	209.045
CD	151.18	161°52'	-143.672	47.052	0.091	0.040	-143.763	47.012
DE	159.10	228°43'	-104.971	-119.557	0.095	0.042	-105.066	-119.599
EA	232.26	300°42'	118.579	-199.709	0.139	0.061	118.44	-199.77
			ΣL=0.51	ΣD=0.224	Σ1=0.51	ΣD=0.224	ΣL=0	ΣD=0

By Transit method

For correction,

CL = ΣL * L / ΣLt

CD = ΣD * D / ΣDt

ΣL = Total error in latitude (Sign consideration)

ΣD= Total error in departure (Sign consideration)

L, D = Latitude, and Departure of any side

ΣDt = Arithmetic sum of Departures (No sign consideration)

ΣLt = Arithmetic sum of Latitudes (No sign consideration)

Correction in latitude of line AB = 0.51×(62.968 / 497.796) = 0.065

Correction in latitude of line BC = 0.51×(67.606 / 497.796) = 0.069

Correction in latitude of line CD = 0.51×(143.672 / 497.796) = 0.147

Correction in latitude of line DE = 0.51×(104.971 / 497.796) = 0.108

Correction in latitude of line EA = 0.51×(118.579 / 497.796) = 0.121

Similarly,

Correction in departure of line AB = 0.224×(63.335 / 638.756) = 0.022

Correction in departure of line BC = 0.224×(209.103 / 638.756) = 0.073

Correction in departure of line CD = 0.224×(47.052 / 638.756) = 0.017

Correction in departure of line DE = 0.224×(119.557 / 638.756) = 0.042

Correction in departure of line EA = 0.224×(199.709 / 638.756) = 0.070

For corrected coordinates,

Corrected latitude= latitude–correction in latitude (since (ΣL) is positive, correction will be negative)

Corrected departure= departure- correction in departure

Line	Length	WCB	Consecutive coordinate		Correction		Corrected coordinate	
			Latitude (Lcosθ)	Departure (Lsinθ)	Latitude	Departure	Latitude	Departure
AB	89.31	45°10'	62.968	63.335	0.065	0.022	62.903	63.313
BC	219.76	72°05'	67.606	209.103	0.069	0.073	67.537	209.03
CD	151.18	161°52'	-143.672	47.052	0.147	0.017	-143.819	47.035
DE	159.10	228°43'	-104.971	-119.557	0.108	0.042	-105.079	-110.5
EA	232.26	300°42'	118.579	-199.709	0.121	0.070	118.458	-199.779
			ΣL=0.51 ΣLt=497.796	ΣD=0.224 ΣDt=638.756	ΣL=0.51	ΣD=0.224	ΣL=0	ΣD=0

Example 1.33:

In a closed traverse, calculate the length and bearing of the missing traverse leg BC.

Line	Length (m)	Bearing
AB	89.31	45°10′
BC	L	θ
CD	151.18	161°52′
DE	159.1	228°43′
EA	232.26	300°42⊠

Solution:

For a closed traverse,

ΣL= 0 (Sum of latitude is zero)

$62.968 + L\cos\theta - 143.672 - 104.971 + 118.579 = 0$

$L\cos\theta = 67.096$ ....(i)

ΣD=0 (Sum of departure is zero)

$63.335 + L \sin\theta + 47.052 - 119.557 - 199.709 = 0$

$L \sin\theta = 208.879$ ....(ii)

Dividing equation (ii) by (i)

$\tan\theta = (208.879 / 67.096)$

$\theta = 72^\circ 11' 31.1''$

Putting value of y on any equation, we get-

$L = 219.391$ m

Example 1.34:

Calculate the missing data in a closed traverse-

Line	Length (m)	Bearing
AB	89.31	45°10′
BC	219.76	72°05′
CD	151.18	161°52′
DE	?	228°43′
EA	232.26	?

Solution:

Consider a closed traverse ABCDA, where the length of line AD be x and bearing θ.

For a closed traverse ABCDA,

$\Sigma L = 0$ (Sum of latitude is zero)

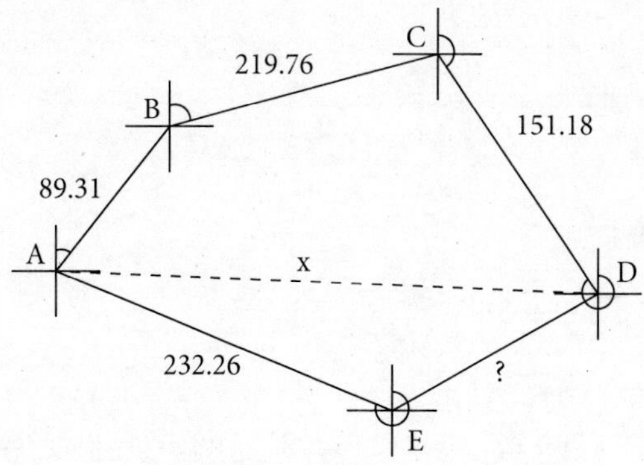

$62.968 + 67.606 + x \cos\theta - 143.672 = 0$

$x \cos\theta = 13.098$ ....(i)

$\Sigma D = 0$ (Sum of departure is zero)

$63.335 + 209.103 + x \sin\theta + 47.052 = 0$

$x \sin\theta = -319.49$ ....(ii)

From eqn (i) and (ii)

$\tan\theta = -24.392$

θ = -87°39′8.58″ (can't be negative)

= 360°–87°39′8.58″

= 272°20′57.42″

x = 319.758 m

Now, in triangle AED, using sine formula;

{(Sin ∠D) / 232.26} = {(Sin ∠A) / ED} = {(Sin ∠E) / 319.758}

Then,

∠D = 272°20′51.42″ – 228°43′

= 43°37′51.42″

((Sin 43°37′51.42″) / 232.26) = ((Sin ∠E) / 319.758)

Sin ∠E = 0.958

∠E=71°47′48.37″

Using cosine law,

Cos D = $((AD^2 + ED^2 - AE^2) / 2(AD)(ED))$

Cos 43°37′51.42″ = $(319.758^2 + ED^2 - 232.26^2) / 2 \times 319.758 \times ED$

ED =158.982 m

Again,

((Sin ∠D) / 232.26) = ((Sin ∠A) / ED)

Sin ∠A= 0.472

∠A=28°11′4.21″

Then,

Bearing of DE = 228°43′

Bearing of ED = 228°43′-180° = 48°43′

Bearing of EA=360°-71°47′48.37″+48°43′= 336°55′11.63″

Exercises for Practice

a. Short Questions

1.35. Write the basic principle of surveying.

1.36. Why is it important to have a knowledge of surveying to a civil engineer?

1.37. Discuss various maps and the scale of topographic maps prepared by Survey of India.

1.38. Draw the symbol and write colours of various features; Railway line, Contours, Landslide, BM, Bridge, Temple, Canal, Road, and Underground tunnel.

1.39. Describe, Survey Station, Survey Lines in a traverse.

1.40. Write the criteria used while making an appropriate selection of traverse stations.

1.41. What are the various methods, employed for distance measurement on the Earth surface?

1.42. What do you understand by the term *Ranging* in surveying? How do you carry out ranging process in the field?

1.43. What is Local Attraction in compass measurement? How do you detect the presence of Local Attraction at a station?

1.44. What is the use of a Levelling Staff? List types of staffs used? What is the least count of levelling staff?
1.45. What are the checks applied to the computed RLs by both the methods?
1.46. Define the following terms; Contour, Contour interval, Horizontal equivalent,
1.47. Draw the contours of a Vertical cliff, Overhanging cliff, Steep slope, and valley.
1.48. What is the use of Plane Table in surveying? List the other accessories and equipment used along with the Plane Table for mapping work.
1.49 Define the terms: Axis of the telescope, Vertical axis of a Theodolite, Trunnion axis of a Theodolite.
1.50. What is a Tacheometry? How do you determine the distance by Tacheometry on a flat ground?
1.51. Write the difference between Triangulation and Trilateration in surveying?

b. Long Questions

1.52. Describe, how the surveying technology has developed in India.
1.53. Describe various types of surveying, based on area covered, based on instruments used, and based on purpose of survey.
1.54. Discuss various sources of errors, likely to be present in survey observations.
1.55. Define the following- True bearing, Magnetic bearing, Whole circle bearing, Quadrantal bearing, Magnetic declination, Fore bearing and Back bearing.
1.56. What is relationship between (i) True bearing and Magnetic bearing, (ii) WCB and QB, (iii) Fore bearing and Back bearing of a line, and (iv) Bearing and included angle.
1.57. Draw neat sketch of Prismatic Compass and explain various components and their function. What is the least count of this compass?
1.58. Define various terms with the help of neat sketches; Levels surface, Level line, Horizontal line, Plumb line, Datum, Mean sea level, Reduced level, Beach mark, Line of sight, Line of collimation, and Height of Instrument, Diaphragm.
1.59. Define Back sight, Intermediate sight, Change point and Fore sight in levelling. Why do you normally keep Back sight and Fore sight distances nearly equal in levelling?
1.60. Discuss various types of levels used in levelling work.
1.61. Write the salient characteristics of an Auto level and a Laser level.
1.62. Explain the steps to be followed for temporary adjustment of a level.
1.63. Discuss the two methods of reduction of levelling observations
1.64. Explain Profile levelling and Cross-section levelling. What is their utility in civil engineering projects?
1.65. Establish a relationship to find the collimation error using Reciprocal Levelling method.
1.66. What are various sources of errors that might be present in levelling observations, and how to minimise them?
1.67. Discuss the factors which decide the contour interval.
1.68. Describe the characteristics of contour maps.
1.69. List various applications of a contour map.
1.70. What are the different method to compute area and volume from a topographic map? Write the relationships.

1.71. List the advantages and disadvantages of Plane Tabling surveying.

1.72. Draw a diagram of Vernier theodolite and write its various parts.

1.73. Differentiate between (i) Transit and Swing of a Theodolite, (ii) Face left and Face right observations, and (iii) Reiteration method and Repetition method of observations.

1.74. Explain the use of a Theodolite (i) For prolonging a straight line, and (ii) as a Level.

1.75. Establish a relationship to compute the distance when the staff is not held vertical.

1.76. What is the purpose of Trigonometrical levelling? Write a simple relationship to determine the height of a building when the base is accessible.

1.77. Determine a relationship to determine the height of a tower when the base is inaccessible and it is not possible to set the instrument in the line of sight of the tower.

1.78. What is the basic purpose of Traversing? Write the steps involved in taking the observations of a traverse and adjusting the observations to compute the coordinates of traverse stations.

1.79. What are the different methods of adjustment of closing error in a closed traverse? Write the relationships.

1.80. Discuss the types of Triangulation schemes.

1.81. Define the terms in triangulation; Centered figure, Base line, Laplace station, Satellite station, Signals, Axis signal correction, Reduction to centre, and Accuracy of triangulation.

c. Unsolved Examples

1.82. A levelling is a carried out to establish the RL of a point C with respect to BM of 100.00 m RL at A. Compute the RL of point C. The staff readings are given below.

Staff station	BS (m)	FS (m)
A	1.545	-
B	-0.860	-1.420
C		0.835

(Ans: RL of C = 101.27 m)

1.83. The staff reading taken on a point of RL 40.500 m below the bridge is 0.645 m. The inverted staff reading taken at the bottom of the deck of the bridge is -2.960 m. Compute the reduced level of the bottom of the deck point.

(Ans: RL of bottom of the deck = 44.105 m)

1.84. During a theodolite survey the following details were noted:

Line	Length (m)	Back Bearing
AB	550	60°
BC	1200	115°
CD	?	?
DA	1050	310°

Calculate the length and bearing of the line CD.

(Ans: Bearing of CD, $\theta = S\ 58^{o}\ 49'\ W = 238^{o}\ 49'$, Length of CD = 855.15 m)

1.85. In a closed traverse ABCDE, it is required to find length and bearing of AE.

Following is the record of readings.

Line	Length (m)	Bearing
AB	130.5	N20°30' E
BC	215.0	N60°15' E
CD	155.5	N30°30' E
DE	120.0	N30°30' E

(Ans: Bearing of EA, θ = S 38° 35′ W = 218° 35′, Length of EA = 596.51 m)

1.86. It is not possible to measure the length and fix the direction of a line AB directly on account of an obstruction between the stations A and B. A traverse ACDB was therefore run and the following data was obtained.

Line	Length (m)	Reduced Bearing
AC	45	N 50° E
CD	66	S 70° E
DB	60	S 30° E

Find the length and direction of line BA

(Ans: Bearing of BA, θ = N 70° 9′ W = 289° 51′, Length of BA = 134.32 m)

References and Suggested Readings

1. Abdalla, Rifaat, (2019), From Land Surveying to Geomatics–Multidisciplinary Technological Trends, IntechOpen, http://dx.doi.org/10.5772/intechopen, 82424.Agor, R., (1980), Surveying, Vol-1 and II, Khanna Publishers, Delhi.
2. Ali, Abdalla Elsadig and Al-garni, Abdullah Mohammed, (1996), Evaluating the Accuracy of Laser Levels for Engineering Surveying, J. King Saud Univ., 8 (1), pp.121-131 (A.H. 1416/1996).
3. Anam, Aysha, (2018), How India was Measured: Story of the Great Trigonometrical Survey of India, India Today Web Desk, New Delhi, July, https://www.indiatoday.in/education-today/gk-current-affairs/story/how-was-india-measured-why-is-it-important-to-measure-land-1281835-2018-07-10
4. Anderson, James M., and Mikhail, Edward M., (2012), Surveying: Theory and Practice Tata McGraw Hill (Indian Edition).
5. Arora, K. R., (2015), Surveying Vol. I & II, Standard Book House.
6. Basak, N.N., (2017), Surveying and Levelling, Tata McGraw Hill.
7. Brock, John, (2019), The Surveyors of Our World -A Journey through History, GIM International, https://www.gim-international.com/content/article/the-surveyors-of-our-world.
8. Chadha, S.M., (1991), Survey of India through the Ages, The Himalayan Journal, Vol. 47.
9. Chandra, A.M., (2005), Plane Surveying, New Age International Publishers.
10. Dangi, Mohan (2020), Theodolite, Digitech Education, https://www.gyanimohan.xyz/2020/04/Theodolite-types-of-theodolite.html
11. Dowson, Emily, (2017), Importance of Laser Levels in the World of Surveying, https://www.howimportant.com/importance-laser-levels-world-surveying/.
12. Duggal, S.K., (2017), Surveying, Vol. I & II, Tata Mc-Graw Hill.

13. Duncan Parnell, (2018), Survey Prism Basics, https://www.duncan-parnell.com/blog/survey-prism-basics.
14. e-Krishi Shiksha (2019), Lesson 17: Land Grading survey and design, http://ecoursesonline.iasri.res.in /mod/page/view.php?id=1985
15. Garg, P.K., (2019), Theory and Principles of Geoinformatics, Khanna Book Publishing House, New Delhi.
16. Garg, P.K. (2020), Introduction to Unmanned Aerial Vehicles, New Age International Publishers Pvt. Ltd., New Delhi.
17. Garg, P.K. (2021), Digital Land Surveying, New Age International Publishers Pvt. Ltd., New Delhi.
18. Gopi Satheesh; Sathikumar, R., and Madhu, N. (2017), Advanced Surveying, Pearson India.
19. Hill, Digital, (2011), Laser Surveying Instruments, https://www.digitalhill.com/blog/laser-surveying-instruments-types-and-uses/.
20. https://geokult.com/2011/09/14/map-symbolisation/
21. https://www.osha.gov
22. https://dreamcivil.com/ranging-in-surveying/
23. Kanetkar, T.P. and Kulkarni, S.V., (2006), Surveying and Levelling, Vol. I and II, Vidyarthi Griha Prakashan, Pune.
24. Keay, John (1983), When Men and Mountains Meet: Explorers of the Western Himalayas, 1820-75 (Traveller's S.), Publisher: Pimlico, New edition (25 August 1983)
25. Mudavath, Kishan, (2018), Digital Level Surveying, We Civil Engineers, https://wecivilengineers.wordpress.com/2018/09/02/features-advantages-of-digital-level-surveying/.
26. One Point Survey (2019), A Buyers Guide to Laser Levels, July 24, https://www.onepointsurvey.com /blogs/guides/buyers-guide-to-laser-levels.
27. Phillimore, R. H., (1945), Historical Records of the Survey of India Eighteenth Century, Vol. I, Survey of India, Dehradun.
28. Punmia, B.C., Jain, Ashok K., Jain, and Arun K., (2016), Surveying, Vol. I & II, Laxmi Publications.
29. Satyaprakash, (2010), Land Surveying in India: Present Status, Geospatial World- July 12, https://www.geospatialworld.net/article/land-surveying-in-india-present-status/.
30. Schofield, W., (1984), Engineering Surveying, Buttersworth.
31. Schultz, R.J. (1987). Leveling. In: Brinker, R.C., Minnick, R. (eds) The Surveying Handbook. Springer, Boston, MA. https://doi.org/10.1007/978-1-4757-1188-2_6
32. Subramanian, R., (2012), Surveying and Levelling, Oxford University Press.
33. Survey Manual, (2014), Chapter 3, Surveying Measurements, Department of Transportation, State of New Jersey, Prepared by Survey Services, https://www.state.nj.us/transportation /eng/documents/survey/Chapter3.shtm.
34. Venkatramaiah, C., (2011), Textbook of Surveying, University Press.
35. Uren, John and Price, Bill, (1994), Surveying for Engineers, Palgrave Macmillan.
36. Wright, John Wilfrid (2019), Surveying Civil Engineering, https://www.britannica.com/technology/surveying/Modern-surveying.
37. Zeiske, Karl, (2004), Surveying Made Easy, Leica Geosystems, Leica Geosystems AG, Switzerland.

02 CURVES

UNIT SPECIFIES

Through this unit, we have discussed the following aspects:

- Horizontal and vertical curves
- Various types of curves
- Various components of curves
- Computation of parameters to set out the curves in the field
- Requirement of Super-elevation
- Computation of parameters for setting out various curves

In this unit, the necessity of curves on roads, and railway lines are discussed. The characteristics of each types of curves are given that would help selecting the right kind of curve for roads, and railways. The relationships given would help in computing various essential parameters of curves. A large number of questions of short and long answer types following lower and higher order of Bloom's taxonomy, assignments through problems, and a list of suggested readings and references are given in the unit so that the students can go through them for practice.

Rationale

This unit on curves helps students to get an idea about the design and establishment of various types of curve in the field. It explains the characteristics of each curve, including the computation of essential components required for layout of curves. All these are discussed at length to develop the understanding of subject matter. Some related problems are given which can help further for getting a clear idea of the concern topics on layout of curves. The provision of curves is to connect two points which can't be otherwise connected by a straight line due to topography of the ground and natural features present. The selection of right type of curve is important to provide a smooth and comfortable ride to the passengers. The super-elevation further helps in smooth riding when a vertical curve is introduced.

Pre-Requisites

Mathematics: geometry and trigonometry; Surveying: Use of theodolite and distance measuring instruments.

Unit Outcomes

List of outcomes of this unit is as follows:

U2-O1: Describe various types of curves

U2-O2: Describe the essential components and characteristics of curves

U2-O3: Realize the role of curves in roads and railways for smooth movement of vehicles

U2-O4: Explain the role of superelevation in vertical curves

U2-O5: Apply the parameters to solve complex problems and layout the curves

Unit-2 Outcomes	Expected Mapping with Programme Outcomes *(1- Weak correlation; 2- Medium correlation; 3- Strong correlation)*					
	CO-1	CO-2	CO-3	CO-4	CO-5	CO-6
U2-O1	3	2	2	2	1	2
U2-O2	3	1	2	1	2	-
U2-O3	2	2	2	3	-	2
U2-O4	3	2	3	2	1	-
U2-O5	2	3	2	1	-	2

2.1 INTRODUCTION

Curves are regular bends provided in the lines of communication, like highways, railways, etc., to make gradual change in the horizontal and vertical directions. Those curves that change the alignment or direction are known as *horizontal curves*, and those that change the slope are called *vertical curves.* Horizontal curves are provided in the horizontal plane to have the gradual change in direction, whereas vertical curves are provided in the vertical plane to obtain the gradual change in the grade. For example, the center line of a road consists of series of straight lines interconnected by curves that are used to change the alignment, direction, or slope of the road. Curves are laid out on the ground along the centre line of the alignment using various surveying equipment, such as theodolite, tape, levels, total station, etc. Various curves are shown in Figure 2.1.

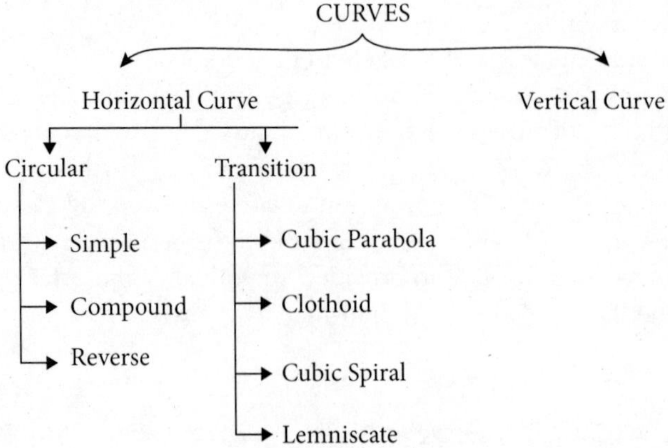

Fig. 2.1: Various types of curves

This unit covers the need of horizontal and vertical. Various types of curves and their various components are explained. Various relationships can be established to compute the essential parameters to set up horizontal and vertical curves. Briefly, super-elevation which is necessary to provide in horizontal curves, and sight distance which is required in vertical curves, are also discussed.

2.2 CLASSIFICATION OF HORIZONTAL CURVES

Circular curves are classified as: (*i*) Simple, (*ii*) Compound, (*iii*) Reverse, and (*iv*) Transition curves.

2.2.1 Simple curves

A simple curve consists of a single arc of a circle connecting two straights. It has the same radius throughout. In Figure 2.2a, a simple curve is shown which is passing through T_1 and T_2 with as OT_1 or OT_2 as its radius.

2.2.2 Compound curves

A compound curve consists of two or more simple curves having different radii bending in the same direction and lying on the same side of the common tangent. Their centres would lie on the same side of the curve. In Figure 2.2b, T_1PT_2 is the compound curve (two curves join at P) with T_1O_1 and T_2O_2 as its different radii.

2.2.3 Reverse curves

A reverse curve is made up of two equal or different radii bending in opposite directions with a common tangent at their junction. Their centres would lie on opposite sides of the curve. In Figure 2.2c, T_1PT_2 is a reverse curve with T_1O_1 and T_2O_2 as its radii. Point P is the junction of two curves in opposite direction. Reverse curves are used when the straights are parallel or intersect at a very small angle. They are commonly used in railway sidings, and sometimes on railway tracks, and is meant for low speeds. They should be avoided as far as possible on main railway lines and highways where speeds are high.

2.2.4 Transition curves

A transition curve is provided in between the straight line and circular curve (Figure 2.2d). It is a curve of varying radius of infinity at tangent point to a circular curve radius provided in between the straight line and circular curve in order to provide gradual centrifugal force. The objectives of providing a transition curve is mainly to gradually introduce the centrifugal force between the tangent point and the beginning of the circular curve, thereby avoiding sudden jerk on the vehicle.

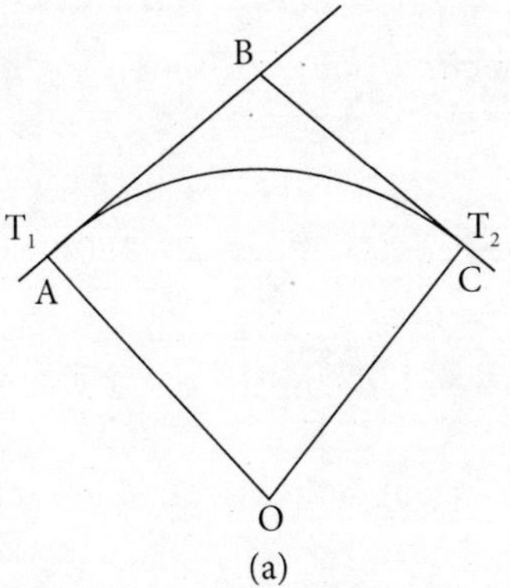

(a)

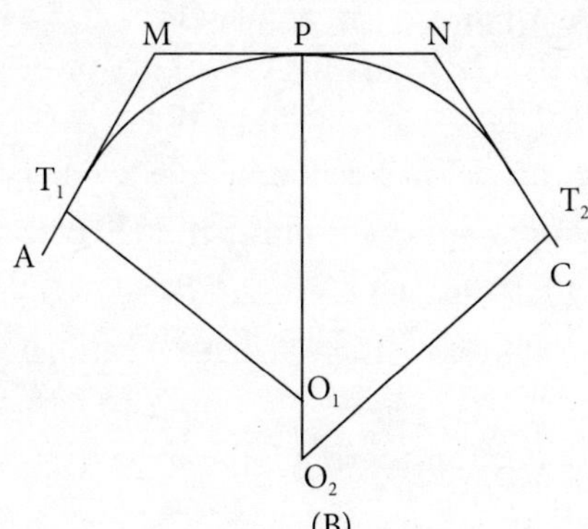

(B)

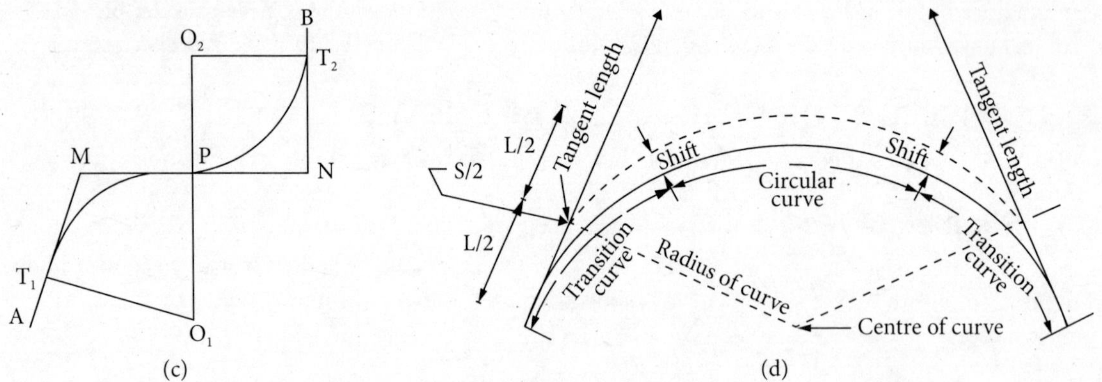

Fig. 2.2: Various curves (a) simple, (b) compound, (c) reverse, and (d) transition curve

2.3 SIMPLE CIRCULAR CURVES

In order to compute various parameters of a simple circular curve for its layout, it is necessary to understand various part of the curve, as shown in Figure 2.3, and explained below.

2.3.1 Various parts of a curve

Tangents or straights: The two straight lines AB and BC, which are connected by the curve, are called the tangents or straights to the curve. The lines AB and BC are tangents to the curve. Line AB is called the first tangent or the rear tangent, and line BC is called the second tangent or the forwarded tangent.

Intersection point: The points of intersection of the two straights (B) is called the intersection or vertex point.

Tangent length: The distance between the two tangent point of intersection to the tangent point (BT_1 and BT_2) is called the tangent length.

Right handed curve: When the curve deflects to the right side of the progress of survey, it is termed as right handed curve.

Left handed curve: When the curve deflects to the left side of the progress of survey, it is termed as left handed curve.

Tangent points: The points (T_1 and T_2) at which the curve touches the tangents are called the tangent points. The beginning of the curve at T_1 is called the tangent curve point and the end of the curve at T_2 is called the curve tangent point.

Long chord: The line joining the two tangent points (T_1 and T_2) is known as the long chord.

Summit or apex: The mid-point (F) of the arc (T_1FT_2) is called summit or apex of the curve.

Length of the curve: The arc T_1FT_2 is called the length of the curve.

Angle of intersection: The angle between the tangent lines AB and BC ($\angle ABC$) is called the angle of intersection (I).

The deflection angle: It is the angle (f) by which the forward tangent deflects from the rear tangent (180°-I) of the curve.

Apex distance: The distance from the point of intersection to the apex of the curve BF is called the apex distance.

Central angle: The angle subtended at the centre of the curve by the arc T_1FT_2 is known as the deflection angle (f).

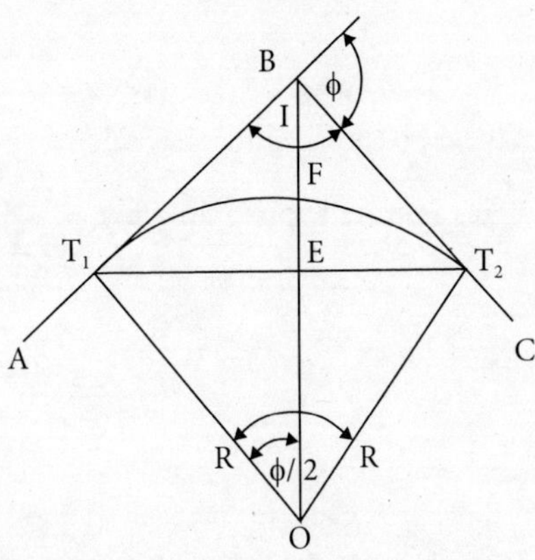

Fig. 2.3: Representations of a simple circular curve

2.3.2 Designation of horizontal curves

A simple circular curve may be designated either by its radius or by the angle subtended at the centre by a chord of a particular length. In India, a simple circular curve is designated by the angle (in degrees) subtended at the centre by a chord of 30 m (100 ft.) length. This angle is called the *degree of curve* (D). The relation between the radius and the degree of the curve may be established as follows (Refer Figure 2.4).

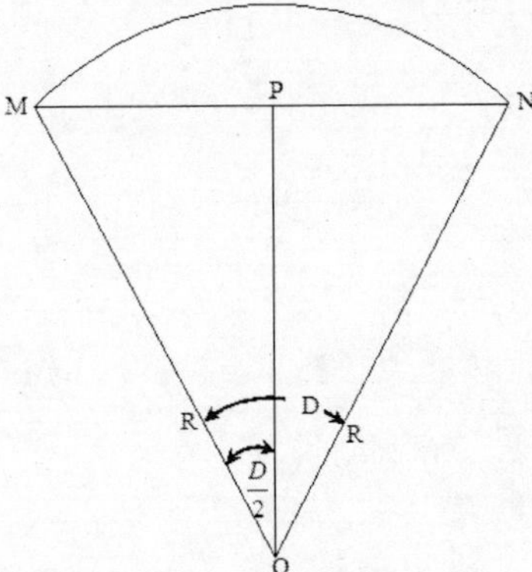

Fig. 2.4: Representation of the degree of curve

If *R* is the radius of the curve in meters, *D* is the degree of the curve, MN is the chord of 30 m length, and P is the mid-point of the chord, then-

In DOMP, OM = *R*

$$MP = \frac{1}{2} MN = 15 \text{ m}$$

$$\angle MOP = \frac{D}{2}$$

Then, $\sin\frac{D}{2} = \frac{MP}{OM} = \frac{15}{R}$

or $R = \frac{15}{\sin\frac{D}{2}}$ (Exact) (2.1)

But when D is small, $\sin\frac{D}{2}$ may be assumed approximately $= \frac{D}{2}$ in radians.

$$R = \frac{15}{\frac{D}{2} \times \frac{\pi}{180^0}} = \frac{15 \times 360}{\pi D}$$

$$= \frac{1718.87}{D}$$

or $R = \frac{1719}{D}$ (approximate) (2.2)

The approximate relation is used for curves up to 50, but for higher degree curves, the exact relation should be used.

2.3.3 Elements of a simple circular curve

From Figure 2.3, $I + \varphi = 1800$

$\angle T1OT2 = 1800 - I = \varphi$ (central angle = deflection angle) (2.3)

Tangent length = BT1 = BT2 = OT1 $\tan\frac{\varphi}{2} = R \tan\frac{\varphi}{2}$ (2.4)

Length of the long chord = $2T_1E = 2 \times OT_1 \sin\frac{\varphi}{2}$

$= 2R \sin\frac{\varphi}{2}$ (2.5)

Length of the curve = Length of the arc $T_1FT_2 = R\varphi$ (in radians)

$= \frac{\pi R\varphi}{180^0}$ (2.6)

Apex distance = BF = BO – OF

$= R \sec\frac{\varphi}{2} - R = R\left(\sec\frac{\varphi}{2} - 1\right)$ (2.7)

Versine of the curve = EF = FO – OE

$= R - R \cos\frac{\varphi}{2}$

$= R\left(1 - \cos\frac{\varphi}{2}\right) = R \text{ versine}\frac{\varphi}{2}$ (2.8)

2.3.4 Methods of horizontal curve setting

A curve may be set out either by (a) linear methods, where chain, tape or EDM is used, or (b) angular methods, where a theodolite with or without a tape or total station is used. Before starting to setting out a curve by any method, the exact positions of the tangent points between which the curve would lie, is to be determined. The steps to be used for setting up the curve between two points A and B (Figure 2.3) are as follows:

1. After fixing the directions of the straights, extend them to intersect at point B.
2. Set up a theodolite at the intersection point B and measure the angle of intersection *I*, using repetition method of angle measurement. Determine the deflection angle φ as (180°-*I*).
3. Calculate the tangent length $\left(\text{length} = R \tan \frac{\varphi}{2}\right)$
4. From the intersection point B, measure the tangent length BT_1 backward along the rear tangent BA to locate T_1.
5. Similarly, locate the position of T_2 by measuring the same distance from B along the forward tangent BC.
6. The chainages of tangent points T_1 and T_2 are determined. The chainage is the distance of point T_1 and T_2 with respect to reference point on the road with known chainage. The chainage of T_1 is obtained by subtracting the tangent length (BT_1) from the known chainage of the intersection point B, while the chainage of T_2 is found by adding the length of the curve (T_1FT_2) to the chainage of T_1.
7. The pegs are then fixed at equal intervals (normally 30 m) on the curve. This distance should actually be measured along the arc, but in practice it is measured along the chord, as the difference is very small and negligible, and the length of the chord should not be more than $(1/20)^{th}$ of the radius of curve.
8. All the pegs on the ground are joined to form the curve.

The distances along the centre line of the curve are continuously measured from the point of beginning of the line up to the end, i.e., the pegs along the centre line of the work should be at equal interval from the beginning of the line to the end. For this reason, the first peg on the curve is fixed at such a distance from the first tangent point T_1 that its chainage becomes the whole number (Integer value). The length of the first chord thus may be less than the peg interval, which is called as a *sub-chord*. Similarly, there may be be a sub-chord at the end of the curve at T_2. Thus, a curve will usually consist of two sub-chords (one at the beginning and another at the end) and all other chords as whole number (equal to peg interval).

a. Linear methods of setting out curves

The following methods of setting out simple circular curve are linear as measurement is doen using chain/tape/distance/EDM:

1. By ordinates from the long chord
2. By successive bisection of arcs
3. By offsets from the tangents
4. By offsets from chords produced

1. By ordinates from the long chord:

In this method, the perpendicular offsets are erected from the long chord to establish points along the curve, as shown in Figure 2.5.

If T_1T_2 is the length of the long chord (L), ED = O_0 which is the offset at mid-point (E) of the long chord (*the versine*), and PQ = O_x which is the offset at distance x from E. Draw a line QQ_1 parallel to T_1T_2 which meets OD at Q_1, and line OQ_1 which cuts T_1T_2 at point E.

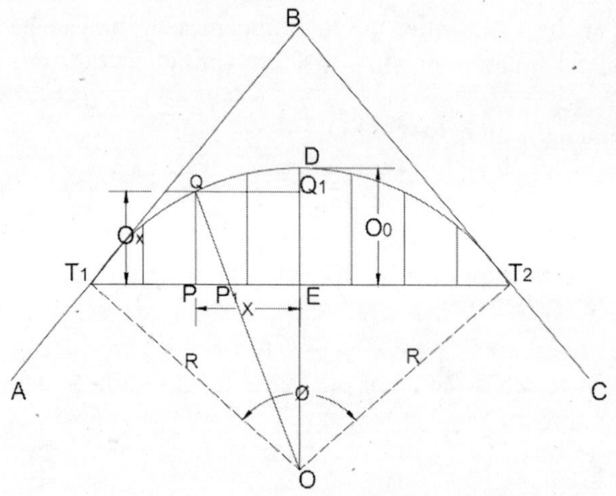

Fig. 2.5: Setting out the curve by ordinates from the long chord

$OQ_1 = OE + EQ_1$

$= (OD - DE) + EQ_1$

$= (R - O_0) + O_x$

From $DOQQ_1$

$OQ^2 = QQ_1^2 + OQ_1^2$

But OQ = R, and $QQ_1 = x$

$R^2 = x^2 + \{(R - O_0) + O_x\}^2$

or $(R - O_0) + O_x = \sqrt{R^2 - x^2}$

Hence $O_x = \sqrt{R^2 - x^2} - (R - O_0)$ (2.9)

$OE = \sqrt{(OT_1^2 - T_1E^2)}$

$= \sqrt{[R^2 - (L/2)^2]}$

Where $O_0 = ED = OD - OE$

But $OE = \sqrt{R^2 - \left(\frac{L}{2}\right)^2}$

So $O_0 = R - \sqrt{R^2 - \left(\frac{L}{2}\right)^2}$ (2.10)

In relationship 2.9, the value of O_0 may be replaced as-

$$\text{Hence } O_x = \sqrt{R^2 - x^2} - [\,R\text{-}\{\,R - \sqrt{R^2 - \left(\frac{L}{2}\right)^2}\,\}] \tag{2.11}$$

The curve is set out as below:

i. Divide the long chord into an even number of equal parts, if possible.
ii. Calculate the offsets using equation (2.11) at each of the points of division.
iii. Set out the offset at respective points on the curve.
iv. Since, the curve is symmetrical about the middle-ordinate, therefore the offsets for the right-half of curve will be same as those for the left–half curve.
v. The method is suitable for setting out short curves e.g., curves for street bends.

2. By successive bisections of arcs:

It is also known as *Versine method.* In Figure 2.6, a curve T_1DT_2 is to be established on the ground by this method. The steps involved are-

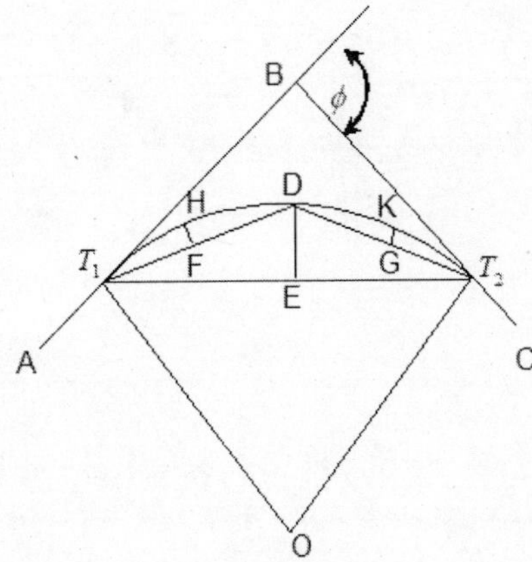

Fig. 2.6: Setting out the curve by successive bisection of arcs

i. Join T_1T_2 and bisect it at E. Set out the offset ED (which is equal to the versine $R\left(1-\cos\frac{\varphi}{2}\right)$, thus a point D on the curve may be fixed.

ii. Join T_1D and DT_2 and bisect them at F and G, respectively. Then set out the offsets FH and KG at F and G, respectively, in the same manner, each equal to $R\left(1-\cos\frac{\varphi}{4}\right)$. Thus, two more points H and K are fixed up on the curve.

iii. Now, each of the offsets can be set out at mid points of the four chords T_1H HD, DK and KT_2 which is equal to $R\left(1-\cos\frac{\varphi}{8}\right)$.

iv. By repeating this process, several points may be set out on the curve, as per the need.
v. This method is suitable where the ground distance outside the curve is not favorable for measurements by tape.

3. By offsets from the tangents:

The offsets may be either redial or perpendicular to the tangents.

a. By radial offsets

In Figure 2.7, if $O_x = PP_1$ which is the radial offset at P from O at a distance of x from T_1 along the tangent AB, then-

$$PP_1 = OP - OP_1 \text{ where } OP = \sqrt{R^2 + x^2} \text{ and } OP_1 = R$$

$$O_x = \sqrt{R^2 + x^2} - R \quad \text{(Exact)} \qquad (2.12)$$

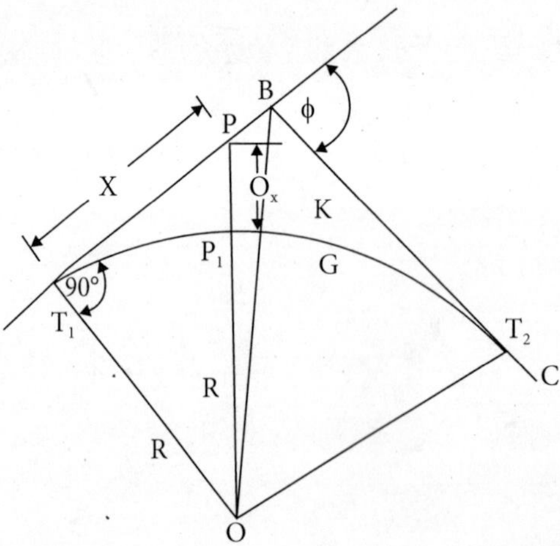

Fig. 2.7: Setting out the curve by radial offsets from the tangents

When the radius of curve is large, the offsets may be calculated by the approximate formula as derived below. Using the property of a circle, we can write;

$$PT_1^2 = PP_1 \times (2R + PP_1)$$

$$x^2 = O_x(2R + O_x) = 2RO_x + O_x^2$$

Since O_x^2 is very small as compared to 2R, it may be neglected. Hence,

$$x^2 = 2R\,O_x$$

$$\text{or } Ox = \frac{x^2}{2R} \quad \text{(approximate) (2.13)}$$

b. By offsets perpendicular to the tangents

In Figure 2.8, $O_x = PP_1$ which is the perpendicular offset at P at a distance of x from T_1 along the tangent AB. Draw P_1P_2 line parallel to BT_1

$$P_1P_2 = PT_1 = x, \text{ and } T_1P_2 = PP_1 = O_x$$

$$\text{Now } T_1P_2 = OT_1 - OP_2$$

$$\text{Where } OT1 = R, \text{ and } OP_2 = \sqrt{R^2 - x^2}$$

$$\text{So } O_x = R - \sqrt{R^2 - x^2} \quad \text{(exact)} \qquad (2.14)$$

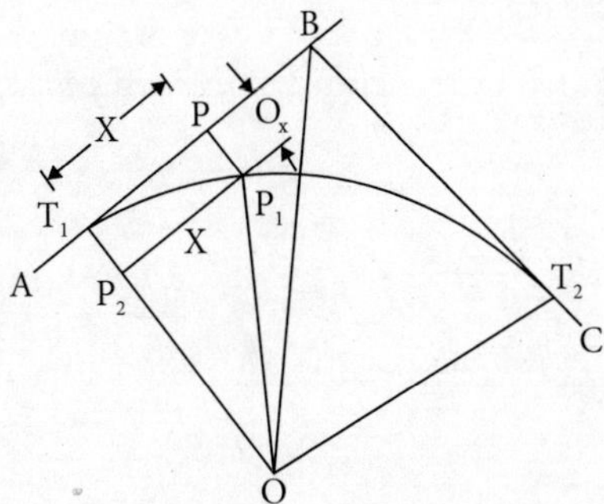

Fig. 2.8: Setting out the curve by offsets perpendicular to the tangents

Approximately, the formula may be obtained similarly as equation 2.13;

$$O_x = \frac{x^2}{2R} \text{ (approximate)} \tag{2.15}$$

Procedure of setting out the curve:

i. Locate the tangent points T_1 and T_2.
ii. Measure equal distances, say 15 or 30 m along the tangent from T_1.
iii. Set out the offsets calculated by any of the above methods at each distance (say x), thus obtaining the required points on the curve.
iv. Continue the process until the apex of the curve is reached.
v. Set out the other half of the curve from the second tangent; being symmetrical in nature.

This method is found to be suitable for setting out sharp curves where the ground outside the curve is favorable for measuring the distance.

4. By offsets from chords produced

In Figure 2.9, if AB is the first tangent, T_1 is the first tangent point, E, F, G etc., are the successive points on the curve. Draw arc EE_1, so $T_1E= T_1E_1= C_1$ which is the first chords. Similarly, EF, FG, etc., are successive chords of length C_2, C_3 etc., each being equal to the full chord length. $\angle BT_1E= \alpha$ in radians (angle between tangent BT_1 and the first chord T_1E). $E_1E= O_1$ which is the offset from the tangent BT_1, and $E_2F= O_2$ which is the offset from the chord T_1E produced. Produce T_1E to E_2 such that $EE_2 = C_2$. Draw the tangent DEF_1 at E meeting the first tangent at D and E_2F at F_1, then-

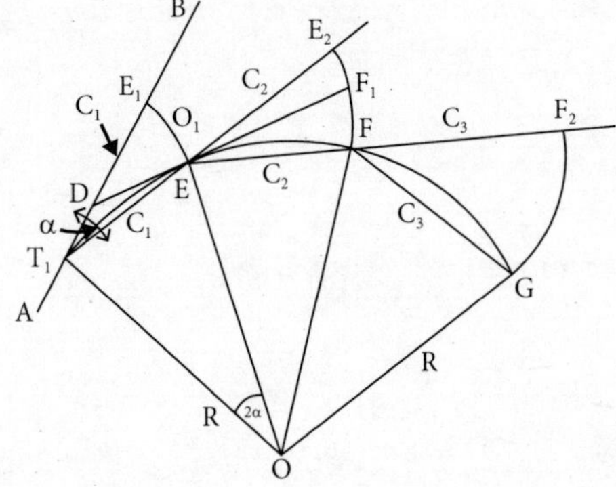

Fig. 2.9: Setting out the curve by offsets from the chord produced

$\angle BT1E = \alpha$

$\angle T1OE = 2\alpha$ (The angle subtended by any chord at the centre is twice the angle between the chord and the tangent

$$\frac{\text{arc}T_1E}{\text{Radius } OT_1} = 2\alpha$$

But arc T_1E may be taken as approximately equal to chord $T_1E = C_1$

$$\text{so, } \frac{C_1}{R} = 2\alpha \text{ or } \alpha = \frac{C_1}{2R} \quad (2.16)$$

$$\text{also } \frac{\text{arc } E_1E}{T.E} = a$$

But arc E_1E is approximately equal to chord $E_1E = O_1$, and $T_1E = C_1$, so

$$O_1 = C_1 \times \alpha$$

Putting there the value of α as calculate above.

$$O_1 = C_1 \times \frac{C_1}{2R} = \frac{C_1^2}{2R} \quad (2.17)$$

Now O_2 = offset $E_2F = E_2F_1 + F_1F$

To find out F_2F_1, consider the two triangles T_1EE_1 and EF_1E_2

$\angle E_2EF_1 = \angle DET_1$ (vertically opposite angles)

$\angle DET_1 = \angle DT_1E$, since $DT_1 = DE$, both being tangents to the circle

$\angle E_1FF_1 = \angle DET_1 = \angle DT_1E$

Both the triangles being nearly isosceles, may be considered as approximately similar. So we may write-

$$\frac{E_2F_1}{EE_2} = \frac{E_1E}{T_1E_1}$$

$$\frac{E_2F_1}{C_2} = \frac{O_1}{C_1}$$

$$\text{or } E_2F_1 = \frac{C_2 \times O_1}{C_1}$$

$$\text{or } = \frac{C_2}{C_1} \times \frac{C_1^2}{2R} = \frac{C_1C_2}{2R}$$

F_1F being the offset from the tangent at E, is equal to

$$\frac{EF_2}{2R} = \frac{C_2^2}{2R}$$

O_2 = offset $E_2F = E_2F_1 + F_1F$

$$O_2 = \frac{C_1C_2}{2R} + \frac{C_2^2}{2R}$$

$$= C_2\left(C_1 + C_2\right) / 2R \quad (2.18)$$

Similarly, the third offset, $O_3 = \dfrac{C_3\left(C_2 + C_3\right)}{2R}$ (2.19)

In the same way, remaining offset O_4, O_5, etc. may be computed using the general relationship.

$$O_n = \frac{C_n(C_{n-1} + C_n)}{2R} \quad (2.20)$$

Since $C_2 = C_3 = C_1$ .... etc., so equations 2.18 and 2.19 may be written as;

$O_2 = \dfrac{C_2^2}{R}$ and

$$O_3 = \frac{C_2^2}{R} \quad (2.21)$$

It is to be noted that first and last offsets may have different length (due to the chainages of their chord lengths), while all intermediate offsets will have equal length.

Procedure of setting out the curve:

i. Locate the tangent points (T_1 and T_2) and find out their chainages. From these chainages, calculate the lengths of first and last sub-chords and find out the offsets by using above equations.
ii. Mark a point E_1 along the first tangent T_1B such that T_1E_1 equals the length of the first sub-chord.
iii. With the zero end of the tape at T_1, swing an arc E_1E equal to radius T_1E_1, and mark point E such that $E_1E = O_1$, thus fixing the first point E on the curve.
iv. Line T_1E is produced and E_2 equal to the second sub-chord is marked, and an arc from E_2 is drawn such that EE_2 = EF to locate the second point F on the curve.
v. Continue this process until the end of the curve is reached.
vi. The last point fixed in this way should coincide with the previously located point T_2. If there is small closing error, all points on the curve are moved sideways by an amount proportional to the square of their distances from the tangent point T_1, but if the error is large, the entire process is repeated.

This method is commonly used for setting out road curves.

a. Angular methods of setting out curves

There are two methods of setting out simple circular curves by angular methods:

1. Rankine's method of tangential angles
2. Two theodolites method

1. Rankine's method of tangential or deflection angles:

In Rankine's method, the curve is set out by the tangential or deflection angles using a theodolite and a tape. The deflection angles are calculated to set out the curve (Figure 2.10).

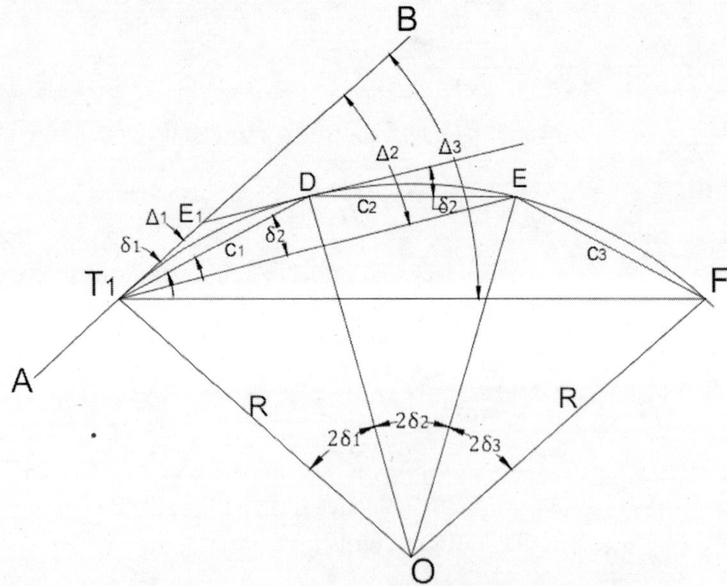

Fig. 2.10: Curve setting by Rankine's method

If T_1 and T_2 are the tangent points and AB the first tangent to the curve, D, E, F etc., are the successive points on the curve, φ is the deflection angle of the curve, R is the radius of the curve, C_1, C_2, C_3, etc., are length of the chords T_1D, DE, EF etc.., δ_1, δ_2, δ_3 etc. are the tangential angles which each of the chords T_1D, DE, EF, etc., makes respectively with the tangents at T_1, D, E, etc., D_1, D_2, D_3 etc., are the total tangential or deflection angles which the chords T_1D, DE, EF, etc. make with the first tangent AB, then-

The chord T_1D can be taken as equal to arc $T_1D = C_1$

$$\angle BT_1D = \delta_1 = \frac{1}{2}\angle T_1OD = 2\delta_1$$

$$\frac{\text{arc } T_1D}{\text{radius } OT_1} = \angle T_1OD \text{ in radians}$$

$$\text{or } \frac{C_1}{R} = 2\delta_1 \text{ radians}$$

$$\text{or } \delta_1 = \frac{C_1}{2R} \text{radians}$$

$$= \frac{C_1}{2R} x \frac{180}{\pi} \text{ degrees}$$

$$\delta_1 = \frac{C_1}{2R} \times \frac{180}{\pi} \times 60 \text{ minutes}$$

$$\delta_1 = 1718.9\, C1 / R \text{ minutes} \tag{2.22}$$

$$\delta_2 = 1718.9 \frac{C_2}{R},\ \delta_3 = 1718.9 \frac{C_3}{R},\ \text{and so on}$$

So we can write a general relation as:

$$\delta_n = 1718.9 \frac{C_n}{R} \text{ minutes} \tag{2.23}$$

Since each of the chord length C_2, C_3, C_4..........C_{n-1} is equal to the length of the full chord, so $\delta_2 = \delta_3 = \delta_4 \ldots\ldots\ldots\ldots = \delta_{n-1}$.

The total tangential angle (D_1) for the first chord (T_1D)

$= \angle BT_1D = \delta_1$

So, $\Delta_1 = \delta_1$

The total tangential angle (D_2) for the second chord (DE) = $\angle BT_1E$

But $\angle BT_1E = \angle BT_1D + \angle DT_1E$

Since, the angle between the tangent and a chord equals the angle which the chord subtends in the opposite segment, so $\angle DT_1E$ is the angle subtended by the chord DE in the opposite segment, therefore, it is equal to the tangential angle (δ_2) between the tangent at D and the chord DE.

$$\Delta_2 = \delta_1 + \delta_2 = \Delta + \delta_2$$

Similarly, $\Delta_3 = \delta_1 + \delta_2 + \delta_3 = \Delta_2 + \delta_3$

A general relationship would be as follows:

$$\Delta_n = \delta_1 + \delta_2 + \delta_3 \ldots\ldots\ldots + \delta n$$

$$\Delta_n = \Delta_{n-1} + \delta_n \tag{2.24}$$

Apply check: The total deflection angle BT1T2 = $\Delta_n = \frac{\varphi}{2}$

If the degree of the curve (D) is known, the deflection angle for a 30 m chord is equal to D/2 degrees, and that for the sub-chord of length C_1, it would be;

$$\delta_1 = \frac{C_1}{30} \times \frac{D}{2} \text{ degrees}$$

$$\delta_1 = \frac{C_1 \times D}{60}$$

$$\delta_2 = \frac{C_2 \times D}{60} \text{ and so on.}$$

$$\delta_n = \frac{C_n \times D}{60} \tag{2.25}$$

Procedure of setting out the curve:

i. Locate the tangent points T_1 and T_2, and find out their chainages. From these chainages, calculate the lengths of first and last sub-chords and the total deflection angles for all points on the curve.

ii. Set up the theodolite at the first tangent point T_1.

iii. Set the initial horizontal circle reading to zero and direct the telescope to the intersection point and bisect it.

iv. Set the first deflection angle D_1 in theodolite, and direct the telescope along T_1D. Along this line, measure T_1D equal in length to the first sub-chord, thus fixing the first point D on the curve.

v. Now set the second deflection angle D_2 in theodolite, and direct the line of sight along T_1E. Hold the zero end of the tape at D and swing the other end until the tape is bisected by the line of sight, thus fixing the second point E on the curve.

vi. Continue the process until the end of the curve is reached.

vii. The end point thus located must coincide with the previously located point (T_2). If not, the distance between them is the closing error. If it is within the permissible limit, only the last few pegs may be adjusted; otherwise the curve should be set out again.

Note: In the case of a left-handed curve, each of the value D_1, D_2, D_3 etc., should be subtracted from 360° to obtain the required value to which the reading in theodolite is to be set i.e., the vernier should be set to (360°-D_1), (360°-D_2), (360°-D_3), etc., to obtain the 1st, 2nd, 3rd etc., points on the curve.

The method is highly accurate, and is most commonly used for railways and other important curves.

2. Two theodolite method

This method is very useful in the absence of distance measurement by tape, and also when the ground is not favorable for accurate distance measurement. It is a simple and accurate method but essentially requires two theodolites to set the curve, so it is not as popular method as the method of deflection angles. In this method, the popular property of a circle "*that the angle between the tangent and the chord equals the angle which that chord subtends in the opposite segment*" is used.

In Figure 2.11, if D, E, etc., are the point on the curve to be established, then-

The angle D_1 between the tangent T_1B and the chord T_1D i.e., $\angle BT_1D = D_1 = \angle T_1T_2D$. Similarly, $\angle BT_1E = D_2 = \angle T_1T_2E$, etc.

The deflection angles D_1, D_2, etc., are calculated to establish the curve using deflection angle.

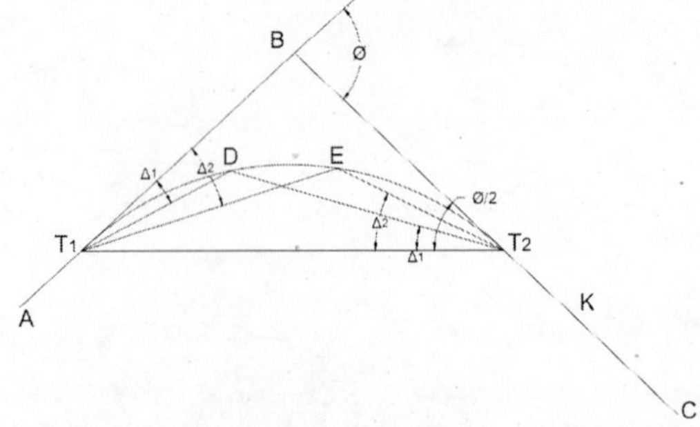

Fig. 2.11: Curve setting by two theodolite method

Procedure of setting out the curve:

i. Set up two theodolites, one at T_1 and the other at T_2.

ii. Set horizontal angle reading of the theodolite at T_1 to zero along T_1B, and similarly the theodolite at T_2 to zero along T_2T_1.

iii. Set both the theodolite at T_1 and T_2 to read the first deflection angle D_1. Now the line of sight of theodolite at T_1 would be along T_1D and that of the theodolite at T_2 along T_2D. The point of intersection of these line of sights is the required point D on the curve. Establish point D on the ground with the help of ranging rods.

iv. Now set both the theodolites to second deflection angle D_2, towards T_1E and T_2E respectively, and proceed as before to establish the second point E on the curve.

v. Repeat the process until the other points on the curve are set out.

Note: If point T_1 is not be visible from the point T_2, in such a case direct the telescope of the instrument at T_2 towards B with reading set to zero. Now set the reading to read an angle of $\left(360^0 - \frac{\varphi}{2}\right)$, directing the telescope along T_2T_1. For the first point D on the curve, set the reading to read $\left(360^0 - \frac{\varphi}{2}\right) + \Delta_1$. Similarly for the second point E, set the reading to read $\left(360^0 - \frac{\varphi}{2}\right) + \Delta_2$ and so on.

2.4 COMPOUND CURVES

A compound curve consists of two different radii, as shown in Figure 2.12, with their centre at O_S and O_L. The radius of curve R_S is smaller than the radius of curve R_L. The two circular curves with different radii meet at a common point O_L. The compound curve is tangential to three straights AB, KM, and BC at T_1, N and T_2, respectively. Points N, O_S and O_L will lie in a straight line. The tangents AB and NK intersect at K, and tangents BC and NM will intersect at M.

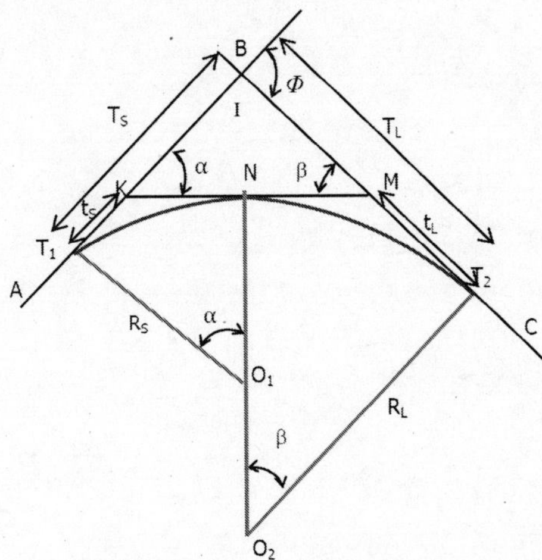

Fig. 2.12: A compound curve

2.4.1 Elements of a compound curve

In Figure 2.12, it is shown that a compound curve has three straights AB, BC and KM which have tangential at T_1,T_2 and N, respectively. The two circular arcs T_1N and NT_2 having centres at O_1 and O_2. The arc having a smaller radius may be first or second curve. The tangents AB and BC intersect at point B, AB and KM at K and BC and KM at M.

If T_1 is the point of curvature, T_2 is the point of tangency, B is the point of intersection, N is the point of compound curve (PCC), T_S is the length of tangent of the first curve, T_L is the length of tangent of the second curve, t_s is the length of tangent to curve T_1N , t_L is the length of tangent to curve NT_2 , K is the vertex of the first curve, M is the vertex of the second curve, R_s is the smaller radius O_1T_1, R_L is the larger radius O_2T_2, Δ is the deflection angle between rear tangent (AB) and forward tangent (BC), α is the deflection angle between rear tangent (AB) and the common tangent (KM), β is the deflection angle between forward tangent (BC) and common tangent (KM), L_1 is the length of first chord, L_2 is the length of second chord, L is the length of long chord from T_1 to T_2, l_S is the length of first arc, l_L is the length of second arc, l is the length of total curve, R_s is the smaller radius O_1T_1, R_L is the larger radius O_2T_2, Φ is the deflection angle between rear tangent (AB) and forward tangent (BC), α is the deflection angle between rear tangent (AB) and the common tangent (KM), and β is the deflection angle between forward tangent (BC) and common tangent (KM), then-

Angle $T_1BT_2 = I = 180° - \Phi$

$$\Phi = \alpha + \beta \tag{2.26}$$

$$KN = KT_1 = t_s = R_s \tan(\alpha/2) \tag{2.27}$$

$$MN = MT_2 = t_L = R_L \tan(\beta/2) \tag{2.28}$$

$$KM = KN + NM = t_S + t_L = R_S \tan\frac{\alpha}{2} + R_L \tan\frac{\beta}{2} \tag{2.29}$$

Applying sine relationship in ΔBKM, we get-

$$\frac{BK}{\sin\beta} = \frac{KM}{\sin I}$$

$$\frac{BK}{\sin\beta} = \frac{KM}{\sin(180^0 - \varphi)}$$

$$\text{or } \frac{BK}{\sin\beta} = \frac{KM}{\sin\varphi}$$

$$\text{so } BK = \frac{KM\sin\beta}{\sin\varphi} = \frac{(t_S + t_L)\sin\beta}{\sin\varphi}$$

But $T_S = BT_1 = BK + KT_1$

$$= \frac{(t_S + t_L)\sin\beta}{\sin\varphi} + t_s \tag{2.30}$$

Similarly, applying sine relationship in ΔBKM

$$BM = \frac{KM\sin\alpha}{\sin\varphi} = \frac{(t_S + t_L)\sin\alpha}{\sin\varphi}$$

$$T_L = BT_2 = BM + MT_2$$

$$= \frac{(KM)\sin\alpha}{\sin\varphi} + t_L \quad (2.31)$$

Length of the first curve = $l_S = R_S\, \alpha\, (\pi / 180)$ (2.32)

Length of the second curve = $l_L = R_L\, \alpha\, (\pi / 180)$ (2.33)

Total length of curve (l) = $l_S + l_L$ (2.34)

2.4.2 Setting out the compound curve

i. The compound curve may be set out by the method of deflection angles from two points T_1 and N; the first curve from point T_1 and the second one from point N.

ii. Locate B, T_1 and T_2, and find out the chainage of T_1 from the known chainage of B and length BT_1.

iii. Find out the chainage of F by adding the length of the first curve to the chainage of T_1, and find the chainage of T_2 by adding the length of the second curve to the chainage of F.

iv. Calculate the deflection angles.

v. Set up the theodolite at T_1, and set out the first curve.

vi. Shift the instrument and set it at point F. With the horizontal angle set to $\left(360^0 - \frac{\alpha}{2}\right)$, take a back sight on T1 and transit the telescope and swing through $\frac{\alpha}{2}$, the line of sight will be directed along the common tangent FE and the reading will read 360°.

vii. Set the vernier to the first deflection angle as calculated for the second curve, thus directing the line of sight to the first point on the second curve.

viii. Process is repeated until the end of the second curve is reached.

Check: Measure the angle T_1FT_2, which must equal $180^\circ - \frac{\varphi}{2}$.

2.5 REVERSE CURVES

A curve consisting of two circular arcs of similar or different radii having their centres on opposite sides of the common tangent at the point of reverse curvature is known as a *reverse curve* (Figure 2.13). It is also known as a *serpentine curve* or *S-curve* due to its peculiar shape. It is generally used when two lines intersect at a very small angle. Reverse curves are used to connect two parallel roads or railway lines. These curves are best suited for hilly terrains and highways for relatively low-speed vehicles. Reverse curves are not advisable to use on the highways and railways which are meant for high-speed traffic movement because of the following reasons:

a. A sudden change in direction can be dangerous for vehicles.

b. A sudden change in curvature and direction increases wear & tear in vehicles, and also provides discomfort to the people traveling along the route.

c. It may cause the vehicle to overturn over a reverse curve, if the vehicle is moving with a greater speed. careless.

d. At the Point of Reverse Curvature (PRC), super-elevation can't be provided.

e. Sudden change in super-elevation from one edge to another edge on reverse curve is required which is difficult to achieve.

f. The curves cannot be properly provided superelevation at the point of reverse curvature

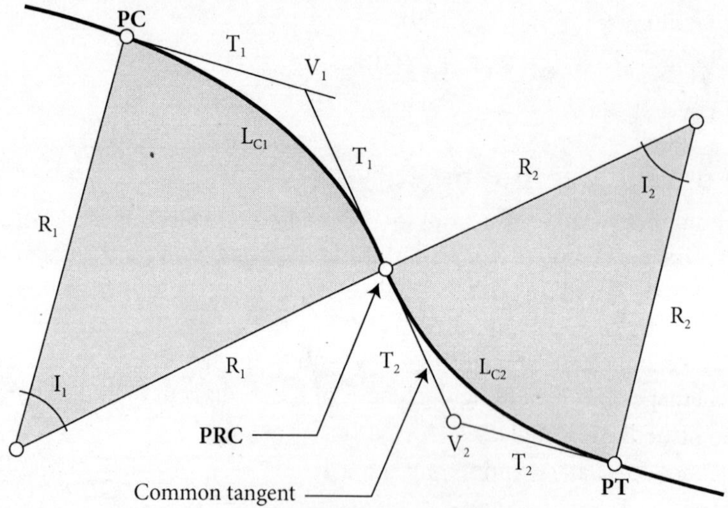

Fig. 2.13: A reverse curve

If PC is the point of curvature, PT is the point of tangency, PRC is the point of reversed curvature, T_1 is the length of tangent of the first curve, T_2 is the length of tangent of the second curve, V_1 is the vertex of the first curve, V_2 is the vertex of the second curve, I_1 is the central angle of the first curve, I_2 is the central angle of the second curve, Lc_1 is the length of first curve, Lc_2 is the length of second curve, L_1 is the length of first chord, L_2 is the length of second chord, and $(T_1 + T_2)$ is the length of common tangent measured from V_1 to V_2, then-

Finding the chainage of PT:

i. Given the chainage of *PC*, then the

Chainage of PT= Chainage of PC+Lc_1+Lc_2 (2.35)

ii. Given the chainage of V_1, then the

Chainage of PT= Chainage of $V_1 - T_1 + Lc_1 + Lc_2$ (2.36)

2.5.1 Elements of a reverse curve

Figure 2.14 shows a reverse curve made up of two different radii. In this Figure, R_1 is the smaller radius ($O_1A=O_1D$), R_2 is the larger radius ($O_2D = O_2B$), Δ_1 is the angle subtended at the centre by the arc of smaller radius R_1, Δ_2 is the angle subtended at the centre by the arc of larger radius R_2, V is the perpendicular distance (AJ=MN) between two straights (parallel tangents) AM and BN, h is the distance between the perpendiculars at A and B, L is the length of the line joining the tangent points A and B.

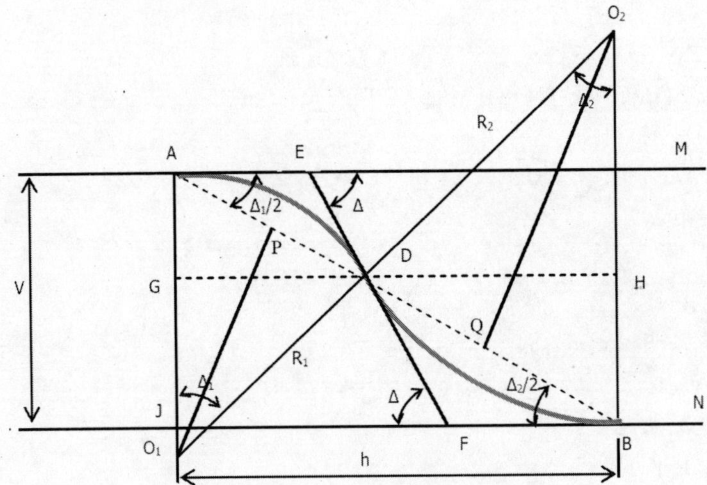

Fig. 2.14: Elements of a reverse curve

D is the point of reverse curvature, and from it a line perpendicular to the straights AO_1 and BO_2 is drawn, cutting these at G and H. Draw perpendicular from O_1 and O_2 at line AB which will cut this line at P and Q, respectively, dividing the angle of deflection into half $\angle AO_1P = \angle DO_1P$, and $\angle BO_2Q = \angle DO_2Q$).

When $\Delta_1 = \Delta_2 = \Delta$

Perpendicular distance (V) = AG + GJ) = (AG + BH)

$V = (O_1A - O_1G) + (O_2B - O_2H)$ (2.37)

Here, $O_1A = R_1$, and $O_2B = R_2$

$\cos \Delta = O_1G / O_1D = O_1G / R_1$

$O1G = R_1 \cos \Delta$

$\cos \Delta = O_2H / O_2D = O_2H / R_2$

$O_2H = R_2 \cos \Delta$

From equation 2.37,

$V = (R_1 - R_1 \cos \Delta) + (R_2 - R_2 \cos \Delta)$

$V = R_1(1 - \cos \Delta) + R_2 (1 - \cos \Delta)$

$= (1 - \cos \Delta)(R_1 + R_2)$

$V = (R_1 + R_2) \text{ versin } \Delta$ (2.38)

Total length (L) = AD + DB

But AD = AP + PD and DB = DQ + QB

In triangle O_1PA

$\sin \Delta/2 = AP / O_1A$

$AP = R_1 \sin \Delta/2$

$AD = 2 R_1 \sin \Delta/2$ (since AP = PD)

Similarly, in triangle O_2QB

Sin $\Delta/2 = BQ / O_2B$

$BQ = R_2 \sin \Delta/2$

$DB = 2R_2 \sin \Delta/2$ (since DQ = QB)

Total length (L) = AD + DB

So, $L = 2R_1 \sin \Delta/2 + 2R_2 \sin \Delta/2$

$L = 2(R_1 + R_2) \sin \Delta/2$ (2.39)

In triangle ABJ

Sin $\Delta/2 = AJ / AB = V / L$

Equations 2.39 can be written as-

$L = 2(R_1 + R_2) V / L$

$L^2 = 2V (R_1 + R_2)$

$L = \sqrt{[2V (R_1 + R_2)]}$ (2.40)

Distance between the end points of the reverse curve measured parallel to the straights (h) = GD + DH

In triangle GO_1D

$\sin \Delta = GD / O_1D$ or $GD = R_1 \sin \Delta$

In triangle BO_2D

$\sin \Delta = DH / O_2D$ or $DH= R_2 \sin \Delta$

So, h = GD + DH

$= R_1 \sin \Delta + R_2 \sin \Delta$

$h = (R_1 + R_2) \sin \Delta$ (2.41)

Length of the first curve AD (l_1) = $R_1 \Delta (\pi /180)$

Length of the second curve DB (l_2) = $R_2 \Delta (\pi /180)$

Total length of the curve ADB ($l_1 + l_2$) = $\Delta (\pi /180) (R_1 + R_2)$ (2.42)

If the radius of the two curves are equal-

From equation 2.38, V = 2R versin Δ (2.43)

From equation 2.39, L = 4R sin Δ/2 (2.44)

or R = L / (4 sin Δ/2)

From equation 2.40, $L = \sqrt{(2V * 2R)} = \sqrt{4 VR}$ (2.45)

or $R = L^2 / 4V$

From equation 2.41, h = 2R sin Δ (2.46)

From equation 2.42, total length of the curve ADB ($l_1 + l_2$) = $2R \Delta (\pi /180)$ (2.47)

2.6 TRANSITION CURVES

A transition curve is a non-circular curve of varying radius which is introduced between a straight and a circular curve for the purpose of giving ease in ride and change of direction along the route (Figure

2.15). The primary purpose of the transition curve is to enable vehicle moving at high speeds to make the change from the tangent section to the curves section, in a safe and comfortable fashion. When a vehicle enters or leaves a circular curve of finite radius, it is subject to an outward centrifugal force which can cause the shifting away of the passengers and the driver.

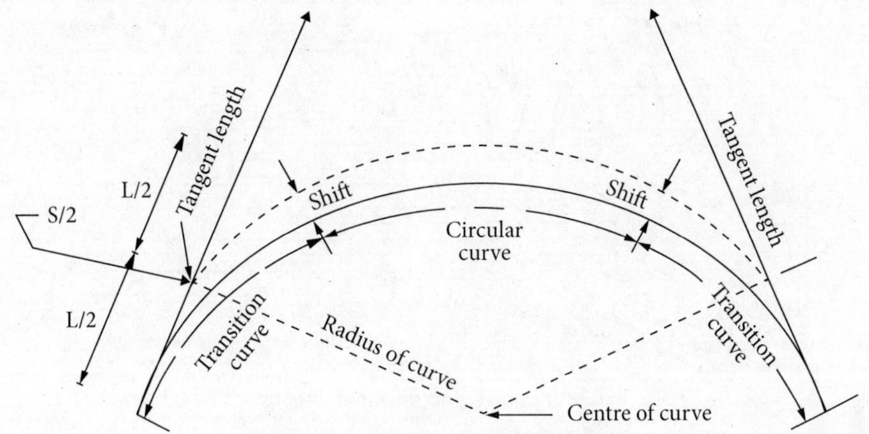

Fig. 2.15: A typical transition curve

The transition curve can also be inserted between two branches of a compound or reverse curve. The transition from the straight line to the tangent to circular curve, and from the circular curve to the straight line should be gradual. The transition curve helps in obtaining a gradual increase of super-elevation from zero on the tangent to the required full amount on the circular curve. It avoids danger of derailment, side skidding or overturning or vehicles while moving, as well as avoids discomfort to the passengers.

The most common types of transition curves are shown in Figures 2.16 and explained below. There are three types of transition curves commonly used: (i) a cubic parabola, (ii) a cubical spiral, and (iii) a lemniscate. The first two are used on railways and highways both, while the third one is used on highways only.

i. **Cubic parabolic curve**–In this curve, the rate of decrease of curvature is much low for deflection angles 4° to 9°, but beyond 9°, there is a rapid increase in the radius of curvature. It is mostly used in railways. The equation is:

$$y = x^3 / (6RL) \tag{2.48}$$

Where y is the coordinate of any point, x is the distance measured along the tangent, R is the radius of curve, and $L$is the length of curve

ii. **Spiral curve**– This an ideal transition curve. It is the most widely used curve as it can easily be set out with its rectangular co-ordinates. The radius of this curve is inversely proportional to length traversed. Hence, the rate of change of acceleration in this curve is uniform throughout its length. It is mostly used in railways.

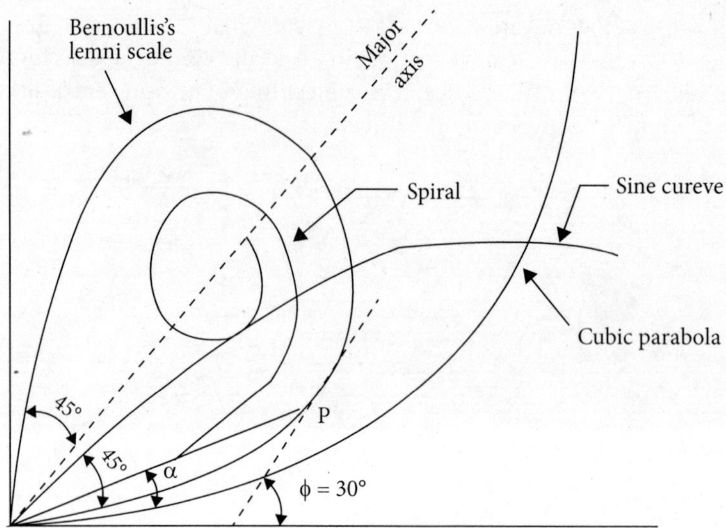

Fig. 2.16: Various types of transition curves

iii. **Lemniscate curve**– In this transition curve, radius decreases as the length increases, and hence there is a slight fall of the rate of gain of radial acceleration. It is mostly used in highways. It can be represented by the Bernoulli's lemniscate curve:

$$L = k \sqrt{(\sin 2\alpha)} \tag{2.49}$$

Where L is the length of polar distance of any point in meters, α is the polar deflection angle of that point in radians, and k is a constant

A transition curve should fulfil the following conditions:

i. If should tangentially meet the tangent line as well as the circular curve.
ii. The curve should have infinite radius (i.e., zero curvature) at the origin.
iii. The rate of increase of curvature along the transition curve should be the same as that of increase of super-elevation.
iv. The length of the transition curve should be such that the full super-elevation is attained at the junction with the circular curve.
v. Its radius at the junction with the circular curve is equal to that of circular curve.

2.6.1 Super-elevation or Cant

When a vehicle passes from a straight line to a curve line, in addition to its own weight, a centrifugal force acts on it, as shown in Figure 2.17. Both the forces act through the centre of gravity of vehicle. The centrifugal force acts horizontally and tends to push the vehicle away from the centre or road. This is because there is no component force to counter balance this centrifugal force. To counteract this effect, outer edge of the curve is elevated or raised by a small amount as compared to the inner one. This raising of the outer edge of curve is called *super-elevation* or *cant.* The amount of super-elevation will depend upon on several factors, such as the speed of the vehicle and radius of the curve.

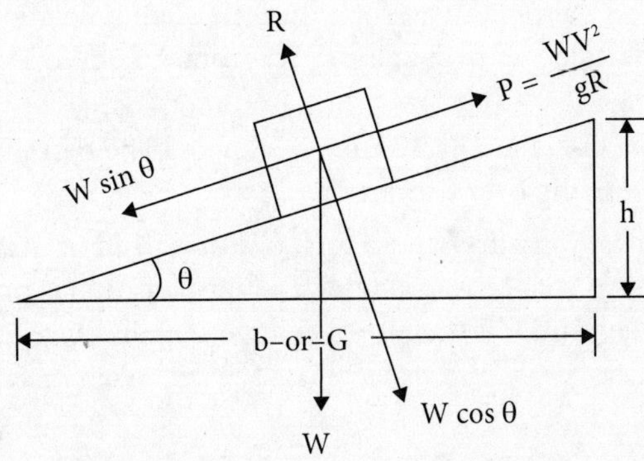

Fig. 2.17: Depiction of super-elevation

If W is the weight of vehicle acting vertically downwards, P is the centrifugal force acting horizontally, v is the speed of vehicle in meters/sec, g is the acceleration due to gravity, 9.81 meters/sec^2, R is the radius of the curve in meters, h is the super-elevation in meters, b or G is the breadth of the road or the distance between the centres of the rails in meters, then-

For maintaining the equilibrium, resultant of the weight and the centrifugal force should be equal and opposite to the reaction perpendicular to the road or rail surface.

The centrifugal force can be represented as, $P = \frac{Wv^2}{gR}$

$$\text{So } W\sin\theta = \frac{Wv^2}{gR}$$

$$\sin\theta = \frac{v^2}{gR}$$

If θ is the inclination of the road or rail surface from ground which is a very small angle, we can write;

$$\sin\theta = \tan\theta = \frac{v^2}{gR}$$

Super-elevation $h = b \tan q$

$$h = \frac{bv^2}{gR} \tag{2.50}$$

2.6.2 Length of a transition curve

The length of the transition curve may be computed from (i) arbitrary gradient, (ii) the time rate, and (iii) rate of change of radial acceleration. It can be determined in the following ways;

1. The length may be already assumed on the basis of experience and judgement as say, 100 m.
2. The length may be such that the super-elevation is applied at a uniform rate of 1 in 300 to 1 in 1200. If h is the amount of super-elevation, then the length of the transition in curve may be from $300h$ to $1200h$.

3. The length of the transition curve may be such that the super-elevation is applied at an arbitrary time rate of '*a*' cm/sec. The value of '*a*' usually varies from 2.5 cm to 5 cm.
4. The length of the transition curve may be such that rate of change of radial acceleration does not exceed a certain value, which is generally 30 cm/sec². It would help in smooth moving of the vehicle and the passengers when moving over the curve.

If L is the length of the transition curve in meters, v is the speed in meters/sec, h is the amount of super-elevation in centimeters, *1 in n* is the rate at which super-elevation is provided, a is the time rate, R is the radius of the curve in meters, and C is the rate of change of radial acceleration in meters/sec². then-

$$L = n\,h$$

$$L = n\,(bv^2 / gR)$$

$$L = nbv^2 / gR \tag{2.51}$$

Other approach could be if the time taken by a vehicle in passing over the length of the transition curve is $\frac{L}{V}$ sec., the super-elevation (h) attained in this time is-

$$h = \frac{L}{V} \times a$$

$$L = \frac{hv}{a} \tag{2.52}$$

The third approach is to compute radial acceleration.

The radial acceleration on the circular curve $= \frac{v^2}{R}$

Time taken by a vehicle to pass over the transition curve $= \frac{L}{V}$ sec.

Radial acceleration attained in this time $= C\frac{L}{V}$ m / sec²

so $\frac{v^2}{R} = C\frac{L}{V}$

$$\text{or } L = \frac{v^3}{CR} \tag{2.53}$$

The length of transition curve can be determined using above relationships (2.51 to 2.53).

2.6.3 Characteristics of a transition curve

In Figure 2.18, a circular curve EE' of radius R, and two transition curves T_1E and $E'T_1$ at the two ends, have been inserted between the two straights. The two straights AB and BC make a deflection angle Δ. It is clear from the figure that in order to fit in the transition curves at the ends, a circular imaginary curve ($T_1F_1T_2$) of slightly greater radius has to be shifted towards the centre as ($E_1EFE'E_1$). The distance through which the curve is shifted is known as shift (S) of the curve, and is equal to $\frac{L^2}{24R}$, where L is the length of each transition curve and *R* is the radius of the desired circular curve (EFE'). The length of shift (T_1E_1) and the transition curve (TE) mutually bisect each other.

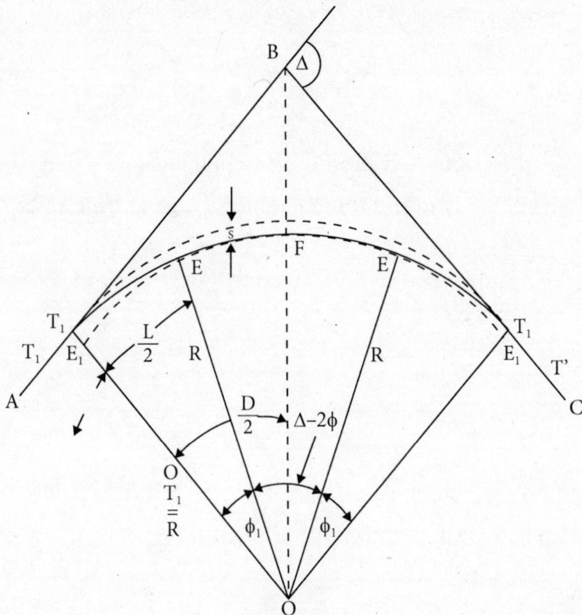

Fig. 2.18:

The tangent length for the combined curve is-

$$BT = BT1 + T1 = OT1\ \tan\frac{\Delta}{2}+\frac{L}{2}$$

$$=(R+S)\tan\frac{\Delta}{2}+\frac{L}{2}$$

The spiral angle $\varphi_1 = \dfrac{L/2}{R} = \dfrac{L}{2R}$ radians (2.54)

The central angle for the circular curve, $\angle EOE' = \Delta - 2\varphi_1$

Length of the circular curve EFE' $= \dfrac{\pi R(\Delta - 2\varphi_1)}{180^0}$

where Δ and f_1 angles are in degrees

Length of the combined curve TEE'T' $= TE + EE' + E'T'$

$$= L + \frac{\pi R(\Delta - 2\varphi_1)}{180^0} + L$$

$$= \frac{\pi R(\Delta - 2\varphi_1)}{180^0} + 2L \qquad (2.55)$$

Chainage of beginning (T) of the combined curve = chainage of the intersection point (B)–total tangent length for the combined curve (BT).

Chainage of the junction point (E) of the transition curve and the circular curve = chainage of T + length of the transition curve (L).

Chainage of the other junction point (E') of the circular curve and the other transition curve = chainage of E + length of the circular curve.

Chainage of the end point (T') of the combined curve = chainage of E' + length of the transition curve.

Check: The chainage of T' thus obtained should be = chainage of T + length of the combined curve.

Note: The points on the combined curve should be marked using chainages so that there will be sub-chords at each end of the transition curves and of the circular curve.

The deflection angle for any point on the transition curve distant l from the beginning of the combined curve (T),

$$\alpha = \frac{l^2}{6RL} \text{ radians} = \frac{1800 l^2}{\pi RL} \text{ minutes}$$

$$= \frac{573 l^2}{RL} \text{ minutes} \tag{2.56}$$

Check: The deflection angle for full length of the transition curve is-

$$\alpha = \frac{l^2}{6RL} = \frac{L^2}{6RL} \quad (l = L)$$

$$= \frac{L}{6R} \text{ radians} = \frac{1}{3}\varphi_1 \text{ (using equation 2.54)} \tag{2.57}$$

The deflection angles for the circular curve are found from, $\delta_n = 1718.9 \frac{C_n}{R}$ minutes.

Check: The deflection angle for full length of the circular curve, $\Delta_n = \frac{1}{2} \times$ central angle

i.e., $\Delta_n = \frac{1}{2}(\Delta - 2\varphi_1)$

The offsets for the transition curve are found from perpendicular offset-

$$y = \frac{x^3}{6RL} \text{ where } x \text{ is measured along the tangent TB.} \tag{2.58}$$

$$\text{Tangent offset, } y = \frac{l^3}{6RL} \text{ where } l \text{ is distance measured along the curve.} \tag{2.59}$$

Checks:

a. The offset at half the length of the transition curve is-

$$y = \frac{l^3}{6RL} = \frac{(L/2)^3}{6RL} \quad (l = L/2)$$

$$= \frac{L^2}{48R} \tag{2.60}$$

b. The offset at junction point on the transition curve is-

$$y = \frac{L^3}{6RL} = \frac{L^2}{6R} \text{ (here } I = L) \tag{2.61}$$

The offsets for the circular curve from chords produced are found from equation 2.20.

$$O_n = \frac{C_n(C_{n-1} + C_n)}{2R}$$

2.7 VERTICAL CURVES

Vertical curves are introduced at changes of gradient to maintain the good visibility as well as avoid any impact while the vehicle is moving along the curve. These are provided to secure safety, appearance and visibility. These curves are set out in a vertical plane to obtain a gradual change of gradient, either higher to lower or lower to higher. These curves may be circular or parabolic, but the latter shape is commonly used. The most common practice has been to use parabolic curves in summit curves, mainly because of the ease of setting them out on the field and the comfortable transition from one gradient to another to provide excellent riding comfort. In case of valley curves, use of cubic parabola is preferred as it closely approximates the ideal transition requirements. Although, for small gradient angles, the difference between a circular and a parabolic curve is negligible.

Designing a vertical curve consists principally of deciding on the proper length of the curve. The longer a curve is, the more gradual the transition will be from one grade to the next; the shorter the curve, the more abrupt the change will be. The change must be gradual enough to provide the required sight distance (Figure 2.19). The sight distance depends on the speed for which the road is designed, the passing or non-passing distance requirements, driver's reaction time, braking time, stopping distance, eye level, and the height of objects. A typical eye level used for designs is between 3.75-4.5 feet; typical object heights are 4 inches to 1.5 feet. For a valley curve, the sight distance (SD) will usually not be significant during daylight, but the nighttime sight distance must be considered when the reach of headlights may be limited by the abruptness of the curve.

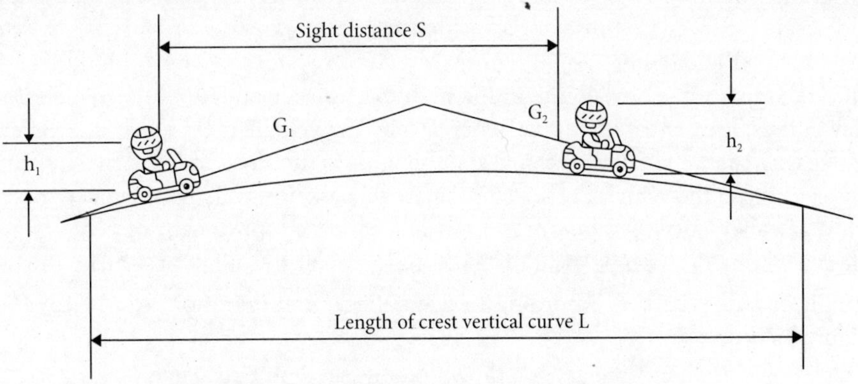

Fig. 2.19: Sight distance in vertical summit curve

The gradient is expressed in two ways:

a. As a percentage, e.g., 1%, 1.5%, etc., and

b. As a ratio, like 1 in n, where n is the horizontal distance and 1 represents vertical distance, e.g., 1 in 100, 1 in 200, etc.

The gradient may be 'rise' or 'fall'. An up gradient is known as 'rise' and is denoted by a positive sign, and a down gradient is known as 'fall' and is indicated by a negative sign. In upgrade curve, the elevation along it increases, whereas in downgrade curve, the elevation decreases.

2.7.1 Types of vertical curves

Vertical curves are usually parabolic, primarily because its shape provides a transition and also computationally efficient. The characteristic of a parabolic curve is that the gradient changes from point to point but the rate of change in grade remains constant. Hence, for finding the length of the vertical curve, the rate of change of grade should be an essential consideration as this factor remains constant throughout the length of the vertical curve. Depending upon the shape of ground profile, a vertical curve may be divided into:

a. **Summit curve:** When two grades meet at the summit and the curve will have convexity upwards, the curve is simply referred as summit curve.

b. **Valley (Sag) curve:** When two grades meet at the valley (sag) and the curve will have convexity downwards, the curve is simply referred as the valley (sag) curve.

Summit curves are usually provided: (i) when an upgrade is followed by a downgrade, (ii) when a steeper upgrade is followed by a milder upgrade, and (iii) when a milder downgrade is followed by a steeper upgrade. Generally, the recommended rate of change of grade is 0.1% per 30 m at summits and 0.05% per 30 m at sags. During the design of a vertical summit curve, the comfort, appearance, and security of the driver is considered. In addition, the sight distances (SD) must be considered during the design. During movement of a vehicle in a summit curve, there is less discomfort to the passengers because the centrifugal force developed by the movement of vehicle on a summit curve act upwards which is opposite to the direction in which its weight acts. This relieves the load on the springs of the vehicle, so stress developed is less. Therefore, the important part in summit curve design is the computation of length of the summit curve which is done by considering the sight distance parameters.

A valley (sag) curve is usually provided; (i) when a downgrade is followed by an upgrade, (ii) when a steeper downgrade is followed by a milder upgrade, and (iii) when a milder upgrade is followed by a steeper upgrade. Valley curves are those curves which have convexity downwards which requires more considerations. During the daytime, visibility in valley curves is not that much hindered as during the night time; and the only source of visibility in the night becomes headlight of the vehicle and the street lights. In valley curves, the centrifugal force generated by the vehicle moving along a valley curve acts downwards along with the weight of the vehicle and this adds to the stress induced in the spring of vehicle which can cause jerking of the vehicle and discomfort to the passengers. Thus, the most important things to consider during valley curve design are: impact and jerk-free movement of vehicles at the design speed, and the availability of stopping sight distance (SSD) under headlight of vehicles during night.

The summit or valley curves, according to geometrical configuration, can be divided into four categories (Figure 2.20):

a. An upgrade (+ive gradient) meets by a downgrade (-ive gradient). Such curve is called a convex curve or a summit curve. The change in gradient would be the algebraic difference of the gradients, i.e., [g1 – (-g2)%] or (g1 + g2).

b. A downgrade (-ive gradient) meets by an upgrade (+ive gradient). Such curve is also called a sag curve or a concave curve. The change in gradient would be the algebraic difference of the gradients, i.e., [(g2 – (-g1)%] = (g1 + g2).

c. An upgrade (+ive gradient) meets by another upgrade (+ive gradient). Such curve is also called a sag curve or a concave curve. The change in gradient would be the algebraic difference between the gradients [i.e., (g2 – g1)%].

d. A downgrade (-ive gradient) meets by another downgrade (-ive gradient). Such curve is also called a summit curve or a convex curve. The change in gradient is the algebraic difference between the gradients, i.e., (g2 – g1)%.

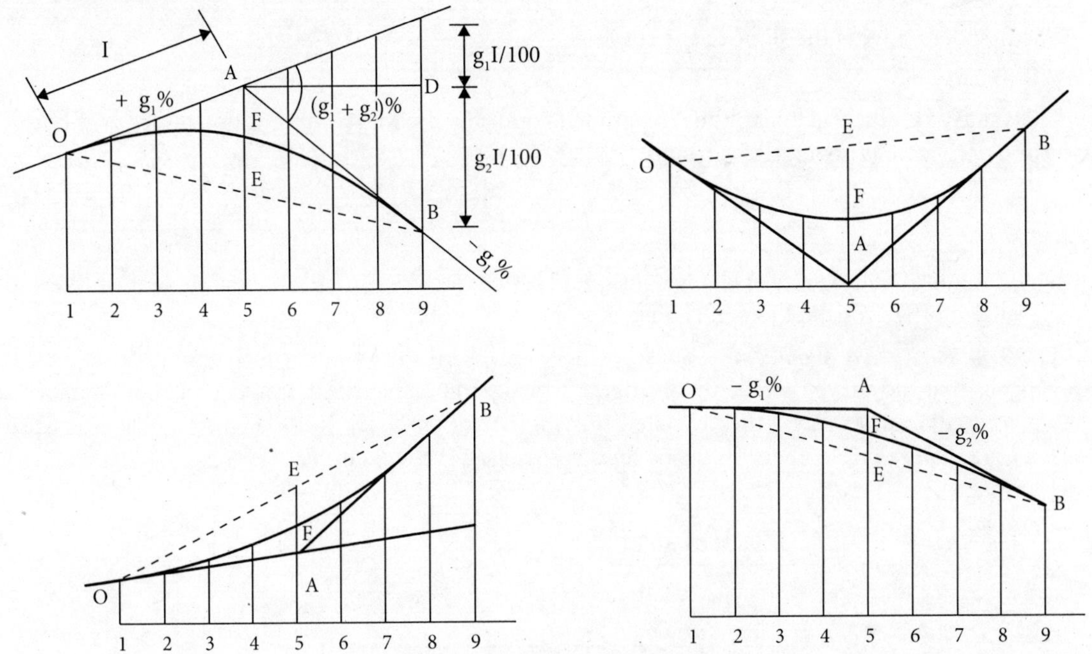

Fig. 2.20: Types of vertical curves

The permissible rate of change in gradient for first class railways is recommended as 0.06% per 20 m station at summits and 0.05% per 20 m station for sags. For second class railways, permissible rate of change of gradient is 0.12% per 20 m station at summits and 0.1% per 20 m station for sags. For small gradients, there is no noticeable difference between a parabola and a circular arc. Suppose the gradient at the beginning of a summit curve is 1.20% and if the rate of change of gradient is 0.05% per 20 m station, the gradients at the various stations will be as given in Table 2.1.

Table 2.1: Distance as per change in 0,05% gradient

Station	Distance from the beginning of the vertical curve (m)	Gradients (%)
1	0	1.20
2	20	1.15
3	40	1.10
4	60	1.05
5	80	1.00

2.7.2 Elements of a vertical parabolic curve

Vertical curves are used to provide gradual change between two adjacent vertical grade lines. The curve used to connect the two adjacent grades is parabola. Parabola offers smooth transition because its second derivative is constant. For a downward parabola with vertex at the origin, the standard equation is-

$$x^2 = -4ay$$

$$\text{or} \quad y = -x^2 / 4a \tag{2.62}$$

The first derivative is the slope of the curve.

$$y' = -x/ 2a$$

The value of y' above is linear, thus the grade (slope) for a summit curve is downward and linear. The second derivative is obviously constant.

$$y'' = -1/ 2a \tag{2.63}$$

which is interpreted as rate of change of slope. This characteristic made the parabola the preferred curve because it offers constant rate of change of slope.

The *symmetrical parabolic curve* does not necessarily mean the curve is symmetrical at $L/2$, it simply means that the curve is made up of single vertical parabolic curve. Using two or more parabolic curves placed adjacent to each other is called *unsymmetrical parabolic curve.* Figure 2.21 is a vertical summit curve. The same elements hold true for a vertical sag (valley) curve.

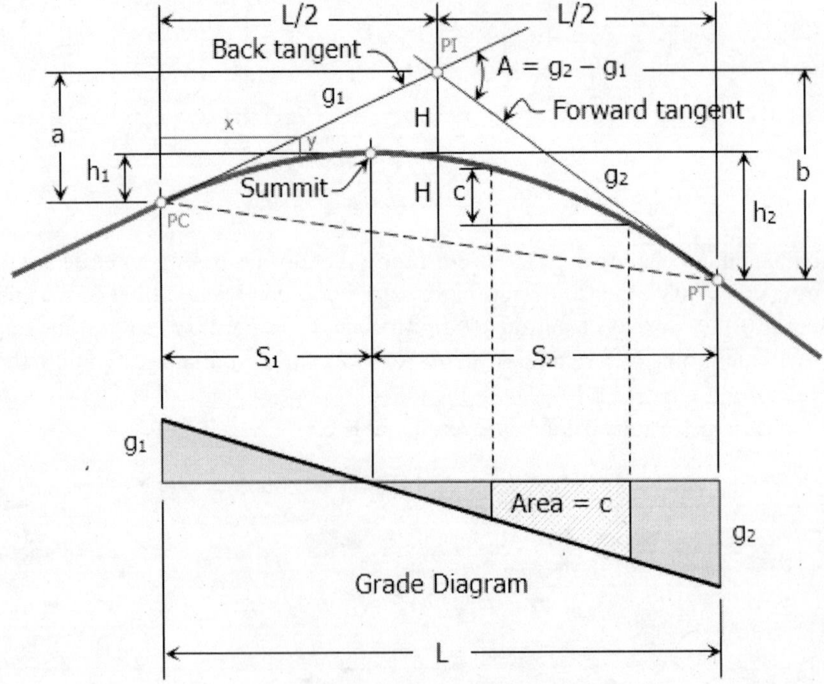

Fig. 2.21: Elements of a summit vertical curve

If PC is the point of curvature, also called as BVC (beginning of vertical curve), PT is the point of tangency, also called as EVC (end of vertical curve), PI is the point of intersection of the tangents, also called PVI (point of vertical intersection), L is the length of parabolic curve, it is the projection of the curve onto a horizontal surface which corresponds to the plan distance, S_1 is the horizontal distance from PC to the highest (lowest in case of sag) point of the summit (or sag) curve, S_2 is the horizontal distance from PT to the highest (lowest in case of sag) point of the summit (or sag) curve, h_1 is the vertical distance between PC and the highest (lowest in case of sag) point of the summit (or sag) curve, h_2 is the vertical distance between PT and the highest (lowest in case of sag) point of the summit (or sag) curve, g_1 is the grade (in percent) of back tangent (tangent through PC), g_2 is the grade (in percent) of forward tangent (tangent through PT), A is the change in grade from PC to PT, a is the vertical distance between PC and PI, b is the vertical distance between PT and PI, H is the vertical distance between PI and the curve, then-

The length of parabolic curve L is the horizontal distance between PI and PT. PI is midway between PC and PT. The curve is midway between PI and the midpoint of the chord from PC to PT. The vertical distance between any two points on the curve is equal to area under the Figure. The vertical distance c is area. The grade of the curve at a specific point is equal to the offset distance in the grade diagram under that point.

Rise = run × slope

$a = g_1L / 2$ (2.64)

$b = g_2L / 2$ (2.65)

Neglecting the sign of g_1 and g_2.

$$S_1 = (g_1L) / (g_1 + g_2) \quad (2.66)$$

$$S_1 = (g_2L) / (g_1 + g_2) \quad (2.67)$$

Vertical distance = shaded area under the diagram

$$h_1 = g_1S_1 / 2 \quad (2.68)$$

$$h_2 = g_2 S_2 / 2 \quad (2.69)$$

Other formulas could be-

$$H = L\,(g1+g2) / 8 \quad (2.70)$$

$$x^2 / y = (L/2)^2 / H \quad (2.71)$$

2.7.3 Characteristics of a vertical curve

The characteristic of a parabolic curve is that the gradient changes from point to point but the rate of change in grade remains constant. Hence, for finding the length of the vertical curve, the rate of change of grade should be an essential consideration as this factor remains constant throughout the length of vertical curve. Generally, the recommended rate of change of grade is 0.1% per 30 m at summits and 0.05% per 30 m at sags.

In Figure 2.22, AB and BC are two gradient lines intersecting at point B. A vertical curve (T_1FT_a) is to be introduced between these two gradients.

If g_1% is the gradient of line AB (In this case, it is +ve), g_2% is the gradient of line BC (In this case, it is -ve), r is the rate of change of grade, T_1 and T_2 are the tangent points at the beginning of the curve and at the end of the curve, T_1T_2 is the line joining the points of tangency or chord of the curve, E is mid-point of the chord of the curve, and F is the mid-point of the curve, then-

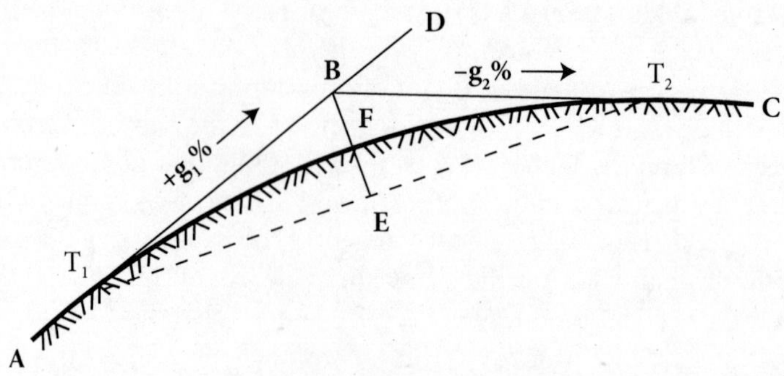

Fig. 2.22: Vertical summit curve

i. It is assumed that length of the vertical curve is equal to the length of the two tangents i.e., $T_1FT_2 = BT_1 + BT_2$

Length of the vertical curve, $L = \dfrac{\text{algabraic difference of two grade}}{\text{rate of change of grade}}$

$$= \frac{g_1 - g_2}{r} \qquad (2.72)$$

The length of the curve on either side of the intersection point is also assumed to be equal i.e., half of the length on one side and the remaining half on other side of the intersection point.

ii. Chainage of beginning of the curve (T_1) = Chainage of B – *L/2* (2.73)

iii. Chainage of end of the curve (T_2)

= Chainage of the intersection point + half the length of vertical curve

= Chainage of B + *L/2* (2.74)

iv. RL of T_1 = RL of B ± level difference between T_1 and B

(+ve sign is used, if g_1 is –ve and vice versa). In this case g_1 is +ve, so -ve sign is to be used.

$$\text{RL of } T_1 = \text{RL of B} - \frac{L}{2} \times \frac{g_1}{100} \qquad (2.75)$$

v. RL of T_2 = RL of B ± level difference between B and T_2

(+ve sign is used if g_2 is +ve, and –ve if g_2 is -ve). In this case g_2 is –ve, so -ve sign is to be used.

$$\text{RL of } T_2 = \text{RL of B} - \frac{L}{2} \times \frac{g_2}{100} \qquad (2.76)$$

vi. $\text{RL of E} = \frac{1}{2}(\text{RL of } T_1 + \text{RL of } T_2) \qquad (2.77)$

vii. It is well known property of parabola that mid-point E of the chord T_1T_2 is situated on the vertical through the point of intersection B of the two tangents, and that mid-point F of the vertical curve is midway between these two points.

BF = FE

$$\text{RL of F} = \frac{1}{2}\ (\text{RL of E} + \text{RL of B}) \qquad (2.78)$$

viii. The curve length of either side of the intersection point is divided into suitable number of parts (15 m, 20 m or 30 m) and the RL of the points on the curve may be calculated by adding algebraically the tangent corrections (tangents offsets) to the RLs of the corresponding points on curve.

The RL of a point at a distance x on the curve = RL of the point at a distance x along the tangent ± tangent correction (y_x).

a. The RLs of points along the tangent can be determined from the known values of RL of the tangent point or intersection point and the gradient of tangent lines.

$$\text{RL of a point at a distance } x \text{ from the tangent point} = \text{RL of the tangent point} \pm \frac{g}{100} \times x \quad (2.79)$$

b. Tangent correction (tangent offset) for a point at a distance x along the tangent.

$$y_x = \frac{g_1 - g_2}{400} \times \frac{x^2}{\text{half the length of the curve}} \quad (2.80)$$

UNIT SUMMARY

This unit discusses two broad type of curves used in roads and railways; horizontal curves and vertical curves. Undwerstanding the importance of curves is neceassry as they are frequently employed due to topography and other features on the ground. Various types of horizontal curves: Simple circular curves, Compound curves, Reverse curves and Transition curves are explained. The characterstics of each curve are given. Mathematical relationships are derived so that the neceassary components of different types of curves are computed and curves established on the ground, mainly using angular and linear measurements. The super-elevation plays an important role while the vehicle is moving a curve, so it is necessary to provide while a curve is laid out. The gradient of vertical curves is quite significant which determines the stopping sight distance at a particular speed of the vehicle. plays an important role.

SOLVED EXAMPLES

Example 2.1:

A circular curve has 300 m radius and 60° deflection angle. Determine its degree by (a) arc definition and (b) chord definition of standard length 30 m. Also compute the (i) length of curve, (ii) tangent length, (iii) length of long chord, (iv) mid-ordinate and (v) apex distance.

Solution:

$R = 300$ m, $D = 60°$

a. Arc definition: s = 30 m,

$R = (s / D_a)(180/ \pi)$

$300 = 30 * 180 / (D_a \pi)$

$D_a = 5.730$ Ans.

b. Chord definition: $R \sin (D_c/ 2) = s /2$

$300 \sin D_c/ 2 = 30 / 2$

$D_c = 5.732$ Ans.

i. Length of the curve: $L = R\,\Delta\,(\pi/180)$

$L = 300 * 60\,(\Delta/180) = 314.16$ m Ans.

ii. Tangent length: $T = R \tan \Delta/2$

$T = 300 \tan 60/2 = 173.21$ m Ans.

iii. Length of long chord: $L_c = 2 R \sin \Delta /2$

$L_c = 2 * 300 \sin 60/ 2 = 300$ m Ans.

iv. Mid-ordinate: $M = R (1- \cos \Delta/2)$

$= 300 (1- \cos 60/2) = 40.19$ m Ans.

v. Apex distance: $E = R (\sec \Delta/2 - 1)$

$= 300 (\sec 60/2) = 46.41$ m Ans.

Example 2.2:

If the approximate perpendicular offset for the mid-point of the circular curve deflecting through 76° 38' is 96.1 m, calculate the radius of the curve.

Solution:

$\Delta = 38$ °76', Ox = 96.1 m

Using Perpendicular offset method:

$O_x = R-\sqrt{[R^2-(x)^2]}$

The distance 'x' from T_1 for locating the apex point = $R \sin \Delta/2$

$= R \sin 76° 38' / 2 = 0.62R$ m

Now

$96.1 = R-\sqrt{[R^2-(0.62R)^2]} = 0.215 R$

$R = 96.1/ 0.215 \doteq 446.98$ m

Example 2.3:

Determine the ordinates of the points on a circular curve having a long chord of 100 m and a versed sine of 5 m. The ordinates are to be measured from the long chord at an interval of 10 m.

Solution:

Length of long chord (L) = 100 m, Versed sine (O_0) = 5 m, Interval = 10 m

The versined or mid-ordinate

$O_0 = R-\sqrt{[R^2-(L/2)^2]}$

$5 = R-\sqrt{[R^2-(100/2)^2]}$

$R = 252.5$ m

Ordinates at intervals of 10 m

$O_0 = \sqrt{[R^2-(x)^2]} - (R-O_0)$

Example 2.4:

Two straight alignments AI and IC of a road intersect at I at a chainage (250 chainage +15 links), angle of deflection being 120°. Calculate the chainages of the point of commencement and the point of tangency if the radius of the right hand circular curve is 200 m. Assume the length of the chain as 30 m.

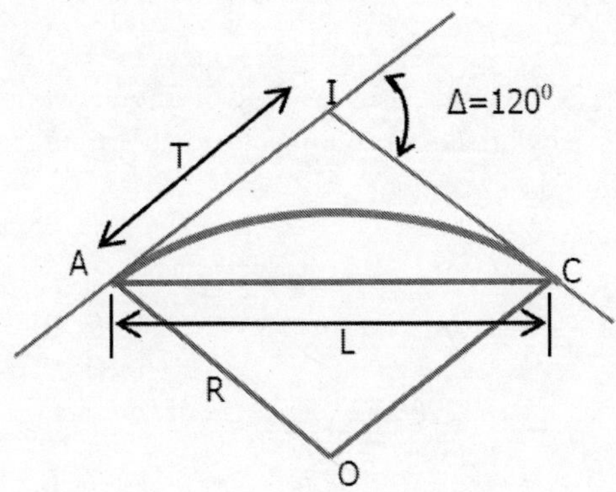

Solution

Chainage at the point of intersection = (250+15) i.e., 250 chains and 15 links.

Length of one chain = 30 m so chainage at the point of intersection = 250 x 30 + 15 x 0.2 = 7503 m

Deflection angle (Δ) = 1200, Radius of curve = 200 m.

Tangent Length T= R tan Δ/2

= 200 tan 120/2 = 346.41m

Length of the curve (l)= πRΔ/ 180 = 200 * 120 π/ 180 = 418.88 m

Chainage at point of curve (T_1) = Chainage at the point of intersection – tangent length

= 7503 – 346.41 = 7156.59 m

Chainage at point of tangency (T_2) = Chainage at point of curve (T_1) + Length of curve

= 7156.59 + 418.88 = 7575.47 m

Example 2.5:

Two straights AC and CB intersect at C at a chainage of 86.22 chains at a deflection angle of 62°. They are to be smoothly connected by a simple curve of radius 12 chains. Find the tangent length, length of curve and the chainages of the starting and end points of the curve. Also, find the length of the long chord.

Solution:

Chainage at the point of intersection (C) = 86.22 chains, Deflection angle (Δ) = 62°, Radius of curve = 12 chains.

Tangent Length T= R tan Δ/2 = 12 tan 62/2 = 7.21 chains.

Length of the curve (*l*) = πRΔ/ 180 = 12 * 62 π/ 180 = 12.985 chains

Chainage at point of curve (T_1) = Chainage at the point of intersection – tangent length

= 86.22 – 7.21 = 79.01 chains

Chainage at point of tangency (T_2) = Chainage at point of curve (T1) + Length of curve

= 79.01+12.985 = 91.995 chains

Length of Long chord (L) = T_1T_2 = 2R sin Δ/2 = 2 * 12 chains sin 62/2 = 12.361 chains

Example 2.6:

Two straights intersect at a chainage 2056.44 m with an angle of intersection is 130°. If the radius of the simple curve is 50 m, set out the curve by offsets from long chord at 5 m interval, and find the following:

i. Chainage of the point of commencement
ii. Chainage at point of tangency
iii. Length of the long chord

Solution:

Chainage at the point of intersection = 2056.44 m, Angle of intersection = 130° , Radius of curve (R) = 50 m

Deflection angle (Δ) = 180°–130° = 50°

Tangent Length T= R tan $\Delta/2$ = 50 tan 50/2 = 23.32 m

Length of the curve (l) = $\pi R\Delta/ 180$ = 50 * 50 π/ 180 = 43.63 m

Chainage at point of curve (T_1) = Chainage at the point of intersection – tangent length

= 2056.44 – 23.32 = 2033.12 m

Chainage at point of tangency (T2) = Chainage at point of curve (T_1) + Length of curve

= 2033.12 + 43.63 = 2076.75 m

Length of Long chord (L) = T_1T_2 = 2R sin $\Delta/2$ = 2 * 50 sin 50/2 = 42.26 m

Starting from centre of long chord, offset at 5 m interval are calculated for half of the long chord, i.e., 42.26/2 = 21.13m

$y = \sqrt{(R^2 - x^2)} - \sqrt{[R^2 - (L/2)^2]}$

$y = \sqrt{(50^2 - x^2)} - \sqrt{[50^2 - (42.26/2)^2]}$

$y = \sqrt{(50^2 - x^2)} - 45.32$

$y_0 = \sqrt{(50^2 - 0^2)} - 45.32 = 4.68$ m

$y_5 = \sqrt{(50^2 - 5^2)} - 45.32 = 4.43$ m

$y_{10} = \sqrt{(50^2 - 10^2)} - 45.32 = 3.67$ m

$y_{15} = \sqrt{(50^2 - 15^2)} - 45.32 = 3.38$ m

$y_{20} = \sqrt{(50^2 - 20^2)} - 45.32 = 0.51$ m

$y_{21.13} = \sqrt{(50^2 - 21.13^2)} - 45.32 = 0$ m

The other half of the curve is symmetrical so there is no need to calculate for the other half.

Example 2.7:

Two roads having a deflection angle of 45° at apex point V whose chainage of apex point is 1839.2 m, are to be joined by a 200 m radius circular curve. Compute the necessary data to set out the curve by: (a) ordinates from long chord at 10 m interval (b) method of bisection to get every 8th point on curve (c) radial and perpendicular offsets from every full 30 m length along tangent. and (d) offsets from chord produced.

Solution:

R = 200 m and D = 45°

Length of tangent = 200 tan 45/2 = 82.84 m

Chainage of T_1 = 1839.2 – 82.84 = 1756.36 m

Length of curve = R * 45 (π/180) = 157.08 m

Chainage of forward tangent T_2 = 1756.36 + 157.08 = 1913.44 m

a. *By offsets from long chord:*

Distance of DT = L/2 = R sin Δ/ 2

= 200 sin 45/2 = 76.54 m

Measuring 'x' from D,

$y = \sqrt{(R^2 - x^2)} - \sqrt{[R^2 - (L/2)^2]}$

At x = 0,

$O_0 = 200 - \sqrt{(200^2 - 76.54^2)}$

= 15.22 m

$O_1 = \sqrt{(200^2 - 10^2)}$ = 14.97 m

$O_2 = \sqrt{(200^2 - 20^2)}$ = 14.22 m

$O_3 = \sqrt{(200^2 - 30^2)}$ = 12.96 m

$O_4 = \sqrt{(200^2 - 40^2)}$ = 11.18 m

$O_5 = \sqrt{(200^2 - 50^2)}$ = 8.87 m

$O_6 = \sqrt{(200^2 - 60^2)}$ = 6.01 m

$O_7 = \sqrt{(200^2 - 70^2)}$ = 2.57 m

At T_1, O = 0.00

b. *Method of bisection:*

Central ordinate at D = R (1–cos Δ/2)

= 200 (1 – cos 45/2) = 15.22

Ordinate at D_1 = R (1–cos Δ/2)

= 3.84 m

Ordinate at D_2 = R (1–cos Δ/2)

= 0.96 m

c. *Offsets from tangents:*

Radial offsets: $O_x = \sqrt{(R^2 + x^2)} - R$

Chainage of T_1 = 1756.36 m

For 30 m chain, it is at = 58 chains + 16.36 m

x_1 = 30 – 16.36 = 13.64

x_2 = 43.64 m

x_3 = 73.64 m and

the last is at x4 = tangent length = 82.84 m

$O_1 = \sqrt{(200^2 + 13.64^2)} - 200$ = 0.46 m

$O_2 = \sqrt{(200^2 + 43.64^2)} - 200$ = 4.71 m

$O_3 = \sqrt{(200^2 + 73.64^2)} - 200$ = 13.13 m

$O_4 = \sqrt{(200^2 + 82.84^2)} - 200 = 16.48$ m

d. Offsets from chord produced:

Length of first sub-chord = 13.64 m = C_1

Length of normal chord = 30 m = C_2

Since length of chain is 157.08 m, so $C_2 = C_3 = C_4 = C_5 = 30$ m

Chainage of forward tangent = 1913.44 m = 63 chains + 23.44 m

Length of last chord = 23.44 m = C_n = C6

$O_1 = C^2_1 / 2R = 13.64^2 / 2 * 200 = 0.47$ m

$O_2 = C_2 (C_1 + C_2) / 2R = 30^2 (13.64 + 30) / 400 = 3.27$ m

$O_3 = C^2_2 / 2R = 30^2 / 400 = 4.5 \text{ m} = O_4 = O_5$

$O_6 = C_n (C_{n-1} + C_n) / 2R = 23.44 (30 + 23.44) / 400 = 3.13$ m

Example 2.8:

Two roads meet at an angle of 127° 30'. Calculate the necessary data for setting out a curve of 15 chains radius to connect the two straight portions of the road if the curve is to be set out the curve by chain and offsets only. Assume the length of chain as 20 m.

Solution:

Angle of intersection = 127° 30', Deflection angle (Δ) = 180°–127° 30' = 52° 30', Length of chain = 20 m, Radius of curve = 15 chains x 20 = 300 m.

Tangent Length T= R tan $\Delta/2$ = 300 tan 52° 30' / 2 = 147.94 m

a. *Radial offset method:*

$O_x = \sqrt{(R^2 + x^2)} - R$

First half of the curve is set from point of curve (T_1)

Assuming interval as 20 m

$O_{20} = \sqrt{(300^2 + 20^2)} - 300 = 0.67$ m

$O_{40} = \sqrt{(300^2 + 40^2)} - 300 = 2.66$ m

$O_{60} = \sqrt{(300^2 + 60^2)} - 300 = 5.94$ m

$O_{80} = \sqrt{(300^2 + 80^2)} - 300 = 10.48$ m

$O_{100} = \sqrt{(300^2 + 100^2)} - 300 = 16.23$ m

$O_{120} = \sqrt{(300^2 + 120^2)} - 300 = 23.11$ m

$O_{140} = \sqrt{(300^2 + 140^2)} - 300 = 31.06$ m

$O_{147.94} = \sqrt{(300^2 + 147.94^2)} - 300 = 34.49$ m

The second half of the curve may be set from point of tangency (T_2)

b. *Perpendicular offset method:*

$O_x = R - \sqrt{[(R^2 - x^2)]}$

First half of the curve is set from point of curve (T_1)

$O_{20} = R - \sqrt{[(R^2 - 20^2)]} = 0.67$ m

$O_{40} = R - \sqrt{[(R^2 - 40^2)]} = 2.68$ m

$O_{60} = R-\sqrt{[(R^2 - 60^2)]} = 6.06$ m

$O_{80} = R-\sqrt{[(R^2 - 80^2)]} = 10.86$ m

$O_{100} = R-\sqrt{[(R^2 - 100^2)]} = 17.16$ m

$O_{120} = R-\sqrt{[(R^2 - 120^2)]} = 25.05$ m

$O_{140} = R-\sqrt{[(R^2 - 140^2)]} = 34.67$ m

$O_{147.94} = R-\sqrt{[(R^2 - 147.94^2)]} = 39.01$ m

The distance 'x' from T_1 for locating the apex point = R sin Δ/2

x = 300 sin 52° 30' = 132.69 m

$O_{132} = 300-\sqrt{[(300^2 - 132.69^2)]} = 30.94$ m

Second half of the curve may be set from point of tangency (T_2)

Example 2.9:

Compute the necessary data for setting out a circular curve of 300 m radius by deflection angle method. The peg interval is 30 m Two tangents intersect at the chainage 1190 m, with the 36° deflection angle.

Solution:

Chainage of apex V = 1190 m, Deflection angle D = 36°, Radius R = 300 m, Peg interval = 30 m.

Length of tangent = R tan Δ/2

= 300 tan 36/2 = 97.48 m

Chainage of T_1 = 1190 – 97.48 = 1092.52 m = 36 chains of 30 m + 12.52 m

C_1 = 30 – 12.52 = 17.48 m, and C_2 = 30

Length of curve = RΔ (π/180)

= 300 * 36 (π/180) = 188.50 m

$C_3 = C_4 = C_5 = C_6 = 30$ m

$C_n = C_7$ = 188.5–17.48– 30*5 = 21.02 m

Chainage of T2 = 1092.52 + 188.50 = 1281.02 m

Ordinates are $O_1 = C^2{}_1/2R = (17.48)^2 /2 *300 = 0.51$ m

$O_2 = C_2 (C_2 + C_1)/ 2R = 30 (30 + 17)/ 2*300 = 2.37$ m

$O_3 = O_4 = O_5 = O_6 = 30^2/ 300 = 3.0$ m

$O_7 = 21.02 (21.02 + 30)/ 2*300 = 1.79$ m

Example 2.10:

Tabulate the data needed to set out a circular curve of radius 600 m by a theodolite and tape to connect two straights having a deflection angle of 18° 24. The chainage of the PI is 2140.00 m and a normal chord length of 20 m is to used.

Solution:

Given Δ =18°24', R = 600 m, normal chord c = 20m, Chainage of PI = 2140.00 m = 21+40.00

Tangent distance, T= R tan (Δ /2) = 600 tan 9°12"= 97.20 m

T=97.20 m = 0 + 97.20

Chainage of PC = 2042.80 m = 20 + 42.80

Next full station on the curve (@20-m intervals) = 20 + 60.00

Therefore, length of initial subchord c = 20 + 60.00 – 20 + 42.80 = 17.20 m

Length of curve = R Δ (π /180) = 600*18.24 (π/180) = 192.68 m

Chainage of PC = 2042.80 m= 20 + 42.80

Length of curve L=192.68 m = 1 + 92.68

Chainage of PT= 2235.48 m = 22 + 35.48

Last full station on the curve (@ 20-m intervals) = 22 + 20.00

Therefore, length of final subchord C_2 = 22 + 35.48 – 22 + 20.00 = 15.48 m

The curve has an initial subchord C_1 of 17.20 m

Eight normal chords of 20 m and a final subchord of 15.48 m

Given the cord length C, the deflection angle from station to station on the is given by Δ =1718.873c / R minutes

Hence, δ_1=1718.873 C_1 / R = 1718.873*17.20 / 600 = 49.27'

δ= 1718.873 C / R = 1718.873*20 / 600 =57.30'

δ_2=1718.873 C_2 /R=1718.873*15.48/600=44.35'

Now, calculate the chainage of each station and the cumulative deflection angles from the back tangent to each station on the curve (tabulated below)

Now, the chainage of each station and the cumulative deflection angles from the back tangent to each station on the curve is computed and tabulated below.

Chainage	Chord (m)	Deflection Angle	Total deflection angle	Total def. angle with 20" theodolite
PC=20+42.80	0	0	0	0
20+60.00	17.20	49.27'	0°49.27'	0°49'20"
+80.00	20.00	57.30'	1°46.57'	1°46'40"
21+00.00	20.00	57.30'	2°43.87'	2°44'00"
+20.00	20.00	57.30'	3°41.17'	3°41'20"
+40.00	20.00	57.30'	4°38.47'	4°38'20"
+60.00	20.00	57.30'	5°35.77'	5°35'40"
+80.00	20.00	57.30'	6°33.07'	6°33'00"
22+00.00	20.00	57.30'	7°30.37'	7°30'20"
+20.00	20.00	57.30'	8°27.67'	8°27'40"
+35.48	15.48	44.35'	9°12.00'	9°12'00"= Δ/2 (check)

Example 2.11:

Tabulate the necessary data for setting out a simple circular curve. The angle of intersection is 144°, chainage of point of intersection is 1390 m, radius of the curve is 50 m. The curve is to be set out by offsets from chords produced with pegs at every 10 m of through chainage.

Solution:

Chainage at the point of intersection= 1390 m, Angle of intersection= 144°, Radius of curve = 50 m.

Deflection angle (Δ)= 180°-144°= 36°

Tangent Length T= R tan Δ /2 = 50 tan 36°/2 = 16.25 m

Length of the curve (l) = π R Δ/ 180= 50 * 36 π/ 180 = 31.42 m

Chainage at point of curve (T_1) = Chainage at the point of intersection – tangent length

= 1390.00 – 16.25 = 1373.75 m

Chainage at point of tangency (T_2) = Chainage at point of curve (T_1) + Length of curve

= 1373.75 + 31.42 = 1405.17 m

Length of initial sub chord (C_1) = 1380 -1373.75 = 6.25 m

Length of final sub chord (C_n) = 1405.17–1400.00 = 5.17 m

Length of normal chord (C) = 10 m

No. of normal chords = (1400-1380) / 10 = 2

Total no of chords = 1 + 2 +1 = 4

Offset for the initial sub-chord O_1= C_1^2 / 2R = 6.25^2 / 2 * 50 = 0.39 m

Offset for the second chord O_2 = C (C + C_1) / 2R = 10 (10 + 6.25) / 100 = 1.625 m

Offset for the third chord O_3 = C_2 / R = 102 / 50 = 2.0 m

Offset for the fourth chord O_4 = C_4 (C + C_4) / 2R = 5.17 (10 + 5.17) / 100 = 0.78 m

Example 2.12:

Two tangents intersect at chainage 1190 m, with 36° deflection angle. Calculate all the required data for setting out a simple circular curve of 60 m radius by the deflection angle method. Take peg interval as 10 m.

Solution:

Tangent Length T= R tan Δ /2 = 60 tan 36 / 2 = 19.50 m

Length of the curve (l) = π R Δ/ 180 = 60 * 36 π / 180 = 37.70 m

Chainage at point of curve (T_C) = Chainage at the point of intersection – tangent length

= 1190.00 – 19.50 = 1170.50 m

Chainage at point of tangency (T_T) = Chainage at point of curve (T_C) + Length of curve = 1170.50 + 37.70 = 1208.20 m

Length of initial sub chord (C_1) = 1180 -1170.50 = 9.5 m

Length of final sub chord (C_n) = 1208.20–1200 = 8.2 m

Length of normal chord (C) = 10 m (peg interval)

No. of normal chords = 1200 – 1180 / 10 = 2

Total no of chords = 1 + 2 + 1 = 4

δ_1 = 90 C_1 / (π R) = 90 * 9.5 / 60 π = 4° 32’ 9.3”

δ_2 = δ_3 = 90 * 10 / 60π = 4° 46’ 28.73”

δ_4 = 90 * C_4 / 60π = 3° 54’ 54.76”

Point	Chainage (m)	Chord Length (m)	Tangential angle	Deflection angle	Actual Theodolite reading
T_C	1170.50		0° 00' 00"	0° 00' 00"	0° 00' 00"
P_1	1180	9.5	δ_1 = 4° 32' 9.3"	Δ_1= δ_1= 4° 32' 9.3"	4° 32' 00"
P_2	1190	10	δ_2 = 4° 46' 28.73"	Δ_2=Δ_1+δ_2= 9° 18' 38.03"	9° 18' 40"
P_3	1200	10	δ_3 = 4° 46' 28.73"	Δ_3=Δ_2+δ_3= 14° 5' 6.76"	14° 05' 00"
T_T	1208.20	8.2	δ_4 = 3° 54' 54.76"	Δ_4=Δ_3+δ_4=18° 00' 1.52"	18° 00' 00"

Check : $\Delta_4 = \Delta/2$ = 36/ 2 = 18° 00' 00"

Example 2.13:

Two straights intersect at a chainage of 1764 m with a deflection angle of 32°, which are to be joined by a 5° curve. Work out the data required to set out a curve by the deflection angle method. Take length of chain as 30 m, peg interval at 30 m, and least count of theodolite as 20".

Solution:

For a 30 m chain length, R = 1719 / Degree of curve of Degree = 343.8 m

Tangent Length T= R tan Δ /2 = 343.8 tan 32 / 2 = 98.58 m

Length of the curve (l) = π R Δ/ 180

= 343.8 * 32 π/ 180 = 192.01 m

Chainage at point of curve (T_C) = Chainage at the point of intersection – tangent length

= 1764 – 98.58 = 1665.42 m

Chainage at point of tangency (T_T) = Chainage at point of curve (T_C) + Length of curve

= 1665.42 + 192.01 = 1857.43 m

Length of initial sub chord (C_1) = 1680 – 1665.42= 14.58 m

Length of final sub chord (C_n) = 1857.43 – 1830.00 = 27.43 m

Length of normal chord (C) = 30 m (peg interval)

5 30 No. of normal chords = (1830 – 1680) / 30 = 5

Total no of chords = 1 + 5 +1 = 7

$\delta_1 = 90\ C_1 / (\pi\ R)$ = 90 * 14.58 / 343.8 π = 1° 12' 53.68"

$\delta_2 = \delta_3 = \ldots\ldots\ldots \delta_6 =$ 90 * 30 / 343.8π = 2° 29' 59.34"

δ_7 = 90 * C_7 / 343.8π = 2° 17' 8.39"

Point	Chainage (m)	Chord Length (m)	Tangential angle	Deflection angle	Actual Theodolite reading
T_C	1165.43		0° 00' 00"	0° 00' 00"	0° 00' 00"
P_1	1680	14.58	δ_1 = 1° 12' 53.68"	Δ_1= δ_1= 1° 12' 53.68"	1° 13' 00"
P_2	1710	30	δ_2 = 2° 29' 59.34"	Δ_2=Δ_1+δ_2= 3° 42' 53.02"	3° 43' 00"

P_3	1740	30	$\delta_3 = 2^\circ\ 29'\ 59.34''$	$\Delta_3=\Delta_2+\delta_3= 6^\circ\ 12'\ 52.36''$	$6^\circ\ 13'\ 00''$
P_4	1770	30	$\delta_4 = 2^\circ\ 29'\ 59.34''$	$\Delta_4=\Delta_3+\delta_4= 8^\circ\ 42'\ 51.7''$	$8^\circ\ 43'\ 00''$
P_5	1800	30	$\delta_5 = 2^\circ\ 29'\ 59.34''$	$\Delta_5=\Delta_4+\delta_5= 11^\circ\ 12'\ 51.04''$	$11^\circ\ 13'\ 00''$
P_6	1830	30	$\delta_6 = 2^\circ\ 29'\ 59.34''$	$\Delta_6=\Delta_5+\delta_6= 13^\circ\ 42'\ 50.38''$	$13^\circ\ 43'\ 00''$
T_T	1857.43	27.43	$\delta_7 = 2^\circ\ 17'\ 8.39''$	$\Delta_7=\Delta_6+\delta_7=15^\circ\ 59'\ 58.77''$	$16^\circ\ 00'\ 00''$

Check : $\Delta_7 = \Delta/2 = 32/\ 2 = 16^\circ\ 00'\ 00''$

Example 2.14:

A right hand compound curve, with radius of the first arc as 200 m and the radius of the second arc as 250 m, has a total deflection angle is 80°. If the chainage of the point of intersection is 1504.80 m and the deflection angle of the first arc is 50°, determine the chainages of the starting point, the point of compound curve and the point of tangency.

Solution:

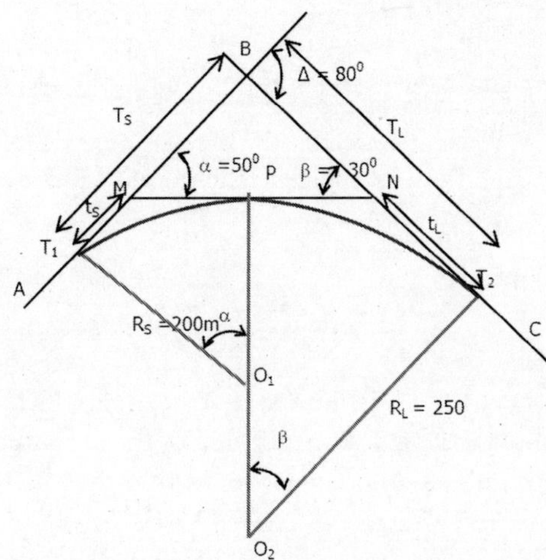

Given- $R_S = 200$ m, $R_L = 250$ m, $\Delta = 80^\circ$, chainage of the intersection point = 1504.80 m, and $\angle BMN = \alpha = 50^\circ$.

Let, BA and BC be the back tangent and forward tangent.

$\Delta = \alpha + \beta = 80^\circ$

Deflection angle of the second arc $\angle BNM = \beta = \Delta - \alpha = 80^\circ - 50^\circ = 30^\circ$

$MP = MT_1 = t_S = R_S \tan \alpha/2 = 200 \tan 50/2 = 93.26$ m

$NP = NT_2 = t_L = R_L \tan \beta/2 = 200 \tan 30/2 = 67.00$ m

MN = MP + NPN = 93.26 + 67.00 = 160.26 m

In triangle BMN,

$\angle BMN + \angle BNM + \angle MBN = 1800$

$\angle MBN = 180^\circ - \angle BMN - \angle BNM = 180^\circ - 50^\circ - 30^\circ = 100^\circ$

Applying sine rule in triangle BMN,

(BN / sin α) = MN / sin MBN

BN = MN sin α / sin MBN

= 160.26 sin 50 / sin 100

BN = 124.66 m

(BM / sin β) = MN / sin MBN

BM = MN sin β/ sin MBN

= 160.26 sin 50 / sin 100

BM = 81.23 m

Tangent length $BT_1 = (T_S) = BM + MT_1 = 81.23 + 93.26 = 174.49$ m

Tangent length $BT_2 = (T_L) = BN + NT_2 = 124.66 + 67.00 = 191.66$ m

Length of the first curve = $R_S\, \alpha\, (\pi / 180) = 200 * 50\, (\pi / 180) = 174.53$ m

Length of the Second curve = $R_L\, \beta\, (\pi / 180) = 200 * 30\, (\pi / 180) = 130.90$ m

Chainage at point of curve (T_1) = Chainage at the point of intersection – tangent length (BT_1)

= 1504.80 – 174.53 = 1330.27 m.

Chainage at point of PCC = chainage at the point of curve + length of first curve

= 1330.27 + 174.53 = 1504.8 m

Chainage at point of tangency (T_2) = chainage at the point of PCC + length of second curve

= 1504.8 + 130.90 = 1635.7 m

Example 2.15:

Two straights AC and BC are intersected by a line MN. The angles AMN and MNB are 150° and 160° respectively. The radius of the first curve is 650 m and that of the second curve is 450 m of a compound curve. If the chainage of point of intersection C is 4756 m, find the chainage of the tangent points and the point of compound curvature.

Solution:

$\Delta_1 = 180 - 150 = 30°$

$\Delta_2 = 180 - 160 = 20°$

$\Delta = \Delta_1 + \Delta_2 = 30 + 20 = 50°$

$t_1 = T_1M = 650 \tan 30/2 = 174.17$ m

Now, TL_1 = T1M + MC

$= R_1 \tan \Delta_1/2 + [R_1 \tan \Delta_1/2 + R_2 \tan\Delta_2/2]\, (\sin\Delta_2/\sin\Delta)$

$TL_1 = 36 \tan \Delta_1/2 + [36\ 2 \tan \Delta_1/2 + 48 \tan (84.5\text{-}\Delta_1)/2\ ]\ [\sin (84.5\text{-}\Delta_1)/ \sin 84.5]$

$38.98 = 36 \tan \Delta_1/2 + 1.0046251 \sin (84.5 - \Delta_1)\ [36 \tan\Delta_1/2 + 48 \tan (84.5\text{-}\Delta_1)/2]$

$F(\Delta_1) = 36 \tan \Delta_1\ 2 + 1.0046251 \sin (84.5 - \Delta_1)\ [36 \tan\Delta_1/2 + 48 \tan (84.5\text{-}\Delta_1)/2] - 38.98$

Solving it by trial and error method, when Δ_1 = °30,

$F(\Delta_1) = -0.0954222$

when $\Delta_1 = °32$, $F(\Delta_1) = -0.123610$.

If $\Delta_1 = °29$, $F(\Delta_1) = -0.08106$

If $\Delta_1 = °28$, $F(\Delta_1) = -0.0665$

If $\Delta_1 = °25$, $F(\Delta_1) = -0.02419195$

If $\Delta_1 = °23$, $F(\Delta_1) = 0.008600$

say $\Delta_1 = 23.5°$ for which $F(\Delta_1) = 0.000914 \approx 0$

Thus, the solution is-

$\Delta_1 = 23.5°$, $\Delta_2 = 84.5 - 23.5 = 61°$

Arc length of first curve = 36 * 23.5 (π/180) = 14.765 chains of 30 m length

Chainage of point of junction of the two curves (C) = 30.5 + 14.765 = 45.265 chains of 30 m length

Length of second curve = 48 * 61 (π/180) = 51.103 chains of 30 m length

Chainage of last tangent point (T_2) = 45.265 + 51.103 = 96.363 chains of 30 m length

For first curve: Length of first sub-chord = 31 – 30.5 = 0.5 chains of 30 m length

Example 2.16:

Two straights BA and AC are intersected by a line EF so that angles BEF and EFC are 140° and 145°, respectively. The radius of the first curve is 600 m and the second curve is 400 m. If the chainages of the intersection point A is 3415 m, compute the chainages of the tangent points and the point of compound curvature.

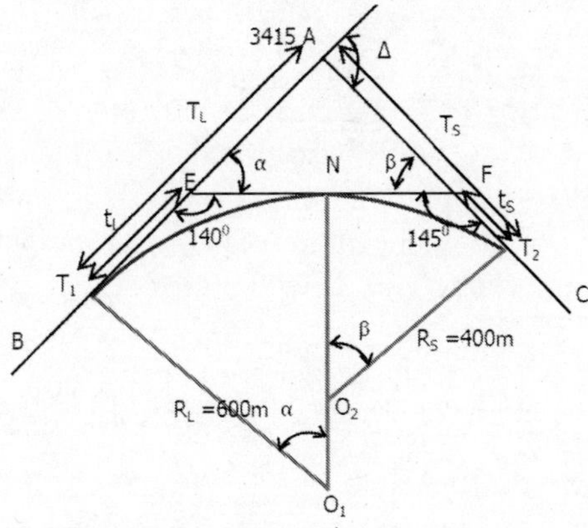

Solution:

In above Figure, R_L = 600 m, R_S = 400m, ∠BEF = 140°, ∠EFC = 140°, Chainage of the intersection point A = 3415 m.

∠AEF = α = 180°–∠BEF =180°–140° = 40°

∠AFE = β= 180°–∠EFC =1800–145° = 35°

$ET_1 = EN = t_t = R_L \tan \alpha /2$ = 600 tan 40/2 = 218.40 m

$FT_2 = FN = t_s = R_S \tan \beta /2$ = 400 tan 35/2 = 126.12 m

EF = EN + FN = 218.40 +126.12 = 344.52 m

Consider, Δ AEF,

$\angle AEF + \angle AFE + \angle EAF = 180°$

$\angle EAF = 180° - \angle AEF - \angle AFE = 180° - 40° - 35° = 105°$

Applying sine rule to the Δ AEF,

(AE / sin β) = (EF/ sin EAF)

AE = EF (sin β/ sin EAF)

AE = 344.52 (sin 35/ sin 105) = 204.60 m

Similarly,

(AF / sin α) = (EF/ sin EAF)

AF = EF (sin α/ sin EAF) = 344.52 (sin 40/ sin 105) = 229.27 m

Tangent length $AT_1 = T_L = AE + t_L$ = 204.60 + 218.40 = 423.40 m

Tangent length $AT_2 = T_S = AF + t_s$ = 229.27 + 126.12 = 355.40 m

Length of the first curve = $\pi R_L \alpha / 180 = 600 * 40 \pi / 180$ = 418.90 m

Length of the Second curve = $\pi R_s \beta / 180 = 400 * 35 \pi / 180$ = 244.35 m

Chainage at point of curve (T_1) = Chainage at the point of intersection – tangent length (T_L) = 3415 – 423 = 2992 m.

Chainage at point PCC (N) = Chainage at the point of curve + Length of first curve

= 2992 + 418.90 = 3410.9 m

Chainage at point of tangency (T_2) = Chainage at the point of PCC + Length of second curve = 3410.9 + 244.35 = 3655.25 m

Example 2.17:

Two straights AC and CB have bearing of 59° 30’ and 127° 30’ intersect at C at a chainage of 1456.5 m. Two points M and N are located on AC and BC so that the bearing of MN is 105° 30’. The straights AC and CB are to be connected by a compound curve consisting of arcs of radii 220 m and 275 m, respectively. If the line MN is the common tangent to the two arcs, compute the chainages of the starting point, point of compound curvature (PCC) and the end point of the curve.

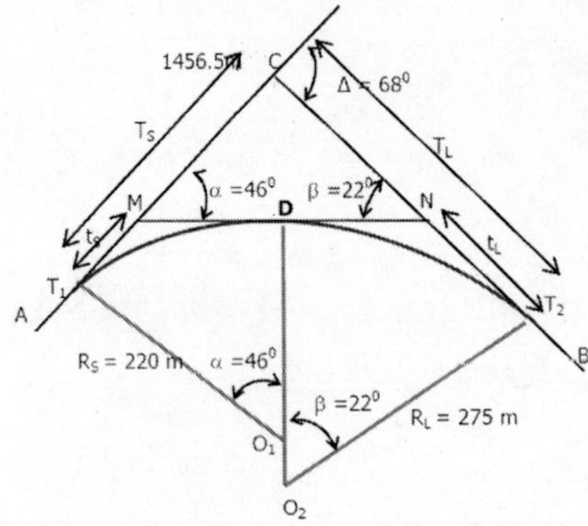

Solution:

Bearing of AC is 59° 30’, of CB is 127° 30’, and of MN is 105° 30’. Chainage of the intersection point is 1456.5 m. Radius of the first arc (R_S) is 220 m, and that of second arc (R_L) is 275 m. Let D be the point on the line MN.

Δ = Bearing of CB – Bearing of AC

$\Delta = 127° 30' - 59° 30' = 68°$

CMN = α = Bearing of MN – Bearing of AC

$= 105° 30' - 59° 30' = 46°$

$\Delta = \alpha + \beta = 68°$

$\beta = \angle CNM = \Delta - \alpha = 68° - 46° = 22°$.

$MD = T_1M = t_S = R_S \tan \alpha/2$

= 220 tan 46/2

= 93.38 m

$ND = T_2N = t_L = R_L \tan \beta/2$

= 275 tan 22/2

ND = 53.45 m

MN = MD + ND = 93.38 + 53.45 =146.83 m

In triangle CMN,

$\angle CMN + \angle CNM + \angle MCN = 180°$

$\angle MCN = 180° - \angle CMN - \angle CNM = 180° - 46° - 22° = 112°$

Applying, sine rule to the triangle CMN,

$(CM / \sin \beta) = MN / \sin MCN$

CM = 146.83 sin 22 / sin 112

CM = 59.32 m

Similarly, $(CN / \sin \alpha) = MN / \sin MCN$

CN = 146.83 sin 46/ sin 112

CN = 113.92 m

Tangent length $CT_1 = t_S = CM + MT_1 = 59.32 + 93.38 = 152.7$ m

Tangent length $CT_2 = t_L = CN + NT_2 = 113.92 + 53.45 = 167.37$ m

Length of the first curve = $R_S\, \alpha\, (\pi / 180) = 220 * 46\, (\pi / 180) = 176.63$ m

Length of the Second curve= $R_L\, \beta\, (\pi / 180) = 275 * 22\, (\pi / 180) = 87.16$ m

Chainage at point of curve (T_1) = Chainage at the point of intersection – tangent length (CT_1)

= 1456.5 – 157.7 = 1298.8 m

Chainage at point of PCC = Chainage at the point of curve + Length of first curve

= 1298.8 + 176.63 =1475.43 m.

Chainage at point of tangency (T_2) = Chainage at the point of PCC + Length of second curve

= 1475.43 + 87.16 = 1562.59 m

Example 2.18:

Two parallel railway lines 12 m apart are to be connected by a reverse curve of same radius throughout. If the maximum distance between tangent points measured parallel to the straight is 48 m. Find the maximum allowable radius. Also calculate the radius of 2nd branch, if both the radii are different, and that of the 1st branch is 60 m. Also calculate the length of both the branches.

Solution:

Let A and B be the points of tangencies and C point of reverse curvature (PRC) in the figure. The distance between the lines V=12 m, and h=48 m.

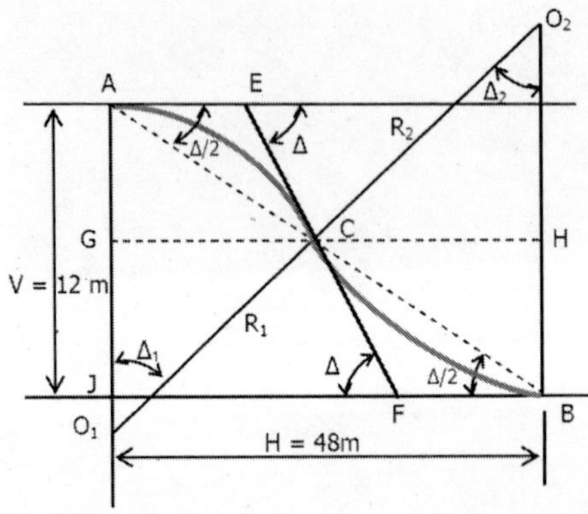

$\text{Tan}\ \Delta / 2 = V / h = 12 / 48 = 0.25$

So $\Delta = 28^\circ\ 04'\ 20.95''$

i. *when the radius is same*

$h = R_1 \sin \Delta + R_2 \sin \Delta$

Here $R_1 = R_2 = R$, so $h = 2 R \sin \Delta$

$R = h / (2 \sin \Delta)$

$R = 48 / 2 \sin 28^\circ\ 04'\ 20.95''$

$R = 51$ m Ans.

ii. *when the radius is different*

Using the above relationship for h

$48 = 60 \sin 28^\circ\ 4'\ 20.95'' + R_2 \sin 28^\circ\ 4'\ 20.95''$

$R_2 = (48 - 60 \sin 28^\circ\ 4'\ 20.95'') / \sin 28^\circ\ 4'\ 20.95''$

$R_2 = 42$ m Ans.

iii. *Length of curves*

Length of first curve $= R_1\ \Delta\ (\pi/180)$

$= 51 * 28^\circ\ 4'\ 20.95''\ (\pi/180) = 29.40$ m Ans.

Length of second curve $= R_2\ \Delta\ (\pi/180)$

$= 42 * 28^\circ\ 4'\ 20.95''\ (\pi/180) = 20.58$ m Ans.

Example 2.19:

Two parallel lines 469 m apart are joined by a reverse curve ABC with two different arcs which deflects to the right by 30° angle from the first straight. If the radius of the first arc is 1400 m, calculate the radius of the second arc. Also calculate the chainage of points B and C, if the chainage of A is 2500 m

Solution:

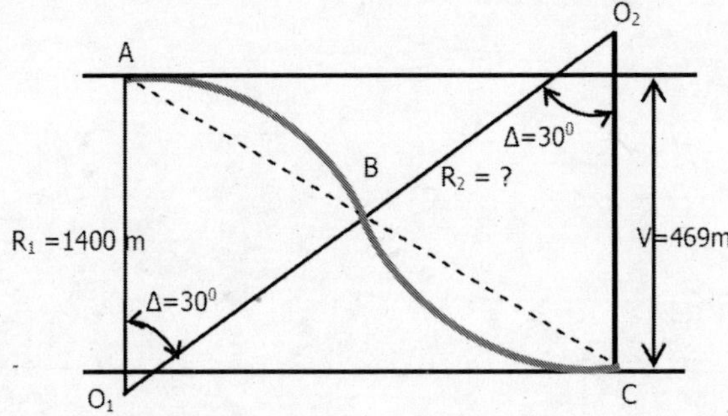

In given Figure, A and C are the points of tangencies and B is the point of reverse curvature (PRC). The distance between the lines V = 469 m

$V = (R_1 + R_2)$ versin Δ

$469 = (R_1 + R_2)(1 - \cos\Delta)$

$469 = (1400 + R_2)(1 - \cos 30^\circ)$

$R_2 = 2100.66$ m Ans.

Length of first curve AB = $R_1 \Delta (\pi/180)$

$= 1400 * 30^\circ (\pi/180)$

$= 733.04$ m

Length of second curve BC = $R_2 \Delta (\pi/180)$

$= 2100.66 * 30^\circ (\pi/180)$

$= 1099.9$ m

Chainage at B = chainage at A + length of first curve

Chainage at B = 2500 + 733.04 = 3233.04 m Ans.

Chainage at C = chainage at B + length of second curve

Chainage at C = 3233.04 + 1099.9 = 4332.94 m Ans.

Example 2.20:

A reverse curve ACB is to be set out between two parallel tangents 30 m apart. The distance between the tangent points A and B is 120 m. Find the radius R, if R_1 is equal to R_2, as well as the radius R_2, if R_1 is 100 m. Also calculate the deflection angle.

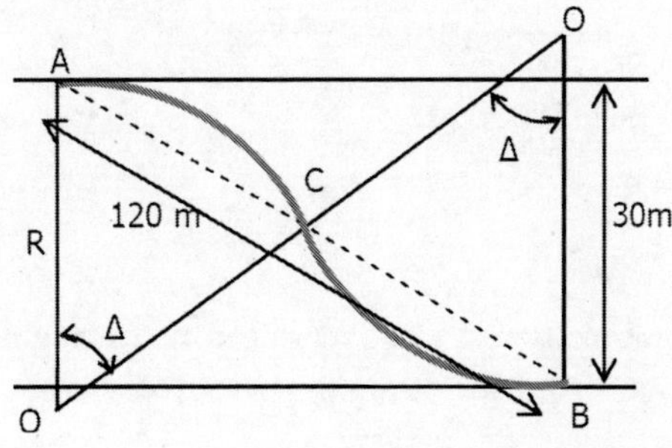

Solution:

In the Figure, the length between tangent points A and B, L = 120 m

V = 30 m.

i. *when the radius is same*

$L^2 = 2V (R_1 + R_2)$

If $R_1 = R_2 = R$, then $L = \sqrt{(4RV)}$

$120 = \sqrt{(4 * 30 * R)}$

R = 120 m Ans.

ii. *when the radius is different*

$L^2 = 2V (R_1 + R_2)$

$120^2 = 2 * 30 (100 + R_2)$

$R_2 = 140$ m Ans.

iii. *The deflection angle*

$V = (R_1 + R_2)$ versin Δ

$30 = (100 + 140) (1 - \cos\Delta)$

$\Delta = \text{Cos}^{-1} (0.875)$

Δ = 18 '57 °28" Ans.

Example 2.21:

A reverse curve is to be set out between two parallel tangents 10 m apart. The distance between the tangent points measured parallel to the tangents is 80 m. If the radius of the first curve is 150 m, calculate the radius of the second curve as well as the lengths of the two branches.

Solution:

v = 10 m, h = 80 m, R1 = 150 m

$\tan \alpha/2 = v / h = 10 / 80$

$\alpha = 14.25°$

$h = R_1 \sin \alpha + R_2 \sin \alpha$

$80 = 150 \sin 14.25 + R_2 \sin 14.25$

$R_2 \sin 14.25 = 43.077$

$R_2 = 175$ m Ans.

Length of first curve = $R_1\ \alpha\ (\pi/180)$

= 150 *14.25 (π/180) = 37.31 m Ans.

Length of second curve = $R_2\ \beta\ (\pi/180)$

= 175 *14.25 (π/180) = 43.52 m Ans

Example 2.22:

A transition curve of the cubic parabola type is to be set out from a straight centreline such that it passes through a point which is 6 m away from the straight, measured at right angles from a point on the straight produced 60 m from the start of the curve. Compute the data for setting out a 120 m length of transition curve at 15-m intervals. Calculate the rate of change of radial acceleration for a speed of 50 km/h.

Solution:

L = 120 m

From expression for a cubic parabola: $y = x^3 / 6RL = cx^3$

$y = 6$ m, $x = 60$ m

$c = y / x^3 = 6 / 60^3$

$= 1 / 36000$

But $c = 1 / 6RL = 1/36000$

$1/ RL = 1/ 6000$

The offsets are now calculated using this constant:

$y_1^3 = 15^3 / 36000 = 0.094$ m

$y_2^3 = 30^3 / 36\,000 = 0.750$ m

$y_3^3 = 45^3 / 36\,000 = 2.531$ m and so on

Rate of change of radial acceleration $q = V^3 / (3.6^3RL)$

$q = 50^3 / (3.6^3 * 6000)$

$q = 0.45$ m/s^3

Example 2.23:

A compound curve AB and BC is to be replaced by a transition curve of 100 m long at each end. The chord lengths *AB* and *BC* are respectively 661.54 m and 725.76 m and radii 1200 m and 1500 m. Calculate the single arc radius: (a) If *A* is used as the first tangent point, and (b) If *C* is used as the last tangent point.

Solution:

$T_1 = AB$, $T_2 = BC$, $R_1 = 1200$ m and $R_2 = 1500$ m

Let I be the intersection point

Chord AB = $2R_1 \sin \Delta_1/2$

Sin $\Delta_1/2 = AB * R_1 /2 = 661.54 / 2*1200$

$\Delta_1 = °32$

Similarly, Sin $\Delta_2/2 = 725.76 / 2*3000$

$\Delta_2 = °28$

Distance $At_1 = t_1B = R_1 \tan \Delta_1/2$

t1B = 1200 tan 16° = 344 m

Similarly, $Bt_2 = t_2C = R_2 \tan \Delta_2/2$

$t_2C = 1500 \tan 14° = 374$ m

$t_1t_2 = 344 + 374 = 718$ m

By sine rule in triangle t_1It_2

$t_1I = 718 \sin 28 / \sin 120$

$t_1I = 389$ m

and $t_2I = 718 \sin 32 / \sin 120$

$t_2I = 439$ m

$AI = At_1 + t_1I = 733$ m

$CI = Ct_2 + t_2I = 813$ m

To find single arc radius:

(a) From tangent point A

$AI = (R + S) \tan \Delta /2 + L/2$

where $S = L^2 / 24R$ and $\Delta = \Delta_1 + \Delta_2 = °60$, and $L = 100$ m

So $733 = [R + (L^2/24R)] \tan 60/2 + 100/2$

$R = 1182$ m

(b) From tangent point C

$CI = (R + S) \tan \Delta /2 + L/2$

$813 = [R + (L^2/24R)] \tan 30 + 50$

$R = 1321$ m

Example 2.24:

Find the length of the vertical curve connecting two grades +0.5% and –0.4% where the rate of change of grade is 0.1%.

Solution:

Length of vertical curve = (0.1 / (30 * (0.4-) -0.5

= ((0.5 + 0.4) * 30 *10) / 1

= 0.9 * 30 * 10

= 270 m

Example 2.25:

Find the length of vertical curve connecting two uniform grades from the following data: (i) + 0.8% and– 0.6%, rate of change of grade is 0.1 per 30 m (ii) – 0.5% and +1%, rate of change of grade is 0.05 per 30 m

Solution:

i. $L = (g_1 - g_2)\, r = (0.8 - (-0.6))\ 0.1 * 30$

= 420 m

ii. $L = (-0.5 - (+1))\ 0.05 * 30$

= 900 m

Example 2.26:

Calculate t6e RL of various stations on a vertical curve connecting two uniform grades of + 0.5% and -0.7%. The chainage and RL of point of intersection is 500 m and 350.750 m, respectively. The rate of change of grade is 0.1% per 30 m.

Solution:

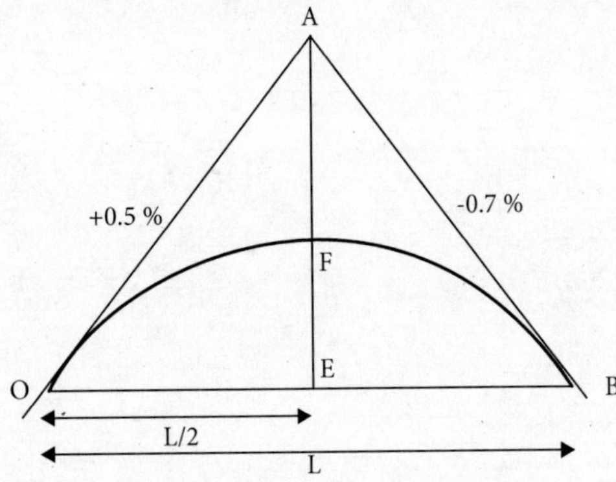

Length of vertical curve L = $(g_1 - g_2)$ r * 30

= (0.5 – (−0.7)) 0.1 * 30 = 360 m

L /2 = 180 m

Chainage of O = chainage of A – L/2 = 500 – 180 = 320 m

Chainage of B = chainage of A + L/ 2 = 500 + 180 = 680 m

RL of point of intersection, A = 330.75 m

RL of O = 330.75–g_1 L / 200

= 330.75–0.5 (360)/ 200

= 329.85 m

RL of B = 330.75–g_2 L/ 200

= 330.75–0.7 (360)/ 200

= 329.49 m

RL of mid-point E of chord OB = ½ (RL of O + RL of B)

= ½ (329.85 +329.49) = 329.67 m

RL of F (vertex of curve) = ½ (RL of B + RL of A)

= ½ (330.75 + 329.67) = 330.21 m

The difference between A and F = 330.75 – 330.21 = 0.54 m

Check: AF = $(g_1 - g_2)$ L/ 800

= [0.5 –(– 0.7)]* 360/ 800 = 0.54 m Checked

1st point on the curve chainage = 350 m , at x= 30 m

RL of 1st point on tangent = 329.85 + (g_1x /100) = 330 m

Tangent correction y = $(g_1 - g_2)$ *x^2/ 200 L

= 0.015 m

RL of 1st point on the curve = 330 m – 0.015 = 329.985 m

Data is shown in table below:

Station	Chainage (m)	Grade Elevation (m)	Tangent Correction (m)	Curve Elevation (m)	
0	320	329.85	0	329.85	Start of V.C
1	350	330.00	– 0.015	329.985	
2	380	330.15	– 0.06	330.090	
3	410	330.30	– 0.135	330.165	
4	440	330.45	– 0.240	330.210	
5	470	330.60	– 0.375	330.225	
6 (F)	500	330.75	– 0.54	330.210	Vertex of V.C
7	530	330.54	– 0.375	330.165	
8	560	330.33	– 0.240	330.090	
9	590	330.12	– 0.135	329.985	
10	620	330.91	– 0.06	329.850	
11	650	330.70	– 0.015	329.685	
12 (B)	680	330.49	0	329.490	End of V.C

Example 2.27:

A 1% grade meets a +2.0% grade at station 470 of elevation 328.605 m. A vertical curve of length 120 m is to be used. The pegs are to be fixed at 10 m interval. Calculate the elevations of the points on the curve by (a) tangent corrections and (b) by chord gradients.

Solution:

Station PVC:

Xo = Sta PVI – L/2 = 470 – (120/2) = 410.0 m

Elevation PVC:

Yo = Elevation at PVI – (g1*L/2) = 328.605 – (-0.01*120/2) = 329.205 m

Equation of Curve

Rate of change of grade, k = (g2-g1)/L = 0.00025

Let x be the distance of any point on the curve, from PVC (BC), and let Y be its elevation.

Equation of an equal tangent vertical parabolic curve is given as,

$Y = Yo + g1*x + k*x^2/2 = 329.205 + (-0.01)*x + (0.00025)*x^2/2$

1st Full Station is at 420.0 m, at x = 420.0 – 410.0 = 10.0 m from PVC

Put x = 10.0 m in above parabolic curve equation-

Elevation @(420.0) = $329.205 + (-0.01)*(10.0) + (0.00025)*(10.0)^2/2 = 329.118$ m

Low Point:

Low point is given at, x = -g1*L/A = 40.0 m from BVC.

hence station (Low point) = Sta_PVC + x = 410.0 + 40.0 = 450.0 m

Elevation at Low point = 329.205 – 0.01* 40.0 + (0.00025)*40.0²/2 = 329.005 m

End Station:

End Station, EVC = PVC + L = 530.0 m at x = L = 120.0 m from PVC

Put 'x' in parabolic curve equation, Elev EVC = 329.805 m

Similarly, put x values of subsequent stations in above parabolic curve equation to get Elevations. Details are tabulated below:

Station	distance x	g1*x	k*x2/2	Elevation(Y)
410.0	0.0	0.0	0.0	329.21
420.0	10.0	-0.1	0.01	329.12
430.0	20.0	-0.2	0.05	329.06
440.0	30.0	-0.3	0.11	329.02
450.0	40.0	-0.4	0.2	329.01
460.0	50.0	-0.5	0.31	329.02
470.0	60.0	-0.6	0.45	329.06
480.0	70.0	-0.7	0.61	329.12
490.0	80.0	-0.8	0.8	329.21
500.0	90.0	-0.9	1.01	329.32
510.0	100.0	-1.0	1.25	329.46
520.0	110.0	-1.1	1.51	329.62
530.0	120.0	-1.2	1.8	329.8

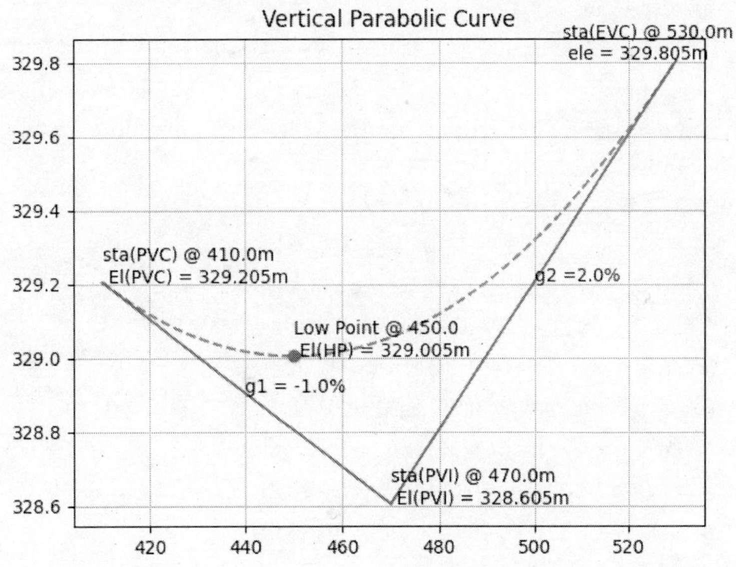

Example 2.28:

Two straights AB and BC falling to the right at gradients 10% and 5%, respectively, are to be connected by a parabolic curve 200 m long. Compute the parameters for a vertical curve for chainage and reduce level of B as 2527.00 m and 56.46 m, respectively. The peg interval is 20 m. Also calculate the sight distance for a car having headlights 0.60 m above the road level, and the headlight beams inclined upwards at an angle of 1.2°.

Solution:

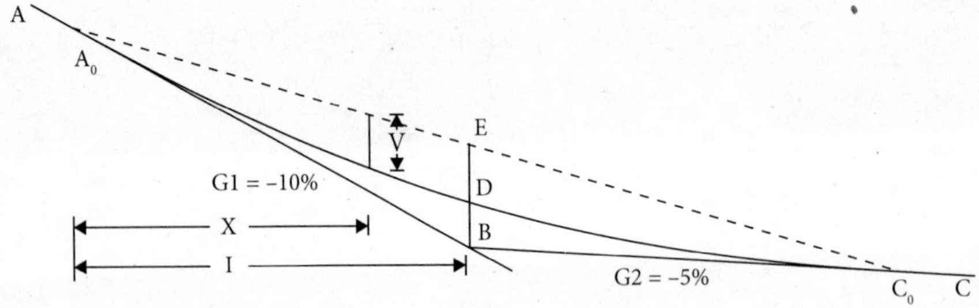

The total number of stations at 20 m interval (2n) = L / 20

2n = 200 / 20 or n = 5

Fall per chord length e1 = g1 * 20 / 100

e1 = -10 * 20 / 100 = -2 m

e2 = g2 * 20 /100 = 20 * 5 / 100 × × = -1 m

Elevation of the beginning of the curve at A0 = Elevation of B – ne1

= 56.46 – 5 × (– 2) = 66.46 m

Elevation of the end of the curve at C0 = Elevation of B + ne^2

= 56.46 – 5 × (–1) = 51.46 m

Tangent correction with respect to the first tangent h = kn^2 where k = e1 – e2 / 4n

k = -2- (-1) / 4 * 5 = – 0.05

RL of the points on the curve = Tangential elevation – tangent correction = H – h (where H is the tangential elevation of a point)

RL on the n'th point on the curve = RL of $A_0 + n' e_1 - kn'^2$

RL of point 1 (A_0) = 66.46 + 1 × (–2) – (– 0.05) × 1^2 = 64.51 m

RL of point 2 = 6.46 + 2 × (– 2) – (– 0.05) × 2^2 = 62.66 m

RL of point 3 = 6.46 + 3 × (– 2) – (– 0.05) × 3^2 = 60.91 m

RL of point 4 = 66.46 + 4 × (– 2) – (– 0.05) × 4^2 = 59.26 m

RL of point 5 = 66.46 + 5 × (– 2) – (– 0.05) × 5^2 = 57.71 m

RL of point 6 = 66.46 + 6 × (– 2) – (– 0.05) × 6^2 = 56.26 m

RL of point 7 = 66.46 + 7 × (– 2) – (– 0.05) × 7^2 = 54.91 m

RL of point 8 = 66.46 + 8 × (– 2) – (– 0.05) × 8^2 = 53.66 m

RL of point 9 = 66.46 + 9 × (– 2) – (– 0.05) × 9^2 = 52.51 m

RL of point 10 (C_0) = 66.46 + 10 × (– 2) – (– 0.05) × 10^2 = 51.46 m (Okay)

Chainage of the intersection point B = 2527.00 m

Chainage of A_0 = Chainage of B – 20n = 2527.00 – 20 × 5 = 2427.00 m

Chainage of C_0 = Chainage of B + 20n = 2527.00 + 20 × 5 = 2627.00 m

hainage of the points and the reduced levels of the corresponding points on the curve are tabulated in Table 7.11.

Point	Chainage (m)	R.L. of points on curve (m)	Remarks
0	2427.00	66.46	A_0 (PC)
1	2447.00	64.51	
2	2467.00	62.66	
3	2487.00	60.91	
4	2507.00	59.26	
5	2527.00	57.71	Apex
6	2547.00	56.26	
7	2567.00	54.91	
8	2587.00	53.66	
9	2607.00	52.51	
10	2627.00	51.46	C_0 (PT)

With the car is at tangent point A0, the headlight beams will strike the curved road surface at a point where the offset y from the tangent at A_0 = (0.60 + x tan 1.2°), where x is the distance from A0.

The offset y at a distance x from A0 is given by y = $(g_1 - g_2)\, x^2 / 400l$

= (-2 + 1) x^2 / (100* 400) = x^2 / 40000 (ignoring the sign)

0.60 + x tan 1.2° = x^2 / 40000

$x^2 - 837.88x - 24000 = 0$

Soling for x, we get Sight distance x = 865.61 m

EXERCISES FOR PRACTICE

a. Short questions

2.29 Show the various elements of a simple circular curve on a neatly drawn sketch. 2.30 Explain the following terms for a simple circular curve: (i) Back and forward tangents, (ii) Point of intersection, curve and tangency, (iii) Deflection angle to any point, and (iv) Degree of curve

2.31 Show the various elements of a compound curve.

2.32 Draw a neat sketch of a reverse curve provided to join two parallel straights.

2.33 Draw the neat sketches to differentiate between simple. compound and reverse curves.

2.34 List the requirements to be satisfied in setting out a transition curve.

2.35 What is a transition curve and where is it used? What are its advantages?

2.36 What is the need of super-elevation and how it is determined?

2.37 Give any five general requirements of a transition curve.

2.38 State the conditions to be fulfilled by a transition curve introduced between the tangent and circular curve.

2.39 What are different types of vertical curves? What is the use of having a vertical curve as parabola and not a circle?

2.40 Why are parabolic curves not generally used for horizontal highway curves?

2.41 What is meant by rate of change of grade on vertical curves and why it is important?

2.42 Explain why the second differences of curve elevations are equal for a parabolic curve.

b. Long answers

2.43 Derive the formulae to calculate various elements to set out a simple circular curve.

2.44 Establish the formulae to calculate various elements to set out the compound curve.

2.45 Derive the relationship between the several elements of the reverse curve.

2.46 Discuss varoious types of transition curves. Derive an expression for the super-elevation to be provided in a transition curve.

c. Unsolved Numerical Problems

2.47 A 4.56° circular curve is to be designed to fit between two intersecting straights. What is the radius of this curve?

(Ans:1256.49 m)

2.48 It was found, while setting a simple circular curve using offsets from the tangents, that for X =16 m, and Y = 0.28 m. Find the radius of the curve.

(Ans: R = 457.28 m)

2.49 A right hand simple circular curve connects two straights AI and IB where A and B are the tangent points. The azimuth of each straight is 50° 43' 12" and 88° 22' 14", respectively. The curve passes through point C of coordinates (321.25, 178.1) m. If the coordinates of A are (240.40, 125.90) m, compute the radius of curve and degree of curve.

Neglecting the existence of point C above and assuming the radius of curve as 450 m and chainage of I as 2140 m, find the (i) chainage of tangents points A and B, (ii) the external distance E and mid-ordinate M, (iii) the length of various chords, and (iv) the corresponding deflection angles.

(Ans: R = 428.87 m, Degree of curvature = 04° 00' 28", Chainage of A = 1986.59 m, Chainage of B = 2282.29 m, E = 25.43 m, M = 24.072 m, C = 20 m, C_1 = 13.10 m, C_2 = 2.296 m, Δ_1 = 00° 51' 13.4", and Δ_2 = 00° 08' 46.2")

2.50 There are two tangents XI and IY for a railroad circular curve where X and Y are the tangent points having coordinates (240.4E, 125.9N) and (253.8E, 218.65N), respectively, and the coordinates of the mod-[point on the curve is (60.13E, 195.89N). Compute the radius of curve, the deflection angle, and the length of the curve.

(Ans: Δ = 270° 57' 56", R = 104.48 m, L = 494.11 m)

2.51 Two straights intersect at a chainage of 2500 m with an angle of deflection as 40°. The straights are to be connected by a simple circular curve. If the coordinates of O and I are (400, 300) m and (680, 460) m, respectively, calculate the radius of curve, tangents lengths, curve length, and the length of long chord.

(Ans: R = 303.04 m, T = 110.3 m, L = 211.56 m, and L_c*= 207.29 m)*

2.52 A right hand circular curve connects two straights AB and BC intersected at point B which has a chainage of 2000 m. If the azimuth of line BC is 120°, and the coordinates of PC and B are (100, 100) m and (200, 200) m, respectively, determine the intersection angle, radius of the curve, degree of curve, length of long chord, length of the circular curve, coordinates of the mid-point, and chainage of PT and O.

(Ans: Δ *= 75 , R = 184.3 m, Degree of curvature = 09°.19' 35.4", * L_c *= 224.39 m, L = 241.25 m, Coordinates of mid-point (X = 152.40 m, Y = 206.26 m), Coordinates of PT (X = 129.29 m, Y = 322.47 m), and Coordinates of centre of curve (X = -30.32 m, Y = 230.32 m))*

2.53 A straight *BC* deflects 24° right from a straight *AB* which are to be joined by a circular curve passing through a point *P*, 200 m from *B* and 50 m from *AB*. Calculate the tangent length, length of curve and deflection angle for a 30 m chord.

(Ans: R = 3754 m, IT = 798 m, curve length = 1572 m, 014°')

2.54 Two straights, which deflect through an angle of 60°00'00", are to be connected by a circular curve of radius 80 m. The curve is to be set out by offsets from its tangent lengths. Calculate the data required: (i) to set out the mid-point of the curve (ii) to set out pegs on the centre line of the curve by offsets taken at 10 m intervals along the tangent lengths.

(Ans: (i) 40 m, 10.72 m (ii) 0 m, 0 m; 10 m, 0.63 m; 20 m, 2.54 m; 30 m, 5.84 m)

2.55 Two straights, which meet at an intersection angle of 135°00'00", are to be connected by a circular curve of radius 60 m. The curve is to be set out by offsets from its long chord. Calculate the data required to set out the: (i) mid-point of the curve (ii) pegs on the centre line of the curve by offsets taken at 5 m intervals along its long chord.

(Ans: (i) 0 m, 4.57 m; (ii) 5 m, 4.36 m; 10 m, 3.73 m, 15 m, 2.66 m; 20 m, 1.14 m)

2.56 Two straights intersect making a deflection angle of 5924'00°", the chainage at the intersection point being 880 m. The straights are to be joined by a simple curve commencing from chainage 708 m. If the curve is to be set out using 30 m chords on chainage basis, by the method of offsets from the chord produced, determine the first three offsets. Also, find also the chainage of the second tangent point.

(Ans: 0.066, 1.806, 2.985 m, 864.3 m)

2.57 A circular curve of radius 900 m is to be constructed between two straights of a proposed highway. The deflection angle between the straights is 14°28'06" and the curve is to be set out by the tangential angles method using a theodolite and a tape. The chainage of the intersection point is 1345.82 m and pegs are inserted at the centre line at 20 m multiples. Tabulate the data required to set out the curve and compute the (i) tangent lengths (ii) length of the circular curve (iii) chainages of the two tangent points.

(Ans: Chainage (m), chord length (m), cumulative tangential angle: 1231.58 (T), 0.00, 00° 00' 00; 1240.00, 8.42, 00° 16' 05"; 1260.00, 20.00, 00° 54' 17"; 1280. 00, 20.00, 01° 32' 29"; 1300.00, 20.00, 02° 10' 41"; 1320.00, 20.00, 02° 48' 53"; 1340.00, 20.00, 03° 27' 05"; 1360.00, 20.00, 04° 05' 17"; 1380.00, 20.00, 04° 43' 29"; 1400.00, 20.00, 05° 21' 41"; 1420.00, 20.00, 05° 59' 53"; 1440.00, 20.00, 06° 38' 05"; 1458.85 (U), 18.85, 07° 14' 05").

2.58 The bearings of three successive intersecting straights AB, BC and CD along the centre line of a proposed highway are 103°29'24", 125°43'22" and 116°12'54", respectively. The horizontal distance BC is 708.32 m. It is proposed to connect AB and BC by a 1500 m radius curve and BC and CD by a 900 m radius curve such that there is an intervening straight on BC between the end of one curve and the start of the other. The through chainage of intersection point B is 1097.65 m and chainage increases from A to D. Calculate the through chainages of the four tangent points on the two curves.

(Ans: 802.52 m, 1384.57 m, 1723.31 m, 1872.66 m)

2.59 A circular curve has to pass through a point *P* which is 70.23 m from *I*, the intersection point and on the bisector of the internal angle of the two straights *AI*, *IB*. Transition curves 200 m long are to be designed at each end and one of these must pass through a point whose coordinates are 167 m from the first tangent point along *AI* and 3.2 m at right angles from this straight. *IB* deflects 37°54' right from *AI* produced. Calculate the radius and tabulate the data for setting out a complete curve.

(Ans: R = 1200 m, AI = IB = 512.5 m, setting-out angles or offsets calculated in usual way)

2.60 A reverse curve is to start at a point *A* and end at *C* with a change of curvature at *B*. The chord lengths *AB* and *BC* are respectively 661.54 m and 725.76 m and the radii as 1200 and 1500 m. Due to irregular topography of the ground, the curves are to be set out using two theodolites method. Calculate the data for setting out the curve.

(Ans: Tangent lengths: 344.09 m, 373.99 m; curve length: 670.2, 733, per 30 m chords: Δ_1 = 054'42°", Δ_2 = 030 '34°")

2.61 A circular curve of 1800 m radius leaves a straight at through chainage 2468 m, joins a second circular curve of 1500 m radius at chainage 3976.5 m, and terminates on a second straight at chainage 4553 m. This compound curve is to be replaced by 2200 m radius transition curves 100 m long at each end. Calculate the chainages of the two new tangent points and the quarter point offsets of the transition curves.

(Ans: 2114.3 m, 4803.54 m; 0.012, 0.095, 0.32, 0.758 m)

2.62 A composite curve consisting of two equal length transition curves and a central circular arc is to be used to connect two intersecting straights TI and IU, which have a deflection angle of 12°24'46". The design speed of the road is to be 70 kmph, the rate of change of radial acceleration 0.30 m/s^3 and the radius of curvature of the circular arc 450 m. The transition curves are to be set out by offsets taken at exact 20 m intervals along the tangent lengths from T and U. The central circular curve is to be set out by offsets taken at exact 20 m intervals from the mid-point of its long chord. Tabulate the data required to set out the three curves.

(Ans: Entry and exit transition curves (starting at T and U, respectively): y = 0.00 m, x = 0.00 m; y = 20.00 m, x = 0.05 m; y = 40.00 m, x = 0.44 m; y = 54.46 m, x = 1.10 m, Central circular arc (measuring along the long chord from its mid-point in both directions): y = 0.00 m, x = 0.51 m; y = 20.00 m, x = 0.07 m; y = 21.505 m, x = 0.00 m)

2.63 A composite curve consisting of entry and exit transition curves of equal length and a central circular arc is to connect two straights on a new road. The design speed for the road is 100 kmph, the radius of the circular arc is 800 m and the rate of change of radial acceleration is 0.30 m/s^3. The through chainage of the intersection point is 3246.28 m and the deflection angle is 15°16'48". (i) Prepare tables for setting out all three curves by the tangential angles method if pegs are required at each 20 m multiples of through chainage (ii) Describe the procedure necessary to establish the common tangent between the entry transition curve and the central circular arc giving the values of any angles required

(Ans: Entry transition curve (chainage, chord, tangential angle from TI): 3094.26 m (T), 0.00 m, 00°00'00"; 3100.00 m, 5.74 m, 00°00'16"; 3120.00 m, 20 m, 00°05'19"; 3140.00 m, 20.00 m, 00°16'47"; 3160.00 m, 20.00 m, 00°34'39"; 3180.00 m, 20.00 m, 00°58'57"; 3183.57 m (T1), 3.57 m, 01°03'58" Central circular arc (chainage, chord, tangential angle from common tangent): 3183.57 m (T1), 0.00 m, 00°00'00"; 3200.00 m, 16.43 m, 00°35'18"; 3220.00 m, 20.00 m, 01°18'16"; 3240.00 m, 20.00 m, 02°01'14"; 3260.00 m, 20.00 m, 02°44'12"; 3280.00 m, 20.00 m, 03°27'10"; 3300.00 m, 20.00 m, 04°10'08"; 3307.61 m (T2), 7.61 m, 04°26'29" Exit transition curve (chainage, chord, tangential angle from UI): 3396.92 m (U), 0.00 m, 360°00'00"; 3380.00 m, 16.92 m, 359°57'42"; 3360.00 m, 20. 00 m, 359°49'04"; 3340.00 m, 20.00 m, 359°34'01"; 3320.00 m, 20.00 m, 359°12'33"; 3307.61 m (T2), 12.39 m, 358°56'02").

2.64 A road 7.30 m wide deflects through an angle of 18°47'26", the through chainage of the intersection point being 1659.47 m. A circular arc of radius 600 m and two equal length transition curves are to be designed for a speed of 85 kph with a rate of change of radial acceleration of 0.30 m/s^3. Calculate: (i) The through chainages of the four tangent points (ii) The theoretical and actual values for the maximum super-elevation on the circular arc, taking g as 9.81 m/s^2 (iii) The data required to set out the entry transition curve and the exit transition curve by the tangential angles method if pegs are to be placed on the centre line at exact 50 m multiples of through chainage.

(Ans: (i) T = 1523.57 m T1 = 1596.70 m T2 = 1720.34 m U = 1793.47 m (ii) Theoretical = 0.69 m; maximum allowable = 0.31 m (iii) Entry transition curve (chainage, chord, tangential angle from TI): 1523.57 m (T), 0.00 m, 00°00'00"; 1550.00 m, 26.43 m, 00°09'07"; 1596. 70 m (T1), 46.70 m, 01°09'50" Exit transition curve (chainage, chord, tangential angle from UI): 1793.47 m (U), 0.00 m, 360°00'00"; 1750.00 m, 43.47 m, 359°35'20"; 1720.34 (T2), 29.66 m, 358°50'10")

2.65 A wholly transitional curve having equal tangent lengths is be designed to connect two intersecting straights which meet at a deflection angle of 07°34'56". If each tangent length is to be 95.38 m and the design speed for the road is 100 kmph, calculate: (i) The minimum radius of curvature (ii) The total length of the curve (iii) The maximum rate of change of radial acceleration

(Ans: (i) R = 719.96 m (ii) Total length of the curve = 190.55 m (iii) Maximum rate of change of radial acceleration = 0.31 m/s^3)

2.66 A 10 m wide road is to be deflected through an angle of 35° 30′. A transition curve is to be used at each end of the circular curve of 500 m radius. It has to be designed for a rate of gain of radial acceleration of 0.2 m/sec^2 /sec and a speed of 60 km/hr. Calculate the suitable length of the transition curve and super-elevation.

(Ans: 46.32 m, 56.6 cm)

2.67 A circular curve of 610 m radius deflects through an angle of 40°30', which is to be replaced by smaller radius transition curves 107 m long at each end. The deviation of this new curve from the old at their mid-points is 0.46 m towards the intersection point. Determine the revised radius assuming that the shift can be calculated with sufficient accuracy on the old radius. Calculate the lengths of track to be lifted and of new track to be laid.

(Ans: R = 590 m, new track = 521 m, old track = 524 m)

2.68 The centre-line of a new road is to be set out through built up area. The two straights of the road T_1I (237.23 m) and T_2I meet giving a deflection angle of 45°, and are to be joined by a circular arc of 572 m with spiral transitions 100 m long at each end. The spiral from T_1 must pass between two buildings, the position of the pass point being 70 m along the spiral from $T1$ and 1 m from the straight measured at right angles. Calculate all the necessary data for setting out the first spiral transition curve at 30-m intervals; and find the (a) first three angles for setting out the circular arc, if it is to be set out by 10 equal chords, (b) design speed and rate of change of centripetal acceleration, given a centrifugal ratio of 0.1, and (c) maximum super-elevation for a road width of 10 m. Given that θ_1 = 9'1", θ_2 = 36'37", θ_3 = 110'40°").

(Ans: (a) 139'14°5 ,"46'29°3 ,"53'44°", (b) 85 km/h, 0.23 m/s^3, (c) 1 m)

2.69 A parabolic vertical curve is to connect a +3.1% gradient to a –2.3% gradient on a single carriageway road having a design speed of 70 kmph. With reference to the current UK Department of Transport design standards, calculate the minimum required length of curve if: (i) The curve is to be designed for overtaking (ii) The curve is to be designed for stopping only.

(Ans: (i) 1080 m (ii) If possible use 162 m; if not, use 91.8 m)

2.70 A parabolic vertical curve is to connect a –2.2% gradient to a +1.9% gradient on a road having a design speed of 85 kmph. Using the current design standards, calculate the minimum required length of the vertical curve.

(Ans: 14.2 82 m)

2.71 A road having an up-gradient of 1 in 15 is to be connected to a down-gradient of 1 in 20 by a vertical parabolic curve 120 m in length. Determine the visibility distance afforded by this curve for two approaching drivers whose eyes are 1.05 m above the road surface. If a new vertical parabolic curve is set out to replace the original curve so that the visibility distance is increased to 210 m for the same height of driver's eye. Determine the (a) length of new curve, (b) horizontal distance between the old and new tangent points on the 1 in 5 gradient, and (c) horizontal distance between the summits of the two curves.

(Ans: 92.94 m, (a) 612 m, (b) 246 m, (c) 35.7 m)

2.72 A vertical parabolic sag curve is to be designed to connect a down-gradient of 1 in 20 with an up-gradient of 1 in 15. The chainage and reduced level of the intersection point of the two gradients is 797.7 m and 83.544 m, respectively. In order to allow for necessary headroom, the reduced level of the curve at chainage 788.7 m on the down-gradient side of the intersection point is to be kept as 85.044 m. Compute the (a) reduced levels and chainages of the tangent points and the lowest point on the curve, and (b) reduced levels of the first two pegs on the curve, the pegs being set at the 30 m chainage.

(Ans: T_1 = 745.24 m, 86.166 m, T_2 = 850.16 m, 87.042 m, lowest point = 790.21 m, 85.041 m, (b) 85.941 m, 85.104 m)

2.73 A proposed road consists of a rising gradient of 2% followed by a falling gradient of 4% with the two gradients joined by a vertical parabolic summit curve of 120 m in length. The two gradients produced meet a reduced level of 28.5 m. Compute the reduced levels of the curve at the ends, at 30-m intervals and at the highest point. What is the minimum distance at which a driver, whose eyes are 1.125 m above the road surface, would be unable to see an obstruction 100 mm high?

References and Suggested Readings

1. Agor, R., (1980), Surveying, Vol-1 and II, Khanna Publishers, Delhi.
2. Anderson, James M., and Mikhail, Edward M., (2012), Surveying: Theory and Practice Tata McGraw Hill (Indian Edition).
3. Arora, K. R., (2015), Surveying Vol. I & II, Standard Book House.
4. Basak, N.N., (2017), Surveying and Levelling, Tata McGraw Hill.
5. Chandra, A.M., (2005), Plane Surveying, New Age International Publishers.
6. Duggal, S.K., (2017), Surveying, Vol. I & II, TataMc-Graw Hill.
7. Gopi, Satheesh; Sathikumar, R., and Madhu, N. (2017), Advanced Surveying, Pearson India.
8. Kanetkar, T.P. and Kulkarni, S.V., (2006), Surveying and Levelling, Vol. I and II, Vidyarthi Griha Prakashan, Pune.
9. Punmia, B.C., Jain, Ashok K., Jain, and Arun K., (2016), Surveying, Vol. I & II, Laxmi Publications.
10. Subramanian, R., (2012), Surveying and Levelling, Oxford University Press.
11. Venkatramaiah, C., (2011), Textbook of Surveying, University Press.
12. Uren, John and Price, Bill, (1994), Surveying for Engineers, Palgrave Macmillan.

03 MODERN FIELD SURVEY SYSTEMS

UNIT SPECIFICS

Through this unit we have discussed the following aspects:

- Principle of Electronic Distance Measurement (EDM)
- Types of EDMs and their working
- Utility of EDMs
- Various components and parts of a Total Station
- Types of Total Stations and their working
- Taking measurements from Total Stations
- GPS and its components
- Working with GPS/GNSS
- Various methods used in GPS/GNSS survey
- Various applications of GPS

In addition to the basic principle of EDM, Total Stations, and GPS/GNSS, their working has been explained. The practical applications of these modern surveying equipment are presented for generating further curiosity and creativity as well as improving working in field data collection. These modern surveying equipment not only save time but they need less manpower in the fieled to collect ground data. Questions of short and long answer types are given following lower and higher order of Bloom's taxonomy, and a list of references and suggested readings are given in the unit so that one can go through them for additional readings.

Rationale

This unit provides principle and theory of using EDM, Total Stations and GPS/GNSS so that the students get familiar about the use of these modern equipment. It explains various components of each of these devices. Working with these instruments are also explained. The advantages and disadvantages of using EDM, Total Station and GPS/GNSS are given to offer students a wide spectrum of uses. Various methods used in the field and accuracies achieved are also given. Various sources of errors and their minimisation are discussed so that the users can minimise the errors from field data. Various applications given in this unit will further enhance the understanding of these equipment.

Pre-Requisite

Mathematics: geometry and trigonometry; Surveying: Theodolite and distance measuring instruments. Electro-magnetic waves.

Unit Outcomes

List of outcomes of this unit is as follows:

U3-O1: Describe various types of EDMs, Total Stations, and GPS/GNSS

U3-O2: Explain the essential components and characteristics of EDMs, Total Stations, and GPS/GNSS

U3-O3: Realize the role of EDMs, Total Stations, and GPS/GNSS for field data collection

U3-O4: Describe various methods of using EDMs, Total Stations, and GPS/GNSS

U3-O5: Apply the parameters collected in the field for various applications

Unit-3 Outcomes	Expected Mapping with Programme Outcomes *(1- Weak correlation; 2- Medium correlation; 3- Strong correlation)*					
	CO-1	CO-2	CO-3	CO-4	CO-5	CO-6
U3-01	2	3	2	2	1	2
U3-02	2	1	3	1	2	-
U3-03	2	2	2	3	-	3
U3-04	3	3	2	1	2	-
U3-05	2	3	2	1	-	2

3.1 INTRODUCTION

Long distance measurements on the ground by conventional method has always been a problem in surveying. The use of electronics in modern surveying equipment has ease the process. Electronic Distance Measuring (EDM) device can be used for measuring the long distances (slope distances) with higher accuracy, thus saving a lot of time (Garg, 2021). The triangulation and trilateration surveys require accurate measurement of long distances. With the advent of EDMs, it has become easy and fast to measure the long distances. Early EDM instruments were large, bulky, complicated and expensive, but the improvements in electronics over a period of time have made them lighter, portable, simpler, and affordable. These EDMs can be mounted on tribrach or standard units or theodolites, and can be used with theodolites (both digital and optical) or as an independent unit (Subramanian, 2012). They provide slope distance, and once the slope distance is known, horizontal distance can be computed trigonometrically by using the vertical angle measured from the theodolite.

Total Station equipment, which is a combination of digital theodolite, an EDM and a small processor to process the data, can be used for taking multiple observations in the field, and thus this equipment eliminates the need of taking several surveying equipment in the field, such as theodolite, level, tacheometer, tape, etc. With the advent of Total Station, angular measurements, distance measurement and height measurements have become quick, easy and accurate (Gopi et.al., 2017). This equipment can be used to measure the horizontal angles, vertical angles, slope distances, and to determine horizontal distances, elevations, coordinates of the points using these measurements. The Total Station has made a revolution in surveying for digital data collection and data analysis, and subsequently creating a digital map.

The Global Positioning System (GPS) traditionally refers to the North American global positioning system, or satellite positioning system. Originally known as Navigation Satellite Timing and Ranging (NAVSTAR), GPS was developed by the US Department of Defense for military use in the 1970s. The Global Navigation Satellite System (GNSS) term refers to the international multi-constellation satellite system, and typically includes GPS, GLONASS, Baidu, Galileo, QZSS, NAVIC, or and any other constellation system. With the advent of Europe's Galileo system and China's BeiDou, users can now use a broader range of signals and get greater reliability as more satellites are available at any given time, greater precision as combinations of signals and frequencies can help to mitigate effects such as atmospheric interference on GNSS precision.

The GNSS based receivers are modern surveying device which are used to collect field data for mapping from all available satellite systems. The multi-GNSS receivers are now very common as they accurately determine their own location by measuring the distance to four or more satellites. The output from the GNSS receivers is 3D coordinates of the point (x, y and z) (Latitude, Longitude and Elevation) as well as time (Garg, 2021). For mapping purpose, we need only 3-D coordinates of various points. The availability of GNSS receivers has further simplified and speeded-up the survey work as there is no need to take angular or linear measurements (unlike Total Station) to compute the 3D coordinates, which are obtained directly as output. The measurements from GNSS can be taken day and night anywhere on the Earth.

The availability of EDM, Total Station and GNSS has eased the data collection and mapping work, and considerable reduced the cost and time to complete the survey work, as compared to conventional surveying equipment (Gopi et. al., 2017). However, today Total Station equipment, which has an in-built EDM, can be used to measure the long distances. So, now-a-days, there is no need to buy the EDM system separately. But Total Station and GNSS are complimentary to each other, so it is better to take both the equipment in the field while taking several observations. Total Station requires horizontal clearance of line of sight and GNSS needs vertical clearance of satellite signals, so field measurement work is not hampered if one of the clearance is not available (Garg, 2019). Looking at the potential use of both (Total Station and GNSS), a combination of both has emerged in the market, called a *Smart Station*. A Smart Station is an instrument which can be used as a Total Station or as a GNSS or in combination on the same system, so instead of carrying two equipment in the field, the surveyor can carry the Smart Station and perform the survey data collection work much faster.

This module will cover the details and working of EDMs, Total Station and GNSS. These electronic devices used for field survey measurements are dependent on the velocity of Electro-magnetic (EM) wave, and the time of travel of wave. Some advantages and disadvantages of each equipment are also given.

3.2 ELECTRONIC DISTANCE MEASUREMENT (EDM) DEVICES

The EDM is an electronic distance measuring device, which measures the distance from the instrument to its target through electromagnetic waves (Garg, 2021). The EDM instruments are highly reliable and convenient, and can be used to measure distances of up to 100 km. Such long distance measurement can be done by instruments, like *geodimeter, tellurometer* or *distomat* etc. The first EDM instrument called '*geodimeter*' was developed in Sweden in the year 1948. It is geodetic distance meter developed based on a modulated light beam (Figure 3.1). The second instrument for EDM was designed and developed in Africa in the year 1957, named '*tellurometer*'. This instrument employs the modulated microwaves. As the technology improved, the EDMs became smaller and light-weight which measure slope distances in digital form and compute horizontal distances.

Fig. 3.1: Old version of a Geodimeter

Source: https://www.aga-museum.nl/aga-geodimeter-3-8

The method of direct distance measurement (e.g., Tape) can't be implemented in difficult terrains or for large distances or where large amount of obstructions exists. The direct method is limited to 15 to 150 m with an accuracy range of 1 in 1000 to 1 in 10000, whereas the EDM with an accuracy of 1 in 10^5 have a distance range of up to 100 km (Gopi, et.al., 2017).

3.2.1 Principle of EDM

The measurement of distances requires the EDM unit (called transmitter) which transmits Electromagnetic (EM) wave, and a reflecting prism (called receiver/reflecting unit). The slope distance is determined between the points on the ground where these two units can be kept apart (Garg, 2021). The EDM is kept at a point known as *master station* and reflecting prism at another point called *remote station*. The general principle used in EDM is that a modulated EM beam is transmitted from a transmitter kept at the master station to a reflector which is kept at the remote station, and receiving the beam back at the master station. The instrument measures the travel time of EM wave from transmitter to receiver and back to transmitter. The slope distance between transmitter and receiver is computed by taking half of travel time multiplied by the velocity of EM wave (i.e., the velocity of light). Since measurement of time requires very precise observation, so in modern EDMs, the distance is determined by modulating the continuous carrier waves at different frequencies, and then measuring the phase difference at the master station between the outgoing and the incoming signals.

The EM wave, generated by the EDM, travels through the atmosphere in a sine wave form, and ultimately strikes to the prism and reflects back to EDM device, as shown in Figure 3.2a. The distance is measured as a function of speed of light and the elapsed time. Frequencies generated within an EDM are used to determine the elapsed travel time of its signal (EDM to prism and prism to EDM). The distance is calculated as:

Distance (D) = (elapsed time/2) x velocity of light (3.1)

However, since the travel time is too small, measuring the precise time by an EDM is a difficult task. The modern EDMs therefore use the phase difference approach where the number of completed waves and incomplete wave is measured. It basically measures the phase changes that occur as wave travels from one end to the other end of a line and returns back (Figure 3.2b). The EDMs that use the phase-shift principle carry out the measurements several times in order to resolve the ambiguity of phase difference.

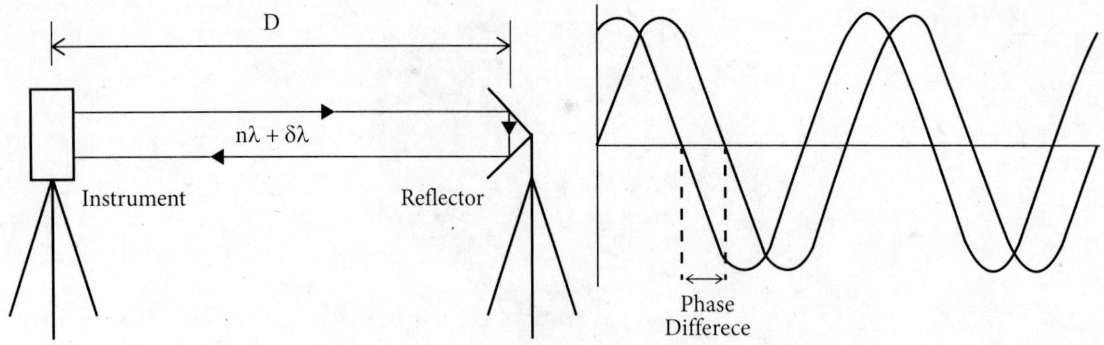

Fig. 3.2: Measurement of (a) time, and (ii) phase difference

The phase shift method is considered to be the most accurate one as it allows a very narrow beam, but its measuring range is limited. Phase shift is typically measured in degrees where a complete cycle is 360°, and the wave form repeats every 360°. As, the EM wave travels in a sinusoidal wave form, the distance between two wave form peak is known as the *wavelength* (λ). The number of cycles a wave completes in per unit time is known as its *frequency* (*f*). The wave leaves the EDM at 0° phase, goes through *n* number of full phases on its path towards the reflector (prism), and returns to the EDM at some angle between 0° and 360°, creating a partial wavelength *p*. The partial wavelength is the difference between the phasing of the emitted and the received signal. When two waves reach at different phase angle at different times, they are out of phase or phase shifted. The EDM compares the phase angle of the returning signal to that of a replica of the transmitted signal to determine the phase shift.

The distance (D) is then determined as one-half of the sum of the number of wavelengths (*n*) in the double path distance multiplied by the wavelength (λ) plus the partial wavelength (*p*) represented by the phase difference. The distance is determined by the equation:

$$D = \frac{1}{2}(n\lambda + p) \quad (3.2)$$

The EDM can very accurately determine the length of the last partial wavelength from its phase. The total EDM to reflector to EDM distance is computed as per example given below:

The wavelength in Figure 3.3 is 6.1 m, and there are 10 full wavelengths before the last partial one.

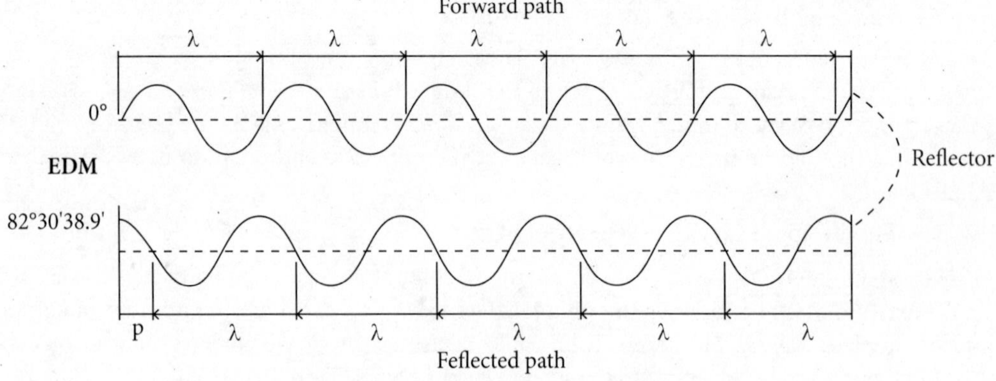

Fig. 3.3: Representation of phase shift

The last partial wave (p) = (82°30'39.9" / 360°00'00") x 6.1 m = 1.40 m. The total distance EDM to reflector to EDM (10 full wavelengths and last partial wave) = 10 x 6.1 m + 1.40 m = 62.40 m. So, the distance between the EDM and reflector is half that 62.40 / 2 = 31.20 m.

In fact, while the EDM can accurately measure the last partial wavelength (p), it doesn't know the number (n) of full wavelengths occurred before it. This is resolved by decreasing the frequency (f) by a factor of 10 or increasing the wavelength (λ) by a factor of 10, and repeating the process of distance measurement. This is repeated a number of times for three or four different frequencies to get the correct value of n or the total distance.

Table 3.1 shows the last partial wavelength (p) given by the EDM for each of 4 corresponding different wavelengths to measure the distance between two fixed points. The digits in bold represent the digits which will be added to the distance as a result of each partial wavelength (p) to get the correct distance.

Table 3.1: Phase shift with different frequencies

λ (m)	*p* (m)	Distance (m)
10	3.68	3.68
100	53.70	53.68
1,000	454	453.68
10,000	8450	8453.68

The total distance is 8453.68 m, and the EDM-reflector distance is 4226.84 = 2/8453.68 m

3.2.2 EDMs classification based on range

The EDMs are also classified on the basis of their range of electro-magnetic (EM) waves (Subramanian, 2012), as:

a. **High range**- radio wave equipment for ranges up to 100 km
b. **Medium range**- microwave equipment with frequency modulation for ranges up to 25 km
c. **Short range**- electro-optical equipment using amplitude modulated infra-red or visible light for ranges up to 5 km

Sun light or electromagnetic spectrum consists of different wavelengths at different frequency. Various wavelength regions of EM waves are shown in Figure 3.4. The EDMs are now-a-days incorporated in electronic theodolites (or Total Stations) that can automatically measure vertical angle to compute horizontal and vertical distances. The present EDM instruments have several other useful features; laser plummet, 30x magnification, high resolution LCD display, upload and transfer of data, data editing and exchange, and bluetooth connectivity.

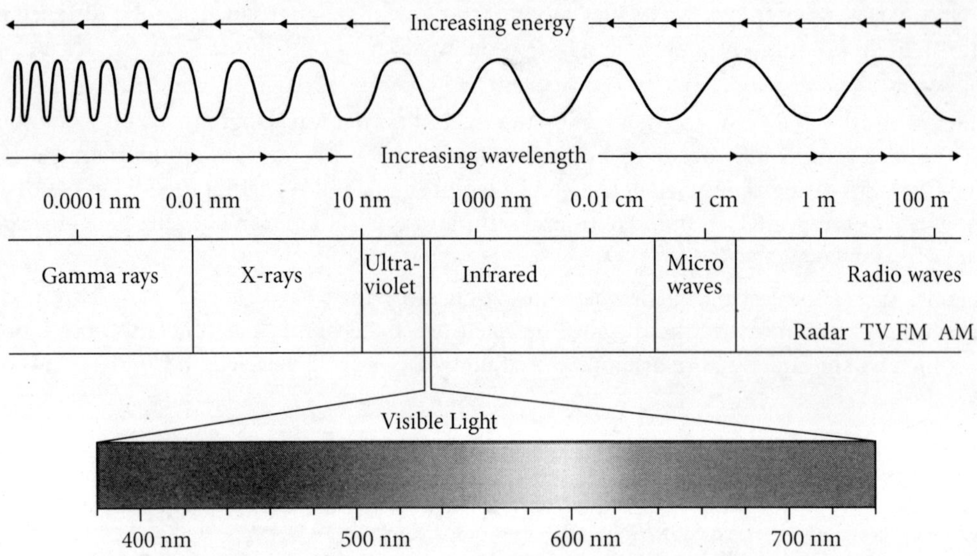

Fig. 3.4: An electromagnetic spectrum (Garg, 2019)

The microwave-based EDMs make use of microwaves, such as Tellurometers, in the wavelength from 10 cm and 3 cm. The range of these instruments is up to 100 km with an error of ±5 to 15 mm/km. They can also be used in night and in adverse weather conditions. They consist of two identical units; where one unit is used as master unit and the other as remote unit. The master unit can be converted into a remote unit and a remote unit into a master unit, if required. Each unit is kept at two ends of a line to determine the distance. For long distance measurement, a communication facility is provided with each unit to interact with survey team members during the measurements.

Visible light or infrared wave-based EDMs rely on propagation of modulated waves, such as Geodimeter or Distomat (Gopi et.al., 2017). The instrument is kept at one end and the reflecting prism at the other end of a line. These instruments are light and economical, and can also be mounted on the top of a theodolite. Such EDMs are very useful for civil engineering project, as the accuracy of these of measurements varies from 0.2 mm/km to 1 mm/km distance. They can be used to measure distance up to 5 km.

3.2.3 Refecting prisms

Reflecting prisms, also known as reflectors, reflect the EM wave back to the Total Station or EDM for computing the distance (Figure 3.5). Prisms are either mounted on a prism pole, or on the top of a tripod. They are used in single or in combination (of 3 or 6 etc.) making a pyramid shape. Large number of prisms are used to get increased accuracy when the distance to be measured is very large. The centre of the prism is bisected with the centre of the diaphragm of the telescope of EDM before taking the measurements. The prisms on top of the tripod are used for back-sighting purposes as well as to get the better accuracy (Garg, 2021).

While procuring a surveying prism, the offset, the thread size, the optics and the size of the prism are to be considered. Relevant factors are the prism construction (glass quality, geometry, coatings), the prism alignment with respect to the line of sight of an instrument and the EDM unit used. There are two main factors for good range measurement: prism diameter and beam deviation. The application requiring high accuracy demands to use circular prisms (Lackner and Lienhart, 2016). For stake out work, mini-prisms which are smaller in diameter, allows the users to get close to a wall or building to get the shot or stake out.

Fig. 3.5: Reflecting prisms

Source: https://www.nsscanada.com/product-category/leica-geosystems/survey-accessories/prisms-reflectors)

3.2.4 Distance measurement

The EDM instruments generate the EM waves which are modulated and propagated. The EDM is kept at one point and the reflecting prism at the other point whose distance is to measured. The EDM is set up properly with temporary adjustment as has been explained for other equipment in Module 1. Many EDMs have optical plummet for accurate centering and auto levelling facility which levels the base within a certain small range. The optical plummet and auto-levelling facility save lot of time in the field for temporary adjustment of EDM. The diaphragm is properly focussed. Now, the prism is bisected accurately at its centre before taking the distance measurements from EDM.

The EDM sends out a laser or infrared beam which is reflected back from prism to the instrument, and the instrument calculates the distance travelled by the beam. Some EDM instruments use pulsed laser emissions, and these instruments determine the distance by measuring the time taken between the transmission of the signal and the reception of the reflected signal, by taking advantage of the pulsed laser beam.

3.2.5 Errors in EDM measurements

The errors in EDM measurements could be due to- Personal errors, Instrumental errors, and Natural errors. A good description of errors in EDM measurements is given by Rajput (2020). Some of these are-

a. Inaccuracy in initial setup of EDMs and the reflectors over the stations
b. Instrument and reflector measurements going wrong
c. Atmospheric pressures, humidity, and temperature errors
d. Atmospheric variations in temperature, pressure as well as humidity, as microwave-based EDMs are more susceptible to these parameters.
e. Calibration errors
f. Errors shown by the reflectors
g. Multiple refraction of the signals
h. Zero error (Additive Constant) in distance measurement which is caused by electrical delays, geometric detours, and eccentricities in the EDM. The additive constant or zero/index correction is added to the measured distances to correct for these differences.

i. Differences between the electronic centre and the mechanical centre of the EDM
j. Differences between the optical and mechanical centres of the reflector. This error may vary with changes of reflector, so only one reflector should be used for EDM calibration.
k. Scale errors are linearly proportional to the measured distance, and can arise from both internal and external sources. Internal sources are ageing, drift and temperature effects (e.g., insufficient warm-up time) of the oscillator (Staiger, 2007).
l. Internal frequency errors, including those caused by external temperature and instrument 'warm-up' effects
m. Non-homogeneous emission/reception patterns from the emitting and receiving diodes (Phase in homogeneities).
n. Cyclic error (Short Periodic error) which is a function of the actual phase difference measurement by the EDM (Staiger, 2007). Phase measurement error is caused by unwanted feed through the transmitted signal onto the received signal (Tulloch, 2012). Cyclic error is usually sinusoidal in nature with a wavelength equal to the unit length of the EDM. The unit length is the scale on which the EDM measures the distance, and is derived from the fine measuring frequency. The stability of the EDM internal electronics can also vary with age, therefore, the cyclic error can change significantly over time. Cyclic error is inversely proportional to the strength of the returned signal, so its effects will increase with increasing distance (i.e., low signal return strength). Calibration procedures exist to determine the EDM cyclic error that consist of taking bench marks measurements through one full EDM modulation wavelength, and then comparing these values to known distances and modeling any cyclic trends found in the discrepancies.

3.3 TOTAL STATIONS

This instrument is an integrated version of an electronic Theodolite and an EDM, as shown in Figure 3.6. It also has a small micro-processor, electronic data collector and storage system. A Total Station is an electronic/optical instrument used to measure sloping distance of object to the instrument, horizontal angles and vertical angles (Garg, 2021). Micro-processor in Total Station processes the collected data to compute the average of multiple angles measured, average of multiple distance measured, horizontal distance, elevation of ground, and 3-D coordinates of the observed points. These data can be stored in the system itself but each Total Station has a limited storage space. The data can be transferred to a laptop and desktop later for its processing.

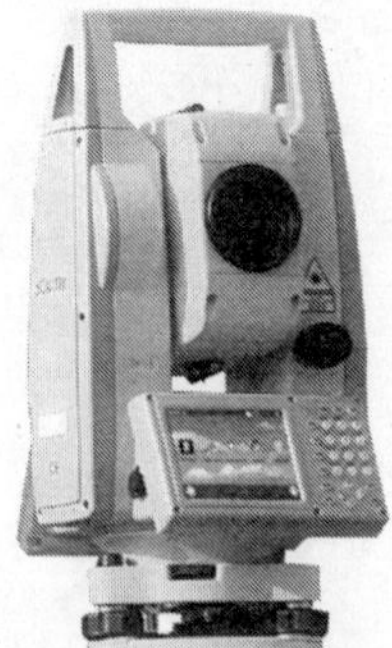

Fig. 3.6: Total Station

Source: https://www.directindustry.com/prod/south-surveying-mapping-instrument-co-ltd/product-160571-1650628.html

The Total Station sends out infrared waves that are reflected by the prism kept at the object. By taking measurements of the prism, the Total Station computes the prism's coordinates as well as reduced level of the ground point. Some of the major advantages of using Total Station over the conventional surveying instruments are; saving in time, ease in working, increase in accuracy, and facility to use computer for storing, processing the data and derive the final result in a desired format (Gopi et.al., 2017).

3.3.1 Various components of a Total Station

Total Station consists of a distance measuring instrument (EDM), an angle measuring instrument (Theodolite) and a simple micro-processor (Figure 3.7). It is a compact instrument, and one person can easily carry it to the field from one point to another. Total Stations with different accuracy, in angle measurement and different range of measurements, are available in the market. The components of a Total Station are as follows:

1. A **tripod** which is used to hold the Total Station
2. A **mainframe device** which is used to record, calculate and even manipulate the field data. Various parts of a Total Station are shown in Figure 3.8.
3. **Prism and prism pole/tripod** which can be used to measure distances up to 6-7 km with triple prism
4. **Battery** is required to provide power source to the instrument.

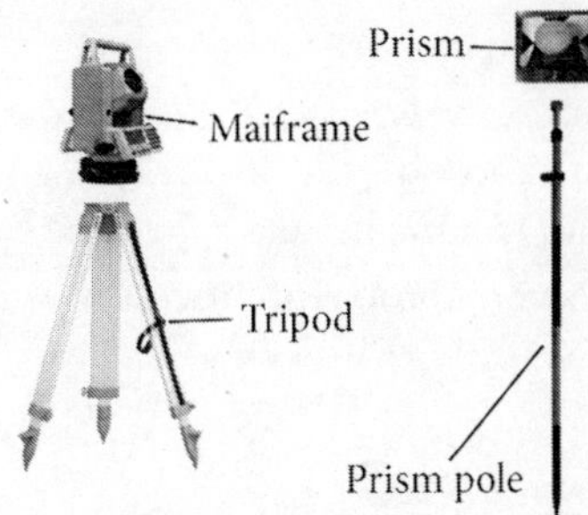

Fig. 3.7: Components of a Total Station

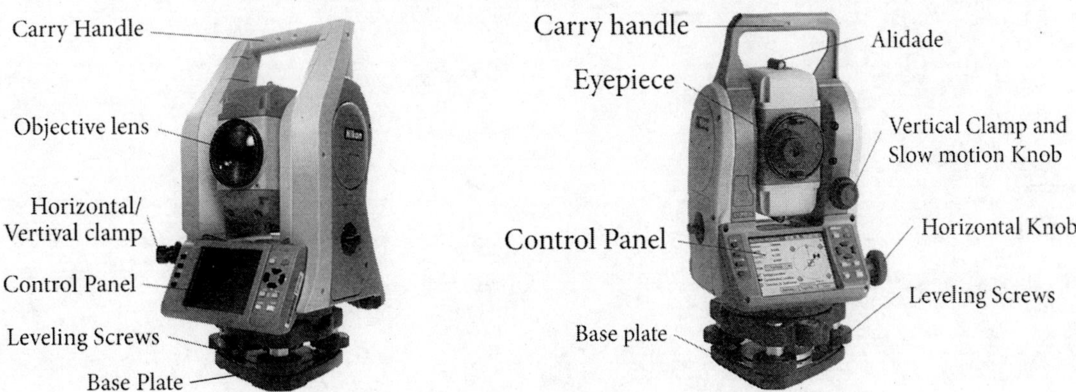

Fig. 3.8: Parts of a Total Station

Source: https://surveyforbeginners.wordpress.com/2019/01/01/total-station

3.3.2 Steps involved in Total Station surveying

a. Fix the instrument on the tripod that has been firmly kept on the ground (Figure 3.9a).

b. Fix the instrument on the tripod with the help of a given screw in tripod (Figure 3.9b).

c. Set up the instrument on the ground point. Some instruments have optical plummet which consists of a sharp laser beam and may be used for centering purpose (Figure 3.9c).

d. Levelling of the instrument approximately with the help of "bull's eye bubble" using eye judgement (Figure 3.9d). Use the standard procedure for levelling (Figures 3.9e & f), and then correct the levelling precisely electronically (Figure 3.9g).

e. Focus the diaphragm (Figure 3.9h).

f. Fix the prism on the prism rod at a known height. Since, the prism rod is graduated so height of the prism is known.

g. Measure the height of instrument with the help of a tape.

h. Set up the working unit in the instrument. Enter the height of prism pole, height of instrument, coordinates of point where instrument is kept (if known), temperature and atmospheric pressure at the site (if known).

i. Focus the prism kept at the other point.

j. Bisect the centre of reflecting prism and with the help of appropriate functional buttons in the instrument. The measurement will be displayed on the display panel. Store all the measurements in the instrument itself.

k. Compute the other data from measured observations.

l. To speed up the work, more than one prism set is used the field. Various persons can set up the prisms at respective locations and observer from the same location of instrument can take observations all around. It saves lot of time in data collection.

m. Transfer the field data into desktop or laptop and process it using the related software.

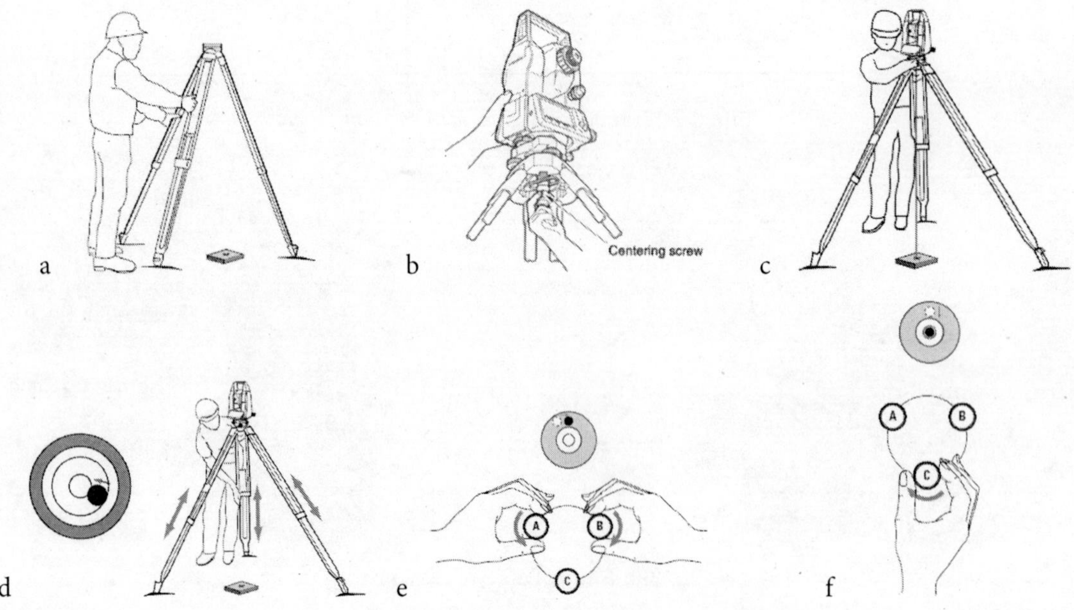

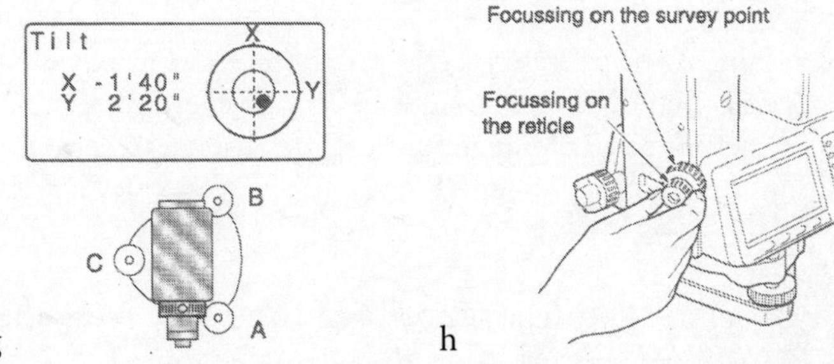

Fig. 3.9: Various steps of Total Station

Sources: https://www.onlinecivilforum.com/site/setting-total-station/ and https://www.southalabama.edu/geography/allison/GY301/Total%20Station%20Setup%20and %20Operation.pdf)

3.3.3 Functions of Total Station

Most Total Stations have a distance measuring accuracy of 2-3 mm at short ranges, which will decrease to about 4-5 mm at 1 km (Mishra, 2014). Although angles and distances can be measured and used separately, the most common applications for Total Stations occur when these are combined to define the position in control surveys.

1. Angle measurement:

To measure horizontal and vertical angles, the Total Station is used with an accuracy of better than one seconds. For horizontal measurement of angles, any direction can be taken as reference. In case of vertical measurement of angles, horizontal direction is taken as reference.

2. Distance measurement:

The EDM is a major part of Total Station. To measure the distance, EDM instrument of Total Station is used with an accuracy of 5-10 mm per km or better. The accuracy varies with each Total Station equipment. They are also available to be used in Robotic mode, with automatic target recognizer. The distance measured is always sloping distance from instrument to the object.

3. Keyboard and display:

Total Station is activated through its control panel, which consists of a keyboard and multiple line LCD (Figure 3.10). A number of instruments have two control panels; one on each face, which makes them easier to use from either side or while changing the face of the instrument. In addition to controlling the Total Station, the keyboard is often used to code the data generated by the instrument; this code will be used to identify the object being measured. In some Total Stations, it is possible to detach the keyboard and interchange them with other Total Stations or GNSS receivers. This is called *integrated surveying.* Electronic display unit is capable of displaying various values when respective key is pressed. The unit is capable of displaying horizontal distance, vertical distance, horizontal and vertical angles, difference in elevations of two observed points and the 3-D coordinates of the observed points. It also displays the graphical figures of the area covered as well as geographical distribution of data points collected in the field.

4. Electronic book:

Each point data can be stored in an electronic note book or stored in a pen drive. The capacity of electronic note book varies up to 10000 data points, depending on the model of instrument. The data from note book can be unload to a computer, and the note book is again empty to re-use in the field.

5. Data processing:

The instrument is provided with an inbuilt microprocessor. The microprocessor averages out the multiple observations. Computation of horizontal distances along with X, Y, Z coordinates is done by the microprocessor of instrument. Hence, if atmospheric temperature and pressure are to be applied, the microprocessor applies suitable corrections to the measurements.

6. Software:

Various softwares are available in the market which can be used to post-process the data from the Total Station. Usually, manufacturers provide their own customised software which allows to export the survey results into other formats. Thus, output can be imported to CAD application or software, like MX Roads, GIS software. Software, like Auto civil and Auto plotter clubbed with AutoCAD can be used for plotting contours at any specified interval and plotting cross-sections along any specified line. Normal survey task, such as traversing, mapping, area, volume, contouring., all are now available in software modules.

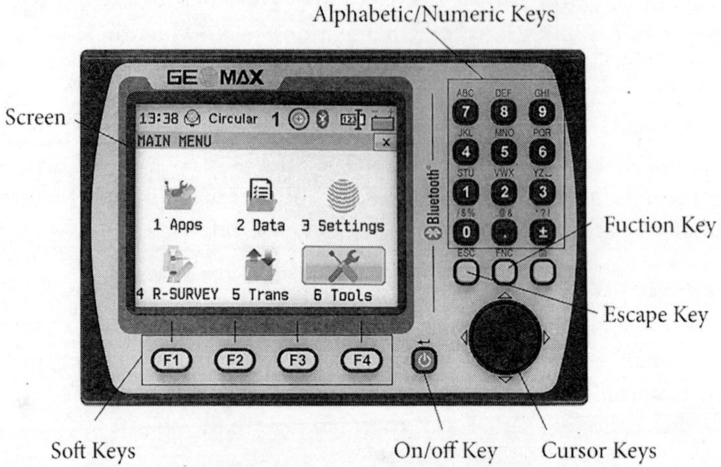

Fig. 3.10: Keyboard and display unit

Source:https://surveyforbeginners.wordpress.com/2019/01/01/total-station)

3.3.4 Relectorless Total Stations

There are two methods of measuring the distance; (i) With reflector and (ii) Reflectorless or prismless modes. The prism method uses a reflective prism at the measurement point, and the non-prism or reflectorless method does not require a reflective prism. Both the methods have been widely used in engineering and industrial surveying systems to measure the distances and angles automatically (Xia et al, 2006). At a small scale and a local coordinate system, the survey by Total Station is more superior and precise as compared to other surveying methods. However, in order to obtain accurate and reliable results, it is necessary to check and adjust the instrument regularly.

With the reflectorless method, the instrument works with a laser (Light Amplification by Stimulated Emission of Radiation) beam (Figure 3.11). The instrument is placed at the measurement point, and the distance is measured using a laser beam (Beshr and Elnaga, 2011). With this method, it is possible to survey the areas of impossible reach, such as disaster areas (e.g., affected by landslides), snow-covered areas, nuclear waste sites, forest fire, etc., safely and efficiently. The laser beam based Total Station can be used in night or underground survey or those locations where it is not possible to keep the prism, such as bottom of the bridge deck, or there is danger to human life to keep the prism. The area to be surveyed may have safety concerns, underground mines, detail surveys of busy road intersections where traffic control is impossible. The main advantage of reflectorless Total Station is the ability to measure inaccessible points to collect data with increasing speed and accuracy.

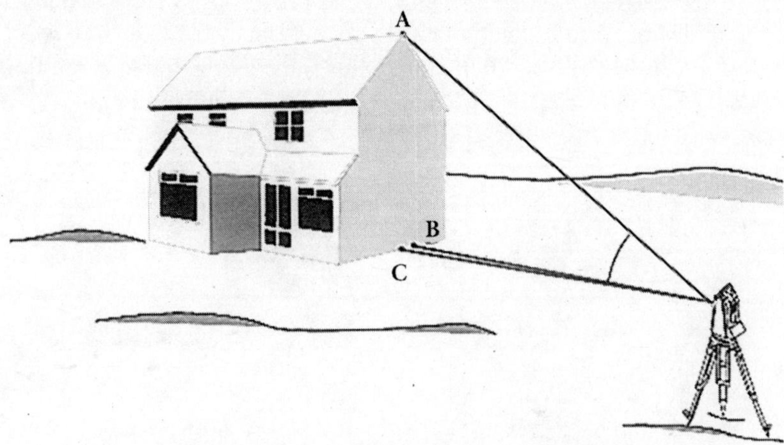

Fig. 3.11: Use of a reflectorless Total Station

Modern Total Stations have the ability to measure inaccessible or hard to reach targets up to 2 km without reflector with an accuracy of 3 mm. It provides the additional advantage of requiring less labour and time as second person is not required to hold the prism at the target point. For example, when surveying busy roads, traffic puts the restrictions to work with a reflective prism. So, the decision to use the prism method or reflectorless method is taken according to the conditions at the survey site. Reflectorless Total Stations are used now-a-days for several applications in geodetic engineering due to their highly accurate and fast measurements in an automated measuring process (Garg, 2021). Structure deformation can also be measured with reflectroless Total Station instrument.

In reflectorless Total Station, the accuracy of measured slope distance for a white surface is higher than the accuracy of any other surface colour; hence this surface has the strongest reflectivity for reflectorless Total Station ray as compared with any other surface. Similarly, the surface of black target has a very low reflectivity, so it absorbs more energy. Increasing the inclination angle of reflecting surface leads to increase in the errors of slope distance measured by the reflectorless Total Station. The reflectorless measurements may have serious errors if the signals have reflected from any surface present in the line between the instrument and the target. For example, the signal may be reflected by a leaf present in between the line of sight. Because the laser pulses reflect-off different surfaces, precaution must be taken when pointing the instrument. This is especially critical when there are multiple surfaces at various orientations near the measurement point. Most reflectorless instruments can also be used with a prism as a conventional Total Station. giving them a greater flexibility of working.

Other main concern with the accuracy of reflectorless Total Stations is divergence of laser beam. The beam divergence of an EM beam is an angular measure of the increase in beam diameter or radius with distance from the optical opening or antenna opening from which the EM beam emerges. As the size of the laser spot increases with distance from the instrument, so the accuracy of the measurement becomes less reliable. The size of the laser beam depends on the distance from the EDM system; the greater the distance, the larger the laser spot size.

3.3.5 Robotic Total Stations

Another advancement in Total Station is the emergence of Robotic Total Station (RTS) which is able to follow a prism horizontally and vertically through servomotor in the instrument. These RTSs are expensive as they have more sensors in the device as well as prism. These sensors make the instrument to work in robotic mode. The servomotor automatically rotates the instrument, as it communicates with an Advanced Tracking Sensor (ATS) fitted in the prism for tracking the movement of the prism (Garg, 2021). A communication link between the RTS and prism allows, controlling the RTS from the prism pole side with a remote controller (Figure 3.12). With this instrument, just one person is required to carry out the entire survey.

Fig. 3.12: Robotic Total Station and prism with remote control unit

Source: https://lasersurvey.com.au/product-category/surveying-equipment/total-stations/robotic-total-stations

Modern Total Stations require only one surveyor to take the field measurements, thus saving time and money. They provide high accuracy, even under low visibility (night) condition. Total Stations are used to increase the productivity for topographic surveying, to set out bridges, dams, canal, houses or boundaries. The RTSs are also used by archaeologists, police, crime scene investigators, insurance companies, for automated guidance of dozers, graders, excavators, harvesters, tractors and scrapers, and for deformation studies, such as dams, towers and plant chimneys. There are many applications of automatic or RTs, such as a stakeout of points, deformation monitoring, cadastral surveys, tunnelling, volume calculations and construction. Its function for automatic tracking survey is very useful especially in the field of dynamic surveying.

3.3.6 Smart Stations

Site surveying is increasingly being carried out using GNNS and Total Station equipment. Integrated survey rovers (called *Smart Stations*) combine the GNSS and Total Station to significantly improve the efficiency of survey work (Figure 3.13). It ensures that in case of obstruction in horizontal visibility, GNSS antenna is connected to the Total Station, and all operations of GNSS observations are performed through the keyboard of the Total Station. Where GNSS does not receive good signals from the satellites due to vertical obstructions, Total Station is used. These latest developments in Total Stations are also capable to provide data for building information modeling (BIM) and virtual design and construction.

Although the use of GNSS is increasing, but the Total Stations are one of the predominant instruments used on site for surveying, and will be used in future along with other surveying equipment to provide accurate and faster data collection approaches. Developments in both technologies (GNSS and Total Station) will find a point where devices can be made that complement both the methods.

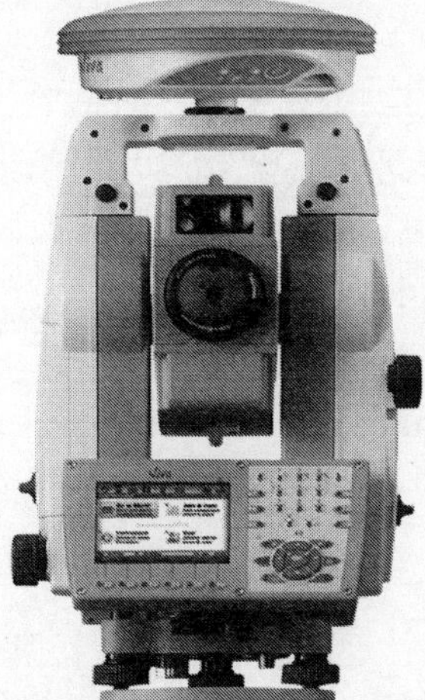

Fig. 3.13: Smart station (Total Station and GNSS combined)

3.3.7 Uses of Total Stations

a. To measure horizontal and vertical angles.
b. To obtain the horizontal distance, slope distance and vertical distance between the points.
c. To get the 3-D co-ordinates (x, y, z) or (northing, easting and elevation) of surveyed points.
d. To locate the points at a pre-determined distance.
e. Plotting of contours
f. Creating detailed maps
g. Carrying out the control surveys
h. To estimate the excavations
i. In crime scene investigations to take measurements of the scene
j. Used to fix the missing pillars
k. Remote Distance Measurement (RDM)
l. Missing Line Measurement (MLM)
m. Remove Elevation Measurement (REM)

3.3.8 Advantages and disadvantages of Total Stations

Advantages:

a. Field work is carried out very fast with a Total Station, saving time in the field.
b. Setting up of the instrument on the tripod can be done easily and quickly by laser plummet and auto levelling facility.
c. The accuracy of measurements is much higher as compared to other conventional surveying instruments.
d. The measured data can be saved and simultaneously transferred to the computer for subsequent use by the software.
e. Since data recording is automatic, no writing or recording errors is committed.
f. Correction for temperature and pressure can also be made in the field data.
g. Accuracy of measurement is high.
h. Calculation of coordinates is very fast and accurate.
i. Contour intervals and scales can be changed in no time.
j. Software can be employed for map making, plotting contour and cross-sections, and 3D models.

Disadvantages:

a. The cost of the instrument is higher than the other surveying instruments.
b. Checking for errors or other things during the operation is slightly difficult.
c. Skilled surveyors are required to handle since it is a sophisticated instrument to operate.

3.3.9 Calibration of Total Stations

Total Stations are worldwide known for their highest precision, reliability and durability. Due to temperature changes, shock or vibrations, instrument precision can change slightly over time. Even the most careful and precise assembly process, however, can result in small deviations which can lead to so called instrument errors. The modern Total Stations, like any other equipment, will provide errors in their measurements due to instrument error which is to be checked on a regular basis using procedures outlined in their manuals. Some instrumental errors are eliminated by taking observation on two faces of the Total Station and taking the average of both, but because one face measurements are common in the field, it is necessary to determine the magnitude of instrumental errors and correct them.

To further minimise these errors, the instrument can be calibrated by the user from time to time (Reda and Bedada, 2012). It is not necessary to send the instruments back to manufacturer for re-calibration. Almost all the modern total stations can be re-calibrated by the users. The user calibration feature is called "Check & Adjust" or just "Adjust" depending on the field software. The different Check & Adjust procedures are very simple to use. If they are followed carefully and precisely, the parameters will be correctly determined by the instrument itself. The instrument does a validity check of the calibration values and warns the user in case of mis-calibration. The calibration values will be stored in the instrument and automatically applied to all subsequent measurements when the tilt compensator and horizontal corrections are activated.

For Total Stations, instrumental errors are measured and corrected using electronic calibration procedures that may be carried out at the site. The electronic calibration procedure is preferred as compared to the mechanical adjustments that used to be done in the labs by trained technicians. Calibration parameters of the instrument usually change because of transportation of equipment, jerking in the field, temperature changes and rough handling, and therefore a high-precision Total Station requires electronic calibration, particularly (i) before using the instrument for the first time, (ii) after long storage periods, (iii) after rough or long transportation, (iv) after long periods of work, and (v) when the temperature changes appreciably.

3.3.10 Errors in Total Station measurements

Like any device, Total Stations also have some sources of error which can affect the surveying observations. All theodolites measure angles with some degree of imperfection, resulting from the fact that no mechanical device can be manufactured with zero error. With the advent of electronic theodolites, mechanical errors are still present but are related in a different way. So, it is important to understand the concepts behind the adjustments for errors that Total Stations now make (Garg, 2021).

1. Circle eccentricity error

Circle eccentricity errors occur when the theoretical center of the mechanical axis of the theodolite does not coincide exactly with the center of the measuring circle. The magnitude of error corresponds to the degree of eccentricity and the part of the circle being read. The circle eccentricity errors appear as a sine curve, graphically. This error in the horizontal circle can always be compensated by taking the measurement on both the faces (opposite sides of the circle) and taking the mean value. However, the vertical circle eccentricity error cannot be compensated this way, since the circle moves with the telescope. More sophisticated techniques are required; (1) Some theodolites are individually tested to determine the sine curve for the circle error in that particular instrument. Then a correction factor is added or subtracted from each angle reading to display corrected measurement, and (2) Other instruments employ an angle measuring system consisting of rotating glass circle that makes a complete revolution for every angle measurement. These are scanned by fixed and moving light sensors. The glass circles are divided into equally spaced intervals which are diametrically scanned by the sensors. The amount of time it takes to input a reading into the processor is equal to one interval, thus only every alternate graduation is scanned. As a result, measurements are made and averaged for each circle measurement. This eliminates the scale graduation and circle eccentricity error.

2. Circle graduation error

In the past, circle graduation error was considered a major source of error. For precise measurements, earlier several set of measurements were repeated beginning with 0°, 90°, 180° and 270°, and mean value adopted to eliminate the circle graduation error. Current technology eliminates the problem of graduation errors by photo-etching the graduations onto the glass circles and making a precise master circle and photographing it. An emulsion is then applied to the circle and a photo-reduced image of the master is projected onto the circle. When the emulsion is removed and the glass circle is etched with very precise graduations.

3. Horizontal collimation (Line of sight) error

The axial error is caused in Total Station when the line of sight is not perpendicular to the tilting axis. It affects all the horizontal circle readings and increases with the steep sight readings. The error can be eliminated by taking observations on both the faces (left and right). For single face measurements, an on-board calibration function is used to determine '*c*', the deviation between the actual line of sight and a line perpendicular to the tilting axis (Figure 3.14a). A correction is then applied automatically for all the horizontal circle readings.

4. Tilting axis or Tilt error

The axial errors occur when the titling axis of the Total Station is not perpendicular to its vertical axis. It has no effect on readings taken when the telescope is horizontal, but introduces errors into horizontal circle readings when the telescope is tilted, especially for steep sights. As with horizontal collimation error, this error is also eliminated by taking two face measurements. Alternatively, a tilting axis error '*a*' is measured in a calibration procedure and a correction is applied to all the horizontal circle readings (Figure 3.14b). If the value of '*a*' is large, the instrument should be returned to the manufacturer for calibration.

5. Compensator index error

If the Total Station is not carefully levelled, this error cannot be eliminated by taking both the face readings. Several Total Stations are equipped with a compensator that will measure the residual tilt of the instrument, and subsequently apply the corrections to both the horizontal and vertical angles. Normally, all the compensators will have a longitudinal error '*l*' and traverse error '*t*' known as *zero point errors* (Figure 3.14c). These are averaged using both the face readings, but for a single face reading, it must be determined by the calibration function of Total Station.

6. Vertical collimation or Vertical index error

A vertical collimation error is present in a Total Station if the 0° to 180° line in the vertical circle does not coincide with its vertical axis. This zero point error is present in all vertical circle readings and similar to horizontal collimation error, and it is eliminated by taking both the face readings. Else, it is eliminated by determining '*i*' (Figure 3.14d).

7. Vertical circle error

It is important to check the vertical circle indexing adjustment on Total Station on a routine basis. When direct and indirect zenith angles are measured to the same point, sum of the two angles should be equal to 360°. Over continuous use, the sum of these two angles may diverge from 360°, and consequently cause errors in vertical angle measurements. While, averaging the direct and indirect zenith angles easily eliminates this error, on many jobs it may not be cost effective to take two readings. Acceptable accuracy may still be maintained for many applications with only a direct reading; as long as the index error is kept to a minimum by periodically performing a vertical adjustment. Most Total Stations are provided with some type of electronic adjustment to correct the vertical circle indexing error. This adjustment takes just a few seconds and will provide good vertical angle readings with just one measurement. The adjustment can be made as per explanation given in manufacturer's manual.

8. Pointing errors

Pointing errors are due to both human's abilities to point the instrument and environmental conditions limiting the clear vision of the observed target. The best way to minimize the pointing errors is to repeat the observations several times and use the average value.

9. Uneven heating

Direct sunlight can heat one side of the instrument enough to cause small errors. For the highest accuracy, the instrument is shaded with an umbrella or a shaded spot is selected to set up the instrument.

10. Vibrations

Avoid setting up the instrument at locations that are subject to vibration, such as the site near airports, railway tracks. Continuous vibrations can cause the compensator unstable.

11. Atmospheric errors

The Total Stations are generally standardized at a specific temperature and pressure. When measurement conditions deviate from either then a proportional correction must be applied. Meteorological data corrections to observed slope distances may be significant for longer distances, so they need to be applied. The refractive indices of electromagnetic waves in air are functions of air temperature, atmospheric pressure and the partial pressure of water vapour. But, light waves and microwaves react somewhat differently to varying atmospheric conditions. These errors can be removed by applying an appropriate atmospheric correction model that takes care of different meteorological parameters from the standard (nominal) one. This is normally done by the Total Station, as it requires some necessary information (temp and pressure) from the operator. For most topographic surveying over short distances, nominal (estimated) temperature and pressure data is acceptable for input into the data collector. Instruments used to measure the atmospheric temperature and pressure, such as thermometers and barometers, must be periodically calibrated.

12. Optical plummet errors

The optical plummet must be periodically checked for misalignment. This would also include Total Stations with laser plummets.

13. Adjustment of prism poles

When using prism poles, precautions should be taken to adjust the leveling bubble. Bubbles tube can be examined by standard tests.

14. Height of standards error

To plunge the telescope of Total Station through a truly vertical plane, the telescope axis must be perpendicular to the vertical axis. The horizontal collimation and height of standards errors are correlated, and can magnify or offset one another. Horizontal collimation error is usually eliminated before checking for height of standards. Height of standards error is checked by pointing to a scale the same zenith angle above a 90° zenith in "face-one" and "face-two." The scales should read the same value in both the faces.

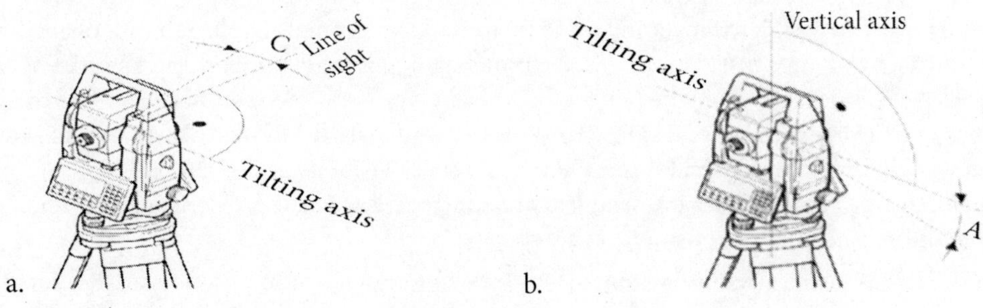

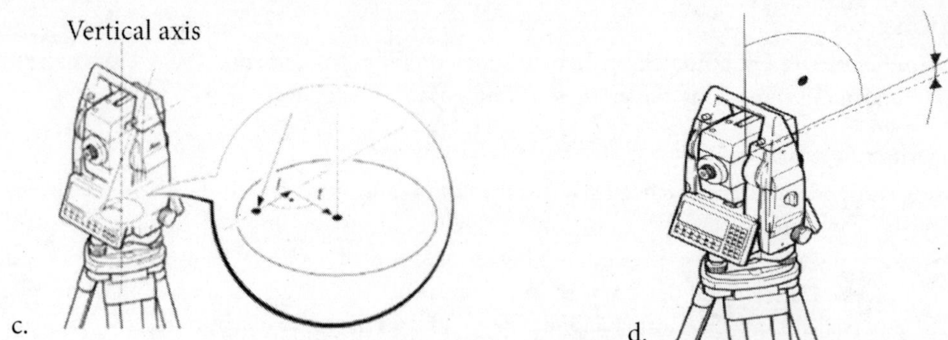

Fig. 3.14: Depiction of various errors from Total Station

Source: https://www.brainkart.com/article/Sources-of-Error-for-Total-Stations_4653

3.4 GLOBAL POSITIONING SYSTEMS (GPS)

The navigation systems in some form or the other have been in use since the civilizations. Human-beings have always been interested to know where they are, where they are going, and how they are going to get there, and back again to their destination, using some appropriate path. So, several crude methods have been developed in the past. In modern era, GNSS (Global Navigational Satellite System), which receives signals from the all the satellites, is used for precise navigation (Garg, 2019). The GNNS broadcasts precise, synchronized timing signals to provide precise position, velocity and time.

The GPS was developed by the Department of Defence (DoD), USA, which is a part of Global Navigation Satellite System (GNSS), primarily to provide precise estimates of position, velocity and time to the U.S. military. In 1973, the US decided to establish, develop, test, acquire, and deploy a first space-borne GPS, resulting in the NAVSTAR (Navigation Satellite Timing and Ranging) GPS. Initial satellites were launched between 1974 and 1977 for use in precision weapon delivery (Herring, 1996). In the early 20th century, ground-based radio-navigation systems were developed. Although, the GPS was initially developed for military applications, but over a period of time, the civil applications of GPS and GNSS have grown at an alarming rate. In the 1980s, the government made the system available for civilian use. It has brought a great technological revolution in surveying where the exact position of any object or phenomena is to be captured.

The European Union developed a system, known as GALELIO navigation satellites. Indian system is known as the IRNSS (Indian Regional Navigation Satellite System), using its seven satellites, which will beam accurate navigation signals over India and up to 1,500 km from its borders. India's own GPS NavIC ('Navigation with Indian Constellation' whose Hindi meaning is '*sailor*' or 'navigator') is the operational name of IRNSS. China has created the BeiDou (Compass) Navigation Satellite System; while the regional service has already been launched. The QZSS (Quasi-Zenith Satellite System) Japan, signal is designed in such a way that it is interoperable with GPS. It improves visibility and DOP (Dilution of Precision) in dense urban area, and provides messaging system during disasters. It provides augmentation data for sub-meter and centimeter level position accuracy.

The GNNS is primarily a navigation system receiving signal from all visible satellites for real-time positioning. The GNSS broadcasts the precise, synchronized timing signals to provide precise estimates of position, velocity and time. With the transformation from the ground-to-ground survey measurements to ground-to-space measurements made possibly by GNSS, this technique overcomes several limitations of ground surveying methods, like the requirement of intervisibility of survey stations, dependability on weather, difficulties in night observations, etc. These advantages over the conventional techniques and the economy of operations make GNSS the most promising surveying technique. With the high accuracy achievable with GNSS in positioning of points, this unique surveying technique has found important applications in diverse fields. Other advantages include (i) increase in usable space vehicles, signals and frequencies, (ii) increase in availability and coverage, (iii) more robust and reliable services, (iv) higher accuracy even in bad conditions, (v) offers less expensive high-end services, (vi) applies better atmospheric correction, and (vi) use in new and applications, such as atmosphere related, short message broadcasting, search and rescue applications, determine soil moisture, wind velocity, sea wave height etc.

Surveying and mapping fields have greatly benefitted with the availability of GPS and GNSS, such as highways, railroads, mining/geology, agricultural, power, telecommunications, health, law enforcement, emergency, crustal movement, etc. The GNSS offers the advantages of accuracy and speed, while collecting data in the field. The other advantage is to provide exact position of objects anytime, anywhere, and in any weather condition. The benefits of using GNSS for mapping and modeling include improved productivity, fewer limitations (such as open to sky requirements), and faster delivery of 3D coordinates. The other technological revolution the GNSS has provided for multiple applications, such as locating a petrol pump, restaurant, cinema hall etc. World-wide, there are a large number of users of GNSS enabled gadgets, like vehicles, mobile phones, wrist watches etc., used for a variety of utility. Today GNSS is considered as a global leader in navigational systems all over the world.

3.4.1 Technical terms in GNSS

Latitude: It is an angular measurement made from the center of the Earth to north or south of the equator. It comprises the north/south component of the latitude/longitude coordinate system, which is used in GNSS data collection.

Longitude: It is an angular measurement made from the center of the Earth to the east or west of the Greenwich meridian. It comprises the east/west component of the latitude/longitude coordinate system, which is used in GNSS data collection.

Datum: The GNSS receivers are designed to collect positions relative to the WGS-84 datum, however the user has the option of defining the datum in which the data will be collected. Users must be familiar with the datum before data collection. For example, for most GIS applications, the WGS-84 datum is similar to the NAD-83 datum, however NAD-27 is significantly different from the NAD-83 datum.

Ellipsoid: An ellipsoid is the 3-D shape that is used as the basis for mathematical modelling of Earth surface. The ellipsoid is defined by the lengths of minor axes (polar axis) and major axes (equatorial axis).

Geoid: A mathematical surface of constant gravitational potential that approximates the sea level, or the equipotential surface of the Earth's gravity field which best fits, in a least squares sense, global mean sea level.

Satellite constellation: The arrangement in space of a set of satellites.

NAVSTAR (NAVigation Satellite Timing And Ranging) System): The formal name given to the United States NAVigation Satellite Timing And Ranging (NAVSTAR) System. which comprises of GPS satellites, monitoring stations, and master control station.

Ephemeris: The current satellite position predictions that are transmitted from a GNSS satellite in the NAVDATA message.

Rover (GNSS) Receiver: Any mobile GNSS receiver and data collector used for determining location in the field. A roving GNSS position can be differentially corrected relative to a stationery base GNSS receiver.

L1 Frequency: The primary L-band carrier used by GNSS satellites to transmit satellite data in frequency 1575.42 MHz. It is modulated by C/A code, P-code and a 50 bit/second navigation

message.

L2 Frequency: The secondary L-band carrier used by GPS satellites to transmit satellite data in frequency 1227.6 MHz. It is modulated by P-code and a 50 bit/second navigation message.

L5 Frequency: The L5 signal operates at 1176.45 MHz which is planned to be used to improve accuracy for civilian use, such as aircraft precision approach guidance. The GNSS has fifteen satellites with L5 available and Galileo with 24 all of which are currently using L5.

Epoch: The measurement interval of a GNSS receiver.

Dual-frequency (GNSS) Receiver: A type of GNSS receiver that uses both L1 and L2 signals from satellites. A dual-frequency GNSS receiver can compute more precise position fixes over longer distances and under more adverse conditions by compensating for ionospheric delays. The receivers are now available which can also receive signal in L5 wavelength, in addition to L1 and L2.

Multi-constellation and Multi-frequency (GNSS) Receiver: Multi-frequency GNSS receivers are used for reliable positioning down to the centimeter level. Among various professional-grade GNSS receivers, there is variability in the satellite constellations and signals that a receiver can access. Those receivers that have access to the highest number of constellations and signals from, for example, Galileo, QZSS, BeiDou, offer the best positioning availability, accuracy and resilience even in challenging environments. A simple GPS receiver only makes use of one global navigation satellite system, while multi-constellation GNSS receivers get information from many such systems at the same time.

Ephemeris: The prediction of current satellite position that is transmitted to the user in the data message.

Data Message: A message included in the GNSS signal, which reports a satellite location, clock correction, and health. It also includes information on other satellites' health and their approximate positions.

Pseudorandom Noise or Number (PRN): A signal that carries a code that appears to be randomly distributed, like noise, but can be exactly reproduced. The PRN codes have a low auto-correlation value for all delays or lags, except when they are exactly coincident. Each NAVSTAR satellite has its own unique PRN code.

Carrier Phase: The difference between the carrier signal generated by the internal oscillator of a roving GNSS receiver and the carrier signal emitted from a particular GNSS satellite.

Coarse or Acquisition (C/A) Code: A pseudorandom noise (PRN) code modulated onto a L1 signal which helps the GNSS receiver to compute the distance from each satellite. Specifically, the difference between the PRN code generated by the GNSS rover software and the PRN code coming in from the satellite is used to quickly compute the distance to a satellite and the position.

P-Code: The precise code transmitted by the GNSS satellites. Each satellite has a unique code that is modulated onto both the L1 and L2 carrier. The P-code is replaced by a Y-code when Anti-Spoofing is active.

Selective Availability (SA): The artificial and deliberate degradation of GPS satellite signals by the United States Department of Defense. Selective Availability was implemented in view of the national security, but was turned off on May 10, 2000 due to the presence of several sources of various differential correction (DGNSS) messages, which rendered SA obsolete. Earlier the potential error due to SA is between 30 to 100 m.

Signal-to-Noise Ratio (SNR): The signal strength of a satellite is a measure of the information content of the signal relative to the signal's noise. The typical SNR of a satellite at 30° is between 47-50 dBHz. The quality of a GNSS position is degraded if the SNR of one or more satellites in the constellation falls below 39. If a satellite's SNR is below the configured minimum SNR, that satellite is not used to compute positions.

Base Station: A base station is comprised of a GNSS antenna and GNSS receiver positioned at a known location specifically to collect data for differential correction. The purpose of the base station is to provide reference data for applying the differential correction on data collected in the field. A base station can be a permanent installation that can be used by multiple users.

VRS (Virtual Reference Station): A VRS system consists of GNSS hardware, software, and communication links. It uses data from a network of base stations to provide corrections to each rover that are more accurate than the corrections from a single base station. To start using VRS corrections, the rover sends its position to the VRS server. The VRS server uses the base station data to model systematic errors (such as ionospheric noise) at the rover position. It then sends correction messages back to the rover.

Initialization of GNSS: Initialization refers to the procedure of telling a GNSS receiver where it is, when it is turned on for the first time. Information required for initialization includes approximate present position in latitude/longitude coordinates, the current local time and date.

Differential Correction: The process of correcting GNSS data collected on a rover with data collected simultaneously at a base station. Because it is on a known location, any errors in data collected at the base station can be measured, and the necessary corrections applied to the rover data. Differential correction can be done in real time, or after the data has been collected by post-processing software.

RTK (Real-Time Kinematic): A real-time differential GNSS method that uses carrier phase measurements for greater accuracy. The RTK measurements can typically yield relative horizontal accuracy of approximately one centimeter.

Accuracy: The degree of conformity with a standard or accepted value. Accuracy relates to the quality of the result, and is distinguished from precision which relates to the quality of the operation by which the result is obtained.

Precision: A measure of the repeatability or uniformity of a measurement. Precision relates to the quality of the operation by which the result is obtained. In order to comply with a specific standard, accuracy results must meet the minimum while complying with the precision required.

Multipath Errors: Errors caused by reflected signals arriving at the GNSS receiver, as a result of nearby structures or other reflective surfaces (e.g., buildings, forest, water). Signals traveling

longer paths produce higher (erroneous) pseudorange estimates and, consequently the positioning errors. Some mapping grade GNSS receivers as well as most or all survey grade GNSS receivers have antennas and software capable of minimizing multipath signals.

Atmospheric Delays: The satellite signal slows down as it passes through the atmosphere. The GNSS system uses a model that calculates an average amount of delay to correct for this type of error.

Receiver Clock Errors: A receiver's built-in clock is not as accurate as the atomic clocks onboard the GNSS satellites, therefore, it may have very slight timing errors.

Orbital Errors: Also known as ephemeris errors, these are inaccuracies of the satellite's reported location.

Dilution of Precision (DOP): It is an indication of the quality of the results that can be expected from a GNSS point position. It is a measure based solely on the geometry of the satellites in the sky.

Geometric Dilution of Precision (GDOP): It deals with the overall accuracy, 3D-coordinates and time.

Positional Dilution of Precision (PDOP): It deals with the position accuracy, and 3D-coordinates. A value expressing the relationship between the error in user position and the error in satellite position. Values considered good for positioning are small, such as 3. The GNSS receiver may be set to collect data at a PDOP level of 6 or less. Values greater than 7 are considered poor.

Horizontal Dilution of Precision (HDOP): It deals with the horizontal accuracy, and 2D-coordinates.

Vertical Dilution of Precision (VDOP): It deals with the vertical accuracy, and height.

Post-processing of Data: The processing of satellite data done after it has been collected in order to eliminate the errors, using. Since the location of base station is known, systematic errors can be detected and removed from the rover data.

SBAS (Satellite Based Augmentation System): This term refers to differential GNSS applied to a specific wide area, such as an entire continent. The SBAS comprised of a series of reference stations that generate GNSS corrections which are broadcast to GNSS rovers *via* geostationary satellites. The WAAS and EGNOS, MSAS, GAGAN are some examples of SBAS network.

3.4.2 Basic principle of GPS

The GPS is based on very simple and quite ancient idea of determining the position, e.g., coordinates (x,y,z), given the distances and directions to other surrounding objects whose positions are known (Garg, 2021). Let us consider three objects whose positions (i.e., their coordinates) as well as their distances from the unknown position of the observer are known. The observer's position can now be related to the known distances and positions of objects by using distance equation. Thus, from the known positions of the objects and their distances from observer's position, we get three distance equations involving three unknowns (x,y,z) corresponding to the position of the observer. These three equations can be solved for the three unknowns i.e., x,y,z. The novelty of GNSS lies in the realization of above concept with the technology of late 20th century in a global navigation system. It has the capability to provide estimates of position, velocity and time to an unlimited number of users instantaneously and continuously.

The accuracy of GNSS height measurements depends on several factors but the most crucial one is the undulating shape of the Earth (Xu, 2010). Height can be measured in two ways. The GNSS uses ellipsoidal height (h) above the reference ellipsoid that approximates the Earth's surface. The traditional, orthometric height (H) is the height above an imaginary surface called the geoid (Figure 3.15), which is determined by the Earth's gravity and approximated by MSL. The orthometric height is often referred as the MSL, and can be obtained by subtracting the geoid height (N) from the GNSS height (h) (i.e., $H = h-N$). The signed difference between the two heights; the difference between the ellipsoid and geoid, is the geoid height (N). Figure 3.15 shows a relationship between the different models; a geoid height (N) is positive when the geoid is above the ellipsoid and negative when it is below.

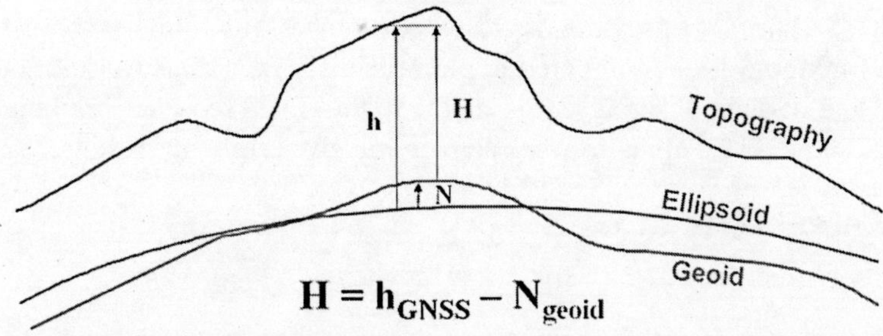

Fig. 3.15: The ellipsoidal height, orthometric height and geoid height

In GNSS based determination of observer's position, satellites act as reference points (i.e., known locations) from which receiver on the ground determines its position. By measuring the travel time of the signals transmitted from at least four satellites, the distances between the receiver and satellite will yield accurate position, velocity and time. Though three-range measurements are sufficient but fourth observation is essential for solving the clock synchronization error between the receiver and satellite. The strength of GPS measurement lies in its ability to measure 1/100 of a cycle of a carrier phase is about 2 to 3 mm in linear distance.

The GNSS satellites cover every corner of the Earth; no matter where you are. If at least four satellites are visible at any time, the GNSS observations can be taken. Each satellite regularly transmits the information about its location and time. These signals that are traveling at the speed of light are intercepted by the GNSS receiver which calculates the distance of each satellite from receiver position based on the measurement of time the signal arrives.

$$\text{Distance} = \text{velocity} * \text{time} \tag{3.3}$$

The GNSS receiver calculates the distance between each one of the satellite and receiver position, the GNSS receiver uses a method called *trilateration* to determine the GNSS receiver position (Figure 3.16). The position thus calculated by GNSS receiver would rely on three accurate measurements; firstly, current time, secondly, position of the satellites, and thirdly, the time delay for the signal. The GNSS time is accurate to about 40 nanoseconds. Higher accuracy can be obtained today by using GNSS in combination with augmentation systems which enable real-time positioning to within a few centimeters.

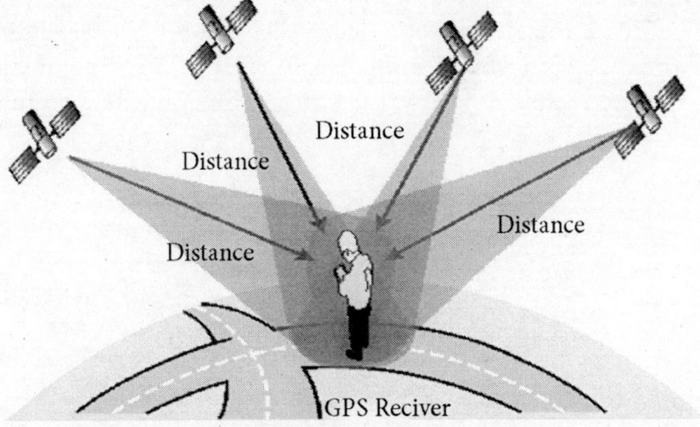

Fig. 3.16: Trilateration principle to compute receiver position

Source: https://owaysonline.com/global-positioning-system-on-ships/)

Even though GNSS technology provides us greater advantages, but it still has some limitations. Since GNSS satellite signals are too weak as compared to phone signals, they don't work indoors, underwater, under trees, etc. The line-of-sight of signal from satellite to the GNSS receiver should not be obstructed by high rise buildings and forested areas, as weak signals will give erroneous results.

3.4.3 Various segments of GPS

There are mainly three segments of GPS (Garg, 2019), as shown in Figure 3.17.

a. Space segment
b. Control segment, and
c. User segment

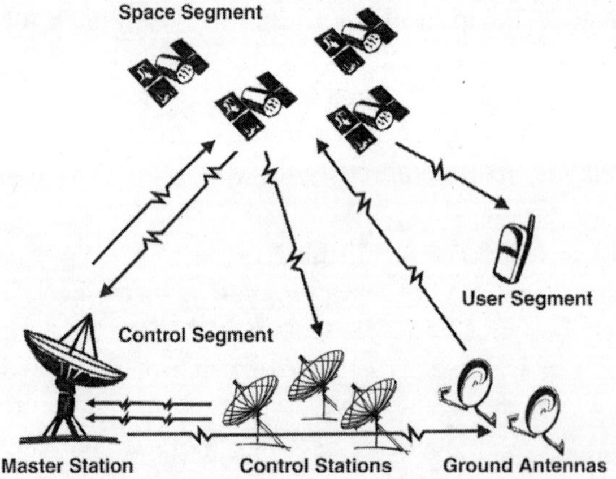

Fig. 3.17: Various segments of a GPS

Source: https://shivkumardas.wordpress.com/agri-tech/an-introduction-to-gps-gis-and-its-uses-in-agriculture/

a. Space segment

The space segment includes constellation of NAVSTAR Earth orbiting satellites, generally 24 satellites for full global coverage. Thus, there are four satellites moving in 6 orbital planes, as shown in Figure 3.18. The orbital planes are inclined at 55° with respect to equator, such that their orbits are separated by 60°. They revolve around the Earth at the altitude of about 12,000 miles (-20,000 km) above Earth's surface with orbital period of approximately 11 hr 55 minutes (Xu, 2010). This type of configuration provides a greater visibility of five to eight satellites at any given time from anywhere on the Earth. Some of the important features of the GPS satellites are given in Table 3.2.

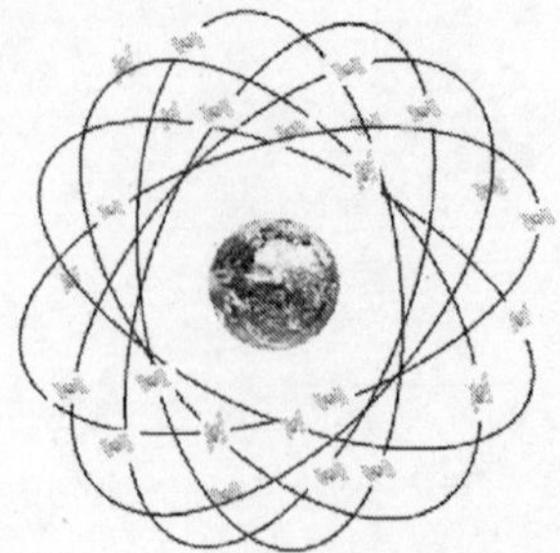

Fig. 3.18: The space segment

Table 3.2: Important features of GPS satellites (as on Sept. 2022)

S.No.	Features	Specifications
1	Design life	5 years (with expendables stored for 7 years)
2	On orbit weight	430 kg
3	End-of-life power	400 W
4	Power Source	$5m^2$ solar arrays tracking the sun and 3 Ni-Cd batteries for eclipse
5	Navigation Pay Load	Pseudo Random Noise (PRN) signal assembly, atomic frequency Standard–Caesium beam atomic clocks accurate to 10-14 sec, processor and L band antenna
6	Codes	(a) Precision (P) Code: generated at GPS clock frequency of 10.23 MHz (equivalent to 30 m in range) interpolated to sub-meter level. Repeats itself after 267 days, resolution=100 nanoseconds. (b) Coarse Acquisition (C/A) Code: code sequence frequency of 1.023 MHz (range 300 m) interpolated to few m. Repeats itself every 1 millisecond, resolution = 1 micro second
7	PRN navigation signals on three frequencies	(a) 1575.42 MHz–L1 band–wavelength 19 cm. (b) 1227.6 MHz–L2 band–wavelength 24 cm. (c) 1176.45 MHz–L5 band–within a 24 MHz bandwidth

Each satellite carries four precise atomic clocks; only one of which is used at a time. It also carries three nickel-cadmium batteries, two solar panels, battery charger, S band antenna-satellite control, and 12 element L band antenna-user control. It has a micro-processor on board for limited data processing. At a given time, several satellites can send their signals to a GPS receiver. Each transmission of signal is time-tagged, and contains the satellite's position. The time-of-arrival of signal at GPS receiver is compared with the time-of-transmission, and this time is multiplied by the speed of light to obtain the corresponding distance between that satellite and GPS receiver. The location of the observer is then determined by the intersection of several of these ranges.

The GPS signals contain three different information, as below:

1. **Pseudo random code (PRC)**– It is simply an ID that identifies which satellite is transmitting the information to receiver. The number attached to each signal bar visible on GPS receiver display identifies the satellite sending the signals.
2. **Almanac data**– It contains details of the orbital path of each satellite. The GPS receiver uses this data to determine which satellites it should track. With almanac data, the GPS receiver can concentrate on those satellites it can clearly see, and neglect the ones that would be over the horizon and out of view.
3. **Ephemeris data**– It provides information to the GPS receiver where each satellite should be at available throughout the day. Each satellite will broadcast its own ephemeris data showing the orbital information. Ephemeris data consists of very precise orbital and clock correction information necessary for precise positioning, but it keeps on changing after short time.

b. Control segment

The control segment is administered by the Department of Defence (DoD), US government, who is responsible for the construction, launching, maintenance and monitoring the performance and health of all GPS satellites launched by them. One master control station is established in USA, and several ground monitoring stations and 4 antenna stations throughout the world, as shown in Figure 3.19. The DoD monitors these stations and tracks all satellites for controlling and predicting their orbits. The master control station is responsible for collecting and tracking the data from the monitoring stations, and computes satellite orbits and clock parameters. Other monitoring stations are responsible for measuring the pseudo range data. This orbital tracking network is used by master control station to calculate the satellite ephemeris and clock correction coefficients, and to forward them to ground antenna. The satellite tracking data from these monitoring stations are transmitted to the master control station for processing.

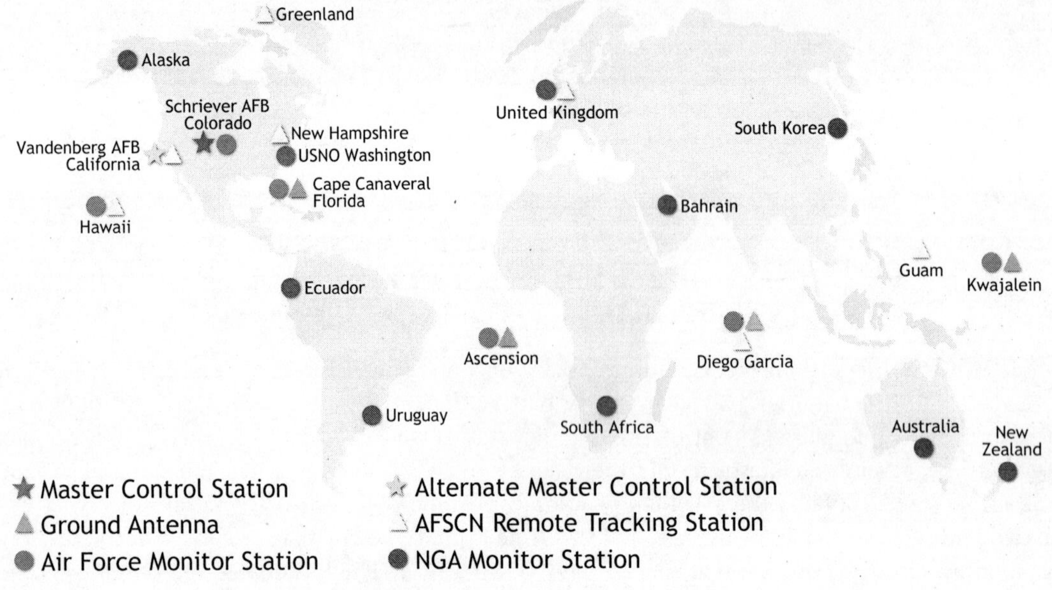

Fig. 3.19: The control segment

Source: https://www.gps.gov/systems/gps/control

c. User segment

It is the segment where observations are taken by the users for various uses. It consists of GPS receiver unit that receives signals from the GPS satellites (Figure 3.20). The typical receiver is composed of an antenna and a pre-amplifier, radio signal micro-processor, control and display device, data recording unit, and power supply. The GPS receivers convert the satellite signals into position, velocity, and time. The GPS receivers are available in different sizes and shapes. A receiver is often described by its number of channels, signifying how many satellites it can monitor simultaneously (Garg, 2021). Presently, the GPS receivers can operate up to 20 channels. The GPS receiver collects two types of data from satellites to get an exact location i.e., almanac and ephemeris data which are continuously transmitted by the GPS satellites. The almanac gives the exact status of satellites and its approximate orbital information which are used to estimate the visibility of satellites. Ephemeris data gives the accurate information about the orbit of satellite which can be used to calculate the location of a satellite precisely. It is updated every two hours and usually valid for 4 hours. The receiver collects and stores the data.

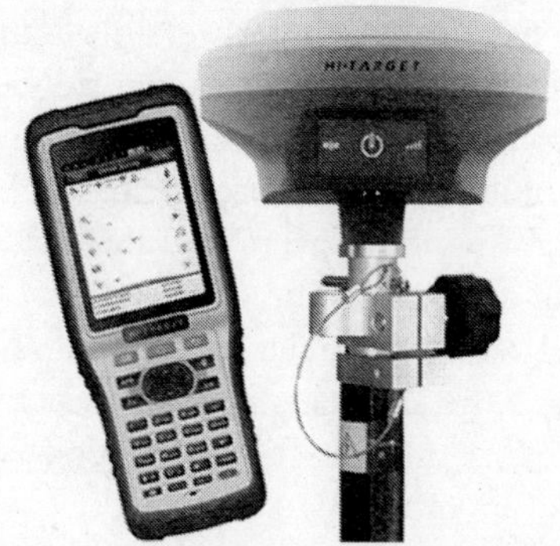

Fig. 3.20: The users segment

Source: https://www.tokopedia.com/aprilianisurvey/gps-geodetik-hi-target-v90-plus-rtk

3.4.4 Signals of GNSS

The GNSS system sends its information through various signals. Broadly, it works with three types of signals; L1, L2, and L5. Figure 3.21 shows the modulation of L1 wave and L2 wave. The L1 signals operate at 1575.42 MHz, and carries both the status message and a pseudo random code (PRC) for timing. The L2 signals operate at 1227.60 MHz, and used for the more precise military work. The L5 signal operates at 1176.45 MHz which was turned on April 2009. The L5 frequency is used to improve accuracy for civilian use, such as aircraft precision approach guidance. The GNSS has fifteen satellites with L5 available and Galileo with 24 all of which are currently using L5. In good satellite visibility the GNSS device with L5 will give a 6 ft accuracy, and 10 ft with GPS only.

A civilian GNSS uses the L1 signal in the UHF band. The L1 frequency carries the navigation message and the standard positioning code signals (SPS). Most receivers are capable of receiving and using the SPS signals, and civilian users world-wide use the SPS without restrictions. The SPS accuracy is intentionally degraded by the DoD by the use of Selective Availability (SA). The SPS provides 100 m horizontal, 156 m vertical, and 340 nanoseconds time accuracy. The signals travelling from satellite to receivers will pass through clouds, dust, gas, particles etc., but will not travel through solid objects, such as buildings and mountains (Garg, 2021).

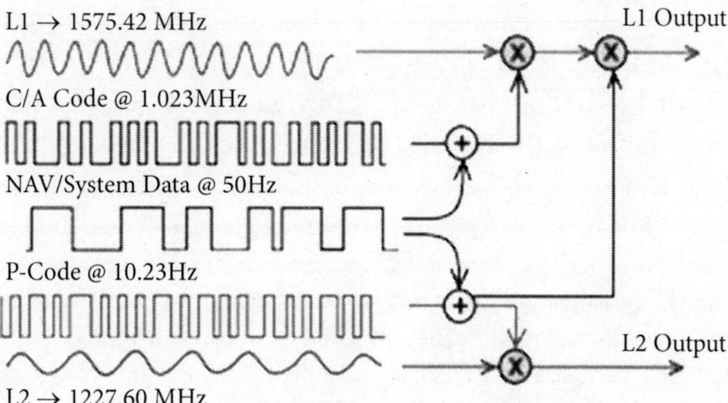

Fig. 3.21: L1 and L2 signals

The pseudoranges, which are derived from signal travel time to the receiver, use two pseudorandom noise (PRN) codes. These codes are modulated onto the carrier frequencies. The carrier frequency with PRN codes is modulated for the real-time positioning through GPS signals. The travel time of a PRN signal code can give pseudo ranges. Dual frequency receivers get signals in L1 and L2 carrier wave at 24.45 cm wavelength. But today many of the multi-band receivers operate in L1, L2 and L5, and they eliminate the ionospheric dispersion which is one of the major sources of systematic range error. The structure of the future operational GPS L5 signal will offer a two carrier components signal. Both components– In-phase (I) and quadrature-phase (Q) will have the same signal power level. The minimum received power is defined with -157.9 dBW, which is 0.6 dB more than the L1 C/A code signal. Both components will carry different but nearly orthogonal and time synchronized PRN Codes. The Q channel of the L5 signal will be a data-less channel, transmitting only a pilot signal modulated with the specific satellite PRN, which is useful for a long coherent integration time. On the I channel the navigation message is modulated with 100-symbol per seconds. In addition, the L5 signal uses a NeumanHoffman synchronization code. The usage of two different PRN Codes helps to prevent possible tracking biases. The two channels are only dependent on the same carrier phase.

The L5 signal uses a chipping rate of 10.23MHz and which is 10 times the rate of the C/A and L2C codes. With this chipping rate the signal has 20.46 MHz null-to-null bandwidth which is exactly the same as the legacy P(Y) code signal. Thus the signal features satisfy the requirements for a new Safety of-Life signal with increased bandwidth, higher signal accuracy und robustness under rough conditions. The L5 I Channel NAV Data is very similar to the L2C channel and includes Space Vehicle ephemerides, system time and clock behavior data, status messages and time information. The L5 band signals have the advantages as they have: (i) a higher power (1.5 dB for the pilot channel of the L5 signal compared to the L1 C/A signal, and 2 dB for the E5a/b signals as compared to the E1 signal for the pilot channels), which means that similar detection performance can be obtained with lower integration times; (ii) a pilot channel, whereas the GPS L1 C/A signal does not have one, therefore this must limit its coherent integration time, which can lead to a longer total integration time; (iii) a lower carrier Doppler and Doppler rate (115/154 ≈ 75%), which can reduce the search space and some constraints for the acquisition architecture; and (iv) a secondary code, which on one side complicates the acquisition, but on another side makes the data synchronization much easier, simplifying the transition to the tracking. The L5 is intended to be a "safety-of-life" signal for aircraft navigation but will be useable for all civil users. This makes L5 to a valuable third civil GPS signal beside the C/A and L2C signal.

A simple GPS receiver will make use of only one global navigation satellite system, while multi-constellation GNSS receivers allows them to "see" much more signals at any given time and get information from many such systems. Each one of the GNSS satellites uses one or more frequencies to transmit ranging signals and navigation data. The more signals the receiver can access, the more information it can collect from the satellites, the more accurate and reliable the computed position will be. Navigation GPS in phones, cars and other consumer devices usually uses GNSS signalsinjustone frequency (L1). Dual-frequency receivers can receive two signals from each satellite system, while multi-frequency receivers receive a multi-signals from any GNSS system. Such multi-frequency receivers enhance the GNSS technology to achieve most accurate, reliable, and robust positioning.

A summary of number of satellites and signals from various GNSS system is given in Table 3.3.

Table 3.3: A summary of number of satellites and signals from various GNSS system

Source: https://www.septentrio.com/en/learn-more/about-GNSS/why-multi-frequency-and-multi-constellation-matters

GNSS	Country	Satellites	Coverage	Signals used
GPS	USA	32	Global	L1CA, L1P, L1C, L2C, L2P, L5
GLONASS	Russia	24	Global	L1CA, L1P, L2CA, L2P, L3CDMA
Galileo	Europe	26+	Global	E1, E1b, E5a, E5b, E6, E5-AltBoc
BeiDou	China	Phase 2: 15+ Phase 3: 25+	Global Phase 2 mostly China regional	B1I, B1C, B2a, B2b, B2I, B3I
QZSS	Japan	4+	Over Japan and Asia Pacific	L1CA, L1C, L1S, L2C, L5, L6
NavIC	India	7+	Over India	L5

Pseudo Random Code (PRC):

The signal that is sent out from GNSS satellites is a random sequence; each part of which is different from every other, called pseudo-random code. It is the prime signal of GNSS, and is physically complicated digital number or complicated sequence of 'on' and 'off' pulses. This random sequence is repeated continuously. All GNSS receivers know this sequence pattern and repeat it internally. Therefore, satellites and the receivers must be in synchronisation. The receiver picks up the satellite's transmission and compares the incoming signal to its own internal signal. By comparing the lag in these signals, the travel time is computed.

There are 2 types of PRC signals generally found.

a. ***Coarse Acquisition Code (C/A):*** The C/A code is made up of sequences called chips, and the sequence repeats itself every millisecond. The C/A code is for the civilians, and is different for every satellite. The L1 code, which is available for civilians, is the C/A-code (Course/Acquisition-code), which has a wavelength of approximately 300 meters. It repeats every 1023 bits and modulates at a 1 MHz rate. The non-availability of C/A-code in L2 allowed the US Government initially to control the level of accuracy available to civilian users.

b. ***Precise Code (P):*** It modulates in both L1 & L2 carries at a 10 MHz rate. The C/A code is made up of sequences called chips, and the sequence repeats itself every millisecond. The C/A code is for the civilians, and is different for every satellite. It is more complicated than C/A code, and is only used by receivers which are designed for PPS (precision positioning service code). The P-code, with a wavelength of approximately 30 meters, is encrypted into the Y-Code in the Anti-Spoofing (AS) mode of operation. The encrypted Y-Code requires a classified AS Module for each receiver channel. The P (Y)-Code is the basis for the PPS. Authorized users with cryptographic equipment & keys, and specially equipped receivers use the PPS. The PPS provides 22 m horizontal, 27.7 m vertical, and 200 nanosecond time accuracy. As the satellite system was fully operational in 1992, access to the P-code was denied to the public by US Government. In order to maintain control over the navigation system, the US military wanted to have most accurate GPS measurements with them only. Therefore, by altering the satellites clocks slightly according to a specific code, civilians used to get some errors in measurements as a result of time error. These signals allowed non-military users to obtain measurements that are accurate to approximately 100 meters.

Later, methods have been developed, where civilians have accurately calculated the receiver's position by comparing the GPS-measured position of a known location with its actual coordinates (e.g., GPS). The corrections were broadcasted to the GPS receiver, and thus it was possible for a civilian to determine the position with an accuracy of millimeters.

3.4.5 Advantages and disadvantages of GNSS

Advantages

- It is easy to navigate with GPS.
- It is available 24 hours anytime
- It has world coverage
- It is independent of weather conditions
- It is independent to visibility conditions: night, fog and dust
- It has open signal
- It requires open to sky clearance
- It is faster and quicker, particularly where Total Station can't be used due to horizontal obstructions.
- Multi-constellations and multi-frequency receivers can provide much accurate results.
- It is independent of visibility of previous points, as all measurements are independent.
- Its data can be easily integrated with Total Station data and GIS.

Disadvantages

- A good visibility for the sky is necessary, as high rise buildings obstruct the signals.
- It can't be used for indoor positioning.
- It can't give precise results in areas like, forest, under a bridge, tunnel, etc.
- Vertical precision is not enough for some applications.
- Extreme atmospheric conditions, storms can contribute to errors in GPS observations.
- With limited number of satellites, accurate results may not be obtained.

3.4.6 Types of GNSS receivers

A wide variety of GPS receivers are commercially available today. Depending upon the type of application, accuracy requirements and cost, the users can select the type of GNSS receiver which best meets the requirements. These receivers cover a wide range from the high-precision receivers with built-in atomic clock, to the hand-held navigation receivers, which can give the precise position to few-metres. Even wrist-watches with built-in GNSS receivers are now commercially available. Three broad categories of GNSS are explained below (Garg, 2021).

a. Navigation receivers:

Navigation in three dimensions is the primary function of GNSS. Navigational receivers are made for aircraft, ships, ground vehicles, and for hand carried by individuals. These are used for navigation, positioning, time dissemination, measuring atmospheric parameters, surveying, geodetic control, and plate tectonic studies. These receivers are normally single-frequency, C/A code, hand-held light weight receivers, which can give the position with a few metres to few tens of metres accuracy. These receivers are very much portable, weighing only few hundred grams, and are fairly cheap. Single channel receivers, which can track 4 or more satellites, are now being replaced by two or five channel receivers. The accuracies in positioning obtained by these type of receivers are in the range of few tens of metres in absolute positioning 10 (in the absence of SA), and few tens of cm in relative positioning, over short baselines of few km.

b. Surveying receivers

The surveying type of receivers are single frequency, multi-channel receivers, which are useful for most surveying applications, including cadastral mapping applications, providing tertiary survey control, engineering surveys, etc. They are more expensive than the navigational receivers, but more versatile. The data from many of these receivers can be directly imported to most commonly used GIS software packages. Most of these receivers can also be used in DGNSS mode.

c. Geodetic receivers

The geodetic receivers are multi-channel, dual-frequency receivers, generally with the capability of receiving and decoding the P-code. They are heavier and more expensive than the navigation and surveying receivers. They are capable of giving accuracies of few cm in absolute positioning with precise post-processed satellite orbit information and of few mm in relative positioning. These receivers are useable for applications related to geodetic, geodynamic, detailed GIS and topographic engineering survey, etc. A modern geodetic receiver should be able to measure accurately and reliably anywhere under any condition.

The GNSS receivers can be classified into two basic types: (i) Code phase receivers, and (ii) Carrier phase receivers (Seeber, 2003; Dhunta, 2001).

I. Code phase receivers

These receivers are also called *code correlating receivers* as they access the satellite navigational P- or C/A-code signal for their operation. They have a unique capability to begin calculations without having an approximate location and time. Code phase receivers provide real-time navigation data using almanac data from satellite message for operation and signal processing. These receivers have anywhere-fix capability as they can synchronize themselves with GNSS time at a point with unknown coordinates. For this purpose, we need to lock the signals of four satellites to start the survey with a quicker start-up time.

In code based receivers, phase position of the received code sequence is compared with the phase of an identical code replica, generated by the receiver (using the same algorithm as used for the code from the satellites) *via* a correlation technique. Hence, the observable is also called the *code phase*. These receivers can be used for the rapid calculation of baselines where high accuracy is not required, for example, in exploration or offshore work. The two code sequences are shifted stepwise in phase until a maximum correlation is obtained. These receivers have a complete code dependent correlation channel which produces: code phase, carrier phase, change of carrier phase (Doppler frequency), and satellite message.

II. Carrier phase receivers

These receivers utilize the actual GNSS signals to calculate a position. Two common types of such receivers are; (i) single frequency, and (ii) double frequency, which are compared together, as given in Table 3.4. The single frequency receivers track L1 frequency signal only, and are cheaper than dual frequency receivers. They can be effectively used in relative positioning mode for accurate baselines of less than 50 km or where ionosphere effects can normally be ignored. The double frequency receivers track both L1 and L2 frequency signals, and are expensive than the single frequency receivers. They can effectively be used to measure longer baselines of more than 50 km where ionosphere effects have a larger impact, as ionosphere effects are eliminated by combining L1 and L2 observations.

Table 3.4: A comparison of single and double frequency receivers

S.No.	Single frequency	Double frequency
1	Access to L1 only	Access to L1 and L2
2	Mostly civilian users	Mostly military users
3	Much cheaper	Very expensive
4	Modulated with C/A and P codes	Not possible for civilian users if Y code is present
5	Affected by ionospheric delay	Almost independent of ionospheric delay
6	Used for short base lines	Used for both long and short base lines
7	Most receivers are coded	Most receivers with dual frequency are codeless

Following factors should be kept in mind while selecting a receiver (Seeber, 2003, Spirent, 2011, https://www.sparkfun.com/GPS_Guide):

i. Capability of tracking all the signals from each visible satellite at any time (as GPS+GLONASS system needs 20 dual frequency channels)
ii. Have low phase and code noise
iii. Contains high data rate (> 10 Hz) for kinematic applications with high memory capacity
iv. Low power consumption, low weight and small size
v. Full operational capability under Anti-Spoofing (AS).
vi. Capability to track weak signals (under foliage, and difficult environmental conditions)
vii. Able to do multipath mitigation, interference suppression, stable antenna phase centre
viii. DGNSS and RTK capability

ix. 1 pps timing output
x. Ability to accept external frequencies
xi. Few or no cable connection
xii. Availability of radio modem
xiii. Can operate over difficult meteorological conditions
xiv. Ease of interfacing with other GNSS systems.
xv. Flexible set up (tripod, pole, pillar, vehicle)

3.4.7 Working of a GNSS

The GNSS operation is based on the concept of ranging and trilateration from a group of satellites, which act as precise reference points. Each of the 24 satellites in GPS emits signals to receivers that determine their locations or ranges by computing the difference between the time that a signal is sent and the time it is received. These signals are captured by a receiver to compute the locations of the satellites and to make accurate positioning. The GNSS satellites carry precise atomic clocks that provide highly accurate time. The time information provided by atomic clocks in a GNSS receiver is placed in the codes broadcast by the satellite so that a receiver can continuously determine the time the signal was broadcasted.

Each satellite broadcasts a navigation message that contains (i) the pseudo-random code, called a Course Acquisition (C/A) code, which contains orbital information about the entire satellite constellation (Almanac), (ii) detail of individual satellite's position (Ephemeris) that includes information used to correct the orbital data of satellites caused by small disturbances, (iii) the GNSS system time, derived from an atomic clock installed on the satellite, with clock correction parameters for the correction of satellite time and delays (predicted by a mathematical ionospheric model) caused by the signal travelling through the ionosphere, and (v) A GNSS health message that is used to exclude unhealthy satellites from the position solution. Once the signals are obtained, the receiver starts to match each satellite's C/A code with an identical copy of the code contained in the receiver's database. By shifting its copy of the satellite's code, in a matching process, and by comparing this shift with its internal clock, the receiver can calculate how long it took the signal to travel from the satellite to the receiver. The distance derived from this method is called a *Pseudo-range* because it is not a direct measure of distance, but a measurement based on time. Pseudo-range is subject to several error sources, including atmospheric delays and multipath errors, but also due to the initial differences between the GNSS receiver and satellite time references. Using trilateration process, the GNSS receiver then mathematically determines its position by using the calculated pseudo-ranges and the satellite position information that has been communicated by the satellites.

Trilateration

The GNSS receivers use a technique called *trilateration* which involves computing the distance with the help of known distances (Garg, 2021). In trilateration, three satellites, each with a known position in space, are required. For example, the first satellite broadcasts a signal to a GNSS receiver, and its angle is not known but the distance is known. In 2D space, this distance forms a circle, but in 3D space it forms a sphere (Figure 3.22). If only one satellite is visible, position location is impossible as the receiver location can be anywhere on the surface of a sphere with the satellite at its centre. So, a large error would be present while estimating the location on the Earth with only one satellite.

If two satellites are visible, the receiver location can be anywhere on a circle where the surfaces of the two spheres intercept. So, position location is still impossible as the GNSS position could be anywhere on that circle. But this time, there are two known distances from two satellites. With two signals, the precise position could be any of the two points where these two spheres intersect. The receivers normally select the point out of two, which is closer to the Earth. But, when a third satellite also becomes visible, the GNSS receiver can establish its position as being at one of two points on the previously derived circle where the third satellite sphere intercepts it. So, now position fixing can be done by trilateration, it is almost certain that only one of the two derived points would be near the surface of the Earth. So fixing of location can be done, but only in two dimensions (latitude and longitude). With at least four satellites visible with their good geometry, four spheres will intersect at only one point in space, so receiver position can be accurately fixed in three dimensions (latitude, longitude and altitude). With five satellites visible, it is possible for the system to automatically detect an erroneous signal. With six satellites visible, it is possible for the system to automatically detect an erroneous signal, identify which satellite is responsible and exclude it from consideration. A modern GNSS receiver will typically track all of the available satellites simultaneously, but only a select of them will be used to calculate the receiver's position.

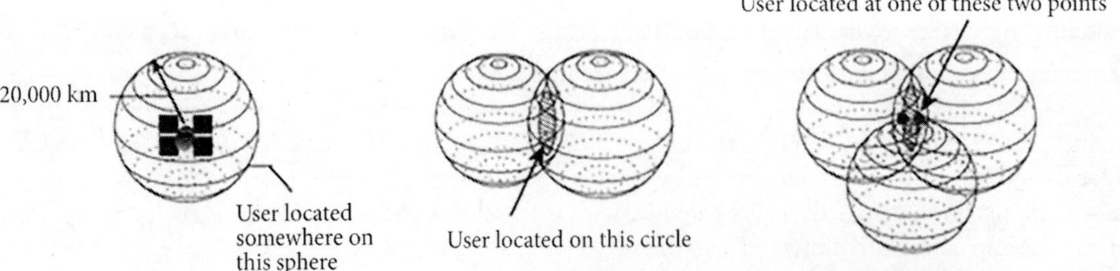

Fig. 3.22: Determination of location from using trilateration method (Langley, 2005)

Altitudes derived from GNSS positions are known as *geodetic altitudes,* and were not initially used for aircraft navigation. Performance-based navigation requires that they, and the navigational information presented by the system, are based on the World Geodetic System (WGS-84) coordinate system, established in 1984.

3.4.8 GNSS surveying techniques

There are various methods available for some specific tasks, which may be kept in mind while collecting the data from GNSS (Garg, 2021). The technique used in a given location, would however depend on the (i) accuracy requirements, (ii) time to complete the work, (iii) local terrain conditions, and (iv) available facility. Advantages and disadvantages of some of the techniques are summarized in Table 3.5. Some of the techniques are given below.

a. Static surveying:

This method is used in surveying that requires reasonable high accuracy, e.g., control surveys from local to state-wide area. It will probably continue to be the preferred method, as the receiver at each point collects data continuously for a defined length of time. The duration of data collection will depend on (i) required precision, (ii) number of visible satellites, (iii) satellite geometry (DOP), (iv) single frequency or dual frequency receivers, and (v) distance between the receivers.

Table 3.5: Advantages and disadvantage of some GPS/GNSS survey methods

Type	Advantages	Disadvantages
Real-time kinematic (RTK)	Real-time corrected positions in a known coordinate plane. Able to navigate to and compute geometries of data points in the field	Significantly increased equipment cost and logistics. Must have radio connection between base and rover. Must set up base on a known position to use advantages
Rapid Static	Reduced equipment expense and complication compared to PPK. Local base stations not necessary	Requires longer occupation times up to 2 hours, with fewer potential measurements. Lower accuracy than RTK
Static	Higher precision than rapid static, less equipment than RTK or PPK Requires long occupation times to reach similar accuracy to RTK or PPK.	Requires more precise mounting and metadata collection than RTK, PPK, and rapid static
Continuous	Highest possible precision and accuracy (mm)	Requires complex infrastructure, precision mounting, and very long occupation times.

Field data are collected from two or more GNSS receivers, and the line between any two receivers is called a *baseline*. The data are collected for a longer duration of time to achieve the higher accuracy of baseline. Figure 3.23 shows the duration of observations for a given areal extent. Multiple baselines can be established simultaneously by using more than two receivers to save time. When the baseline between a known point and a new point is measured, the new point can be used as a known point for other baselines, and this process continues. To strengthen the network of baselines, control points should be situated at commanding locations. The number and location of control points will depend on the size and shape of network. The larger the constellation of satellites, the better the available geometry, the lower the positioning dilution of precision (PDOP), and the shorter the time of observation needed to achieve the required accuracy. To achieve higher accuracy, the post-processing software is used.

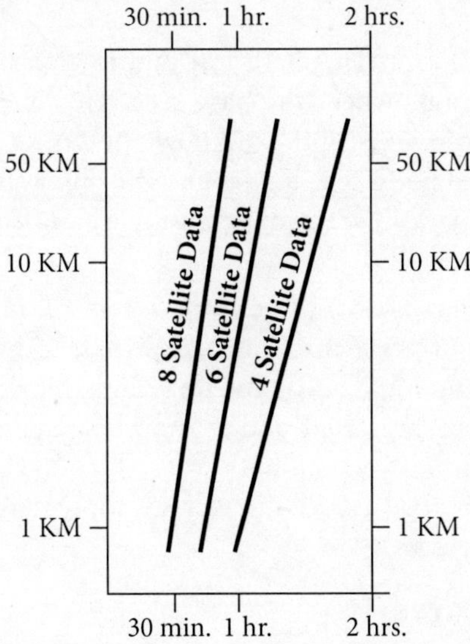

Fig. 3.23: Static sessions lengths (Sickle, 2015)

b. Rapid static surveying:

The rapid static method is a popular method of GNSS surveying where short-time measurements are taken. As the name suggests, it is easy, quick and efficient. The rapid static technique is well suited for short range applications, such as control densification and engineering surveys or surveying many points. It is developed for duel frequency phase measurements or pseudo range measurements. The main basis of this technique is the ability of the software to resolve the ambiguities using a very short observation period.

The required observation time of rapid static observations is mainly a function of baseline length, number of satellites and ionospheric disturbance. Ionospheric disturbance varies with time, day/night, month, year, position on Earth's surface. As ionospheric disturbance is much lower at night, night observation for rapid static technique can often be halved, or the baseline range doubled. Thus at night, it might be advantageous to measure the baselines of 20 to 30 km. Observations with a minimum of 5 satellites above 15° and a good geometric dilution of precision (GDOP < 8) should be used. Table 3.6 provides an approximate guide to baseline length and observation time for mid-latitudes under the ionospheric activity when a dual frequency receiver is used.

Table 3.6: Use of static and rapid static methods (Garg, 2021)

Obs. method	No. of satellites GDOP ≤ 8	Baseline length (km)	Approximate time observation	
			By day	By night
Static	4 or more	15 to 30	1 to 2 hours	1 hour
	4 or more	Over 30	2 to 3 hours	2 hours
Rapid Static	4 or more	Up to 5	5 to 10 mins	5 mins
	4 or more	5 to 10	10 to 20 mins	5 to 10 mins
	5 or more	10 to 15	Over 20 mins	to 20 mins

c. Kinematic surveying

Kinematic surveying uses differential carrier phase tracking to record observations simultaneously. It is used in most surveying applications where the base receiver remains stationary and placed at the known point, while the rover receiver will visit the unknown points for a very short time. The recording interval for static observations could be 10 sec., for rapid static observations 5-10 sec, and for kinematic observations 0.2 sec or more. Kinematic GNSS can use multiple bases and/or multiple rovers in the same survey, if necessary.

Kinematic GNSS surveying is generally suitable for any type of surveying or mapping in areas with no high rise buildings, overhanging trees, dense forest, over-passes or such structures in rover's route. Possible errors in kinematic surveying could be; (i) antenna height may change between points, especially if a prism pole with a sliding mechanism is used, and (ii) improper centering the antenna over the point. This method is undergoing rapid improvement, and OTF-AR (On-The-Fly–Ambiguity Resolution) is making it ideal for surveys, such as road centre line survey, topographic survey, hydrographic survey, airborne applications and many more.

d. Stop and go kinematic surveying

It is known as stop-and-go technique because only the coordinates of the receiver are used when it is stationary ('stop' part) but the receiver continues to function while it is moving ('go' part) from one stationary station up to the next station. The base receiver remains stationary, while the rover receiver will visit the unknown points for a very short time (< 2 min). This is the kinematic technique because the user's receiver continues to track the satellites while in motion. The software sorts out the recorded data for different points, and differentiates the kinematic or 'go' data (not of interest) from the static or `stop' data (of interest).

The initial ambiguity must be resolved by the software before this survey starts. Once the ambiguities are resolved, the user's receiver is moved from point to point, collecting data just for a minute or so. It is vital that the antenna continues to track the satellites while collecting the data. This method is suitable for details surveys as topographic mapping or boundary survey work when many points close together have to be surveyed, and the terrain poses no problems in terms of signal disruption. This may require special antenna mount on vehicles if the survey is to be carried out over a large area. Speed is the main advantage of the kinematic survey. The technique can also be implemented in real-time if a communication link is provided to transmit the data from the reference receiver to rover receiver. This method also has a limitation that the lock on the same satellites must be maintained during the entire survey. If for any reason a cycle slip occurs, the rover must return to any previous point which had been determined without cycle slip.

e. Real-time kinematic (RTK) surveying

The RTK method is preferred for many survey applications as it provides positioning in real-time. There is no post-processing of the carrier phase data required. The RTK method provides positional accuracy nearly as good as static positioning method using carrier phase, but it is much faster. It involves the use of at least one stationary receiver (the reference), and at least one moving receiver (the rover), as well as the data link (Figure 3.24). All the receivers involved observe the same satellites (five satellites minimum), simultaneously. Once set up, the reference receiver will continuously transmit its carrier phase measurements to the roving receiver. The rover keeps updating the coordinates of the locations while moving as long as lock on satellites is maintained. The RTK method is most suitable for stakeout surveys, and can provide 0.3-0.5 m accuracy.

Table 3.7 presents the advantages and disadvantages of various GPS survey methods.

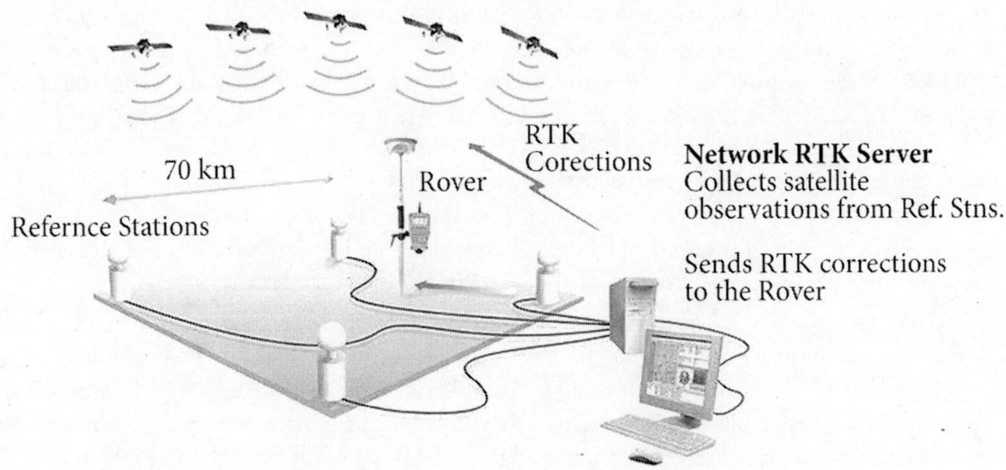

Fig. 3.24: Real-time kinematic (RTK) surveys

Source: https://www.smartnetna.com/hiw03_nrtk.cfm

Table 3.7: Advantages and disadvantages of various methods (Garg, 2021)

Conventional static GNSS		Rapid-static, stop and go, kinematic GNSS	
Advantages	**Disadvantages**	**Advantages**	**Disadvantages**
Highest accuracy Robust technique Ambiguity resolution not critical Minor effect on multi path Undemanding on hardware/software	Long observations session Inappropriate for engineering and cadastral application	Higher accuracy than pseudo range solution Appropriate for many survey applications High productivity Similar procedures to modern terrestrial surveying	Special hardware & software Susceptible to atmospheric and multi path biases Higher capital costs Ambiguity fixed or continuous lock required

f. Pseudo-kinematic surveying

Pseudo-kinematic GNSS surveying is similar to stop-and-go technique, and is a combination of both static and kinematic methods. The method in pseudo kinematic surveys is somewhat similar to the kinematic surveys, except the process of initialization as in stop-and-go method. This feature offers a more favourable positioning technique for the areas having obstructions to signals, such as bridge overpasses, tall buildings, and overhanging vegetation. It is also suitable for lower order control, such as photogrammetric control etc. Loss of lock of satellites that may result due to these obstructions is more tolerable when pseudo-kinematic techniques are employed. Kinematic method has its advantage of speed, but there is no need to lock 4 satellites (Xu, 2010). This method is less precise of all, but highly productive. Each point is occupied for 5-10 minutes for baselines up to 10 km or less. These points are revisited multiple times; may be after 1 hour but not more than 4 hours. Multiple observations of same point with different times will resolve the integer ambiguity (Tusuat and Turgut, 2004).

g. Differential GNSS (DGNSS) surveying

The DGNSS technique requires two identical GNSS units, and is used to improve the accuracy of a standard GNSS. It works by placing a GPS receiver at a known location, called a reference station, and another GNSS unit, known as rover station which is kept at unknown points to determine the coordinates (Figure 3.25). The reference station may be a bench mark whose exact location is already known. Thus, the difference between the GNSS derived position and the true position is determined at the bench mark. The reference station actually calculates the difference between the measured and actual ranges for each of the satellites visible from that station. This calculated difference is called the "*differential correction*" for that group of satellites. At rover station, the correction in the observed value can be applied using the differential correction.

The DGNSS technique is based on pseudo ranges and code phase. Although the accuracy of code phase applications has improved a lot with the removal of Selective Availability (SA) in May 2000, yet reduction of the effect of correlated ephemeris and atmospheric errors by differential corrections requires achieving the accuracy better than 2 to 5 m. The DGNSS based on C/A code SPS signals can however offer meter or even submeter accuracy. Pseudo-range formulations can be developed from either the C/A-code or the more precise P-code. These formulations with the C/A-code can provide a precision of around 3 m with a range measurement. Point positioning accuracy for a differential pseudo-range formulated solution is generally found to be in the range of 0.5-10 m.

Differential corrections may be used in real-time, with post-processing techniques. Usually, pseudo-range corrections are broadcasted from the reference to the rover or rovers for each visible satellite. Using the computed satellite position and its own known location, it computes the range to the satellite. The real-time signal is sent to the receiver over a data link. Some systems require two-way, some one-way communication with the base station. Radio systems, internet, radio signal, or cell phone, geostationary satellites, low-earth-orbiting satellites and cellular phones are some of the options available for two-way data communication.

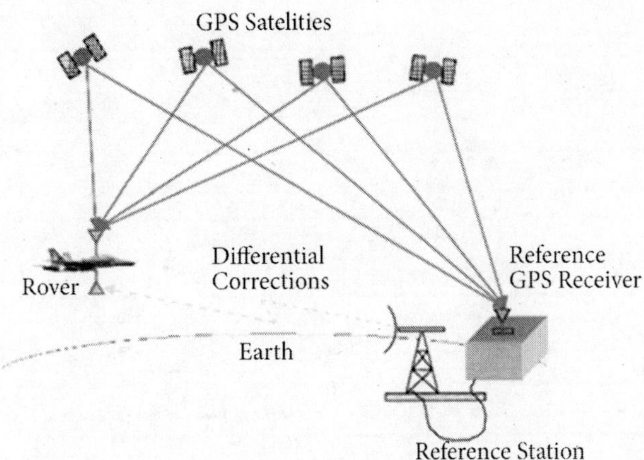

Fig. 3.25: The concept of DGNSS survey

Source: http://emitech-infosystems.com/our-services/gps-dgps

3.4.9 Other satellite based augmentation systems (SBAS)

Most GPS units today are SBAS-enabled. The SBAS are used to augment the GNSS data, and provide higher accuracy, integrity, continuity and availability. Some correction data, like satellite orbit, satellite clock and atmospheric data are broadcasted from communication satellites to get accurate results. It is like a highly advanced real-time DGNSS. There are different types of SBAS, launched by various countries; (i) WAAS, USA, (ii) MSAS, Japan, (iii) EGNOS, Europe, (iv) GAGAN, India, and (v) SDCM, Russia.

1. Wide Area Augmentation System (WAAS) survey:

The WAAS is a combination of ground-based and space-based navigation systems that augments the GNSS Standard Positioning Service (SPS) to improve the accuracy and reliability of location data (Figure 3.26). The WAAS can potentially improve the horizontal accuracy from 5-30 m to 1-5 m. It currently augments the GPS L1 signal providing improved accuracy and integrity. Originally designed for high quality air traffic management in 2003, WAAS signals are broadcasted by geostationary satellites (they remain at the same point over the Earth at all times), and require a clear view of the horizon at higher altitudes. The WAAS is designed to enhance and improve the satellite navigation over the continental United States, and portions of Mexico and Canada, and uses its own geostationary satellites in fixed orbit. There are 25 ground reference stations positioned across the US that monitor the GPS satellite signals. These stations continuously receive and correct the GPS satellite information against their own known precise positions (Garg, 2019).

The WAAS base stations transmit their measurements to a master station where corrections are calculated and then uplinked to two geosynchronous satellites. The WAAS satellite then broadcasts differentially corrected signals at the same frequency as GPS signals. The WAAS signals compensate for position errors measured at WAAS base stations as well as clock error corrections and regional estimates of upper atmosphere errors (Yeazel, 2006). The WAAS-enabled receivers devote one or two channels to WAAS signals, and are able to process the WAAS corrections.

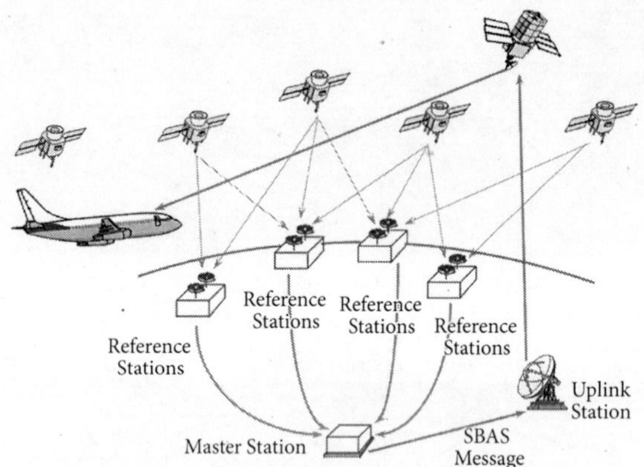

Fig. 3.26: Concept of WAAS GNSS survey

Source: https://ifreebsd.ru.com/product_tag/59755354_.html

2. MSAS Japan:

The MSAS (MTSAT (Multi-functional Satellite) Satellite-based Augmentation System) Japan decided to implement in 1993. Its operation started from September 2007 with the goal of improving its accuracy, integrity, and availability. It augments GPS L1 signals; Two GEOs- MTSAT-1R (PRN129), MTSAT-2 (PRN137), Ground Facility- 2 Master Control Stations (MCSs), 6 Ground Monitoring Stations (GMSs) (Two of them are with the MCSs), 2 Monitoring and Ranging Stations (MRSs), as shown in Figure 3.27. The SBAS signal that is made by Ministry of Land, Infrastructure, Transport and Tourism (MLIT) is now transmitted from the QZS-3 GEO satellite using the QZSS SBAS transmission service since April 2020 (https://gssc.esa.int/navipedia/index.php/MSAS_General_Introduction).

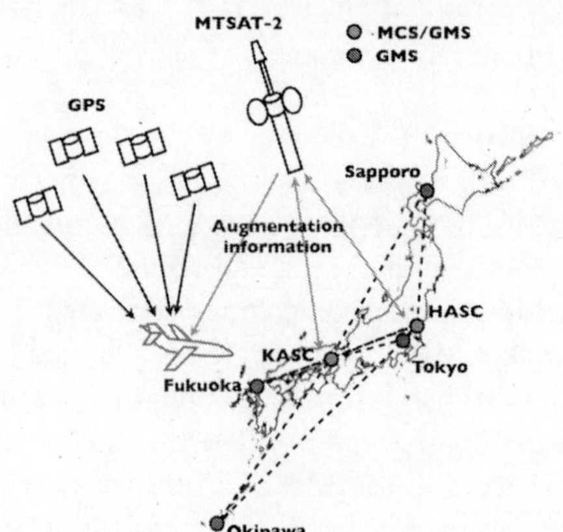

Fig. 3.27: Concept of MSAS survey

Source: https://www.icao.int/APAC/APAC-RSO/GBASSBAS%20 *Implementation%20Workshop/1-6_MSAS%20 System%20Development_Rev2%20(S%20Saito).pdf*

3. EGNOS, Europe:

The European Geostationary Navigation Overlay Service (EGNOS) is Europe's SBAS that is used to improve the performance of GNSSs, such as GPS and Galileo. It will be used to provide the safety of life navigation services to aviation, maritime and land-based users over most of Europe (https://www.euspa.europa.eu/european-space/egnos/what-egnos). The EGNOS augments the GNSS L1 Coarse/Acquisition (C/A) civilian signal by providing corrections and integrity information for space vehicles (ephemeris, clock errors) and most importantly, information to estimate the ionosphere delays affecting the observations (https://egnos-user-support.essp-sas.eu/new_egnos_ops/egnos-system/about-egnos).

The EGNOS uses GNSS measurements taken by accurately located reference stations installed across Europe. All measured GNSS errors are transferred to a central computing centre, where differential corrections and integrity messages are calculated, and results broadcasted over the covered area using geostationary satellites that serve as an augmentation to the original GNSS message. As a result, the EGNOS improves the accuracy and reliability of GNSS positioning, while also providing a crucial integrity message regarding the continuity and availability of a signal. In addition, the EGNOS also transmits an extremely accurate universal time signal. The services will be progressively extended to the European neighbourhood countries.

4. GAGAN, India:

The GAGAN (GPS Aided GEO Augmented Navigation) is a SBAS, jointly developed by ISRO and AAI to provide the best possible navigational services over Indian FIR (Flight Information Region) with the capability of expanding to neighbouring FIRs. It has is a system of geostationary satellites and ground stations that provide signal corrections to give better position accuracy for several services, such as aviation, forest management, railways signalling, scientific research for atmospheric studies, natural resource and land management, Location based services, Mobile, Tourism. The GAGAN corrects for GPS signal errors caused by ionospheric disturbances, timing and satellite orbit errors and also it provides vital information regarding the health of each satellite (https://gagan.aai.aero/gagan/).

The GAGAN consists of set of ground reference stations positioned across India, called Indian Reference Station (INRES), which collect GNSS satellite data. A master station collects the data from reference stations and creates GNSS correction messages. The corrected differential messages are broadcasted on a signal from three geostationary satellites (GSAT-8, GSAT-10 and GSAT-15). The information on this signal is compatible with basic GPS signal structure, which means any SBAS enabled GNSS receiver can read this signal (https://www.asimindia.in/what-is-gagan-waas/). The system is interoperable with other international SBAS systems, such as the WAAS, EGNOS), and MSAS, and provides seamless air navigation across regional boundaries.

5. SDCM, Russia:

The System for Differential Corrections and Monitoring (SDCM) is the SBAS currently being developed by the JSC (Russian Space Systems), as a component of GLONASS (https://gssc.esa.int/navipedia/index.php/SDCM). The main differentiator of SDCM with respect to other SBAS systems is that it is conceived as an SBAS augmentation that would perform integrity monitoring of both GPS and GLONASS satellites, for providing high accuracy solutions, updated and reliable integrity data. The SDCM is expected to provide horizontal accuracy of 1.0-1.5 m and vertical accuracy of 2.0-3.0 m (https://www.unoosa.org/documents/pdf/icg/activities/2008/expert/2-5b.pdf).

The GLONASS and GPS system signals measurement data from measurement data collection stations is transferred to system processing center, where associated correction data and navigational field integrity data are generated, and which are delivered to users via geosynchronous communications satellites or via ground communication channels. User's receiver performs overlapped processing of this data and GLONASS and GPS system signals, which allows solving navigational tasks with improved precision and reliability characteristics.

3.4.10 Accuracy of GNSS observations

The GPS is a pervasive and wonderful technology. It is the first positioning system to offer very high accuracy in most surveying and navigational applications at low-cost and with high efficiency. There are different levels of accuracy that can be achieved using appropriate receiver and technique. There are two main factors that determine the accuracy of a GNSS position (Garg, 2021): (i) error on range measurement (Noise+Systematic), and (ii) geometry of the satellites. The first factor has many components that are controlled by the receiver and the local environment of the system. The second factor is not under the control of the user, except by using a receiver with large channels. Accuracies routinely achieved in measurement of baseline lengths in relative mode, using high precision geodetic instrumentation, with many hours of observations and scientific data processing, are as follows:

i. 0.1-4 mm in local surveys (10 m-100 km baseline lengths)
ii. 4-10 mm in regional surveys (100-1000 km baseline lengths), and
iii. 1-2 cm in global surveys (1000-10000 km baseline lengths)

Most GPS units in standalone mode may have an accuracy of about ±10 m. The DGNSS technique further improves that accuracy better than ±1 m by adding ground-based reference station. For standalone users, the extent use of phase is a major factor in achieving the accuracy. The noise on the phase is typically 1 mm to provide the range 10 cm to 1 m. Therefore, the use of phase data in a way can improve the accuracy. The multipath also significantly reduces the phase measurement. In point positioning mode, accuracy within meter with 1 epoch solution and 24 hours of observations may be possible (depends on Selective Availability). In Differential mode, accuracy in cm (in P Code) and 1-5 m (in CA code) may be obtained, while accuracy of 5 mm ± 1 ppm may be obtained on Differential phase mode. Special techniques (called Kinematics) make use of the phase to achieve higher accuracies. However, they are normally used to range 25 to 50 km baselines. On May 1 2000, the US government made the SA available to all civil and commercial users world-wide (Garg, 2019).

Figure 3.28 presents the accuracy of some available GPS systems. The vertical axis is the expected accuracy or error level, while the horizontal axis is the distance along the Earth's surface between the reference station and the user. If there is no reference station, the line is drawn all the way to 10,000 km, all over the Earth.

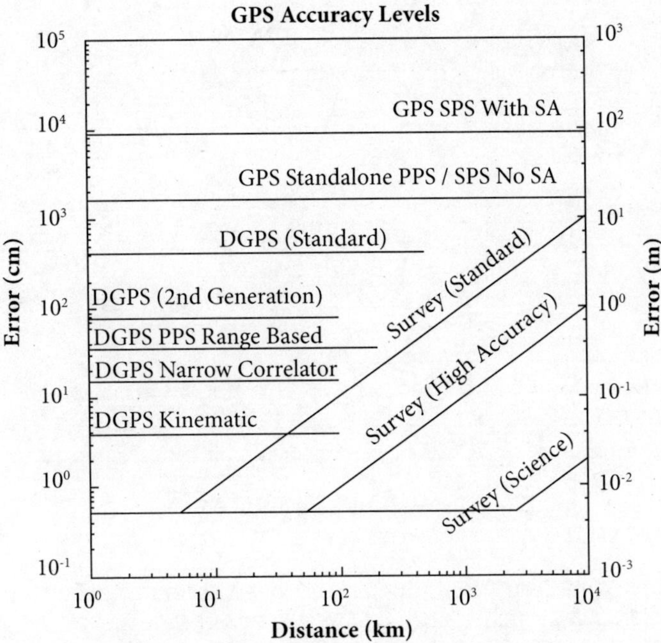

Fig. 3.28: Various accuracy levels using GPS

Source: https://www.oc.nps.edu/oc2902w/gps/gpsacc.html

3.4.11 Errors in GNSS observations

The process of transmitting, receiving and detecting the GNSS signals may contain certain errors. The GNSS positioning accuracy is affected due to (Garg, 2021): (i) satellite signal blockage owing to high-rise buildings, bridges, dense forest trees, etc., (ii) indoor or underground use of GNSS, and (iii) signals reflected-off buildings or walls ("multipath"). Other causes may include: (i) radio interference or jamming, (ii) major solar storms, (iii) satellite maintenance/ manoeuvres creating temporary gaps in coverage, and (iv) improperly designed devices that do not comply with GNSS interface specifications. Due to mapping software, the errors may include, such as, (i) incorrectly drawn maps, (ii) mislabelled businesses and other points of interest, and (iii) missing features, etc. The major sources of errors and their magnitudes are given in Table 3.8. These are shown in Figure 3.29, and explained below-

Table 3.8: The GPS error budget (at 1σ value) (Rathore, 2017)

Error type	Error (m)	Segment
Ephemeris	3.0	Signal-in-space
Clock	3.0	Signal-in-space
Ionosphere	4.0	Atmosphere
Troposphere	0.7	Atmosphere
Multipath	1.4	Receiver
Receiver	0.8	Receiver

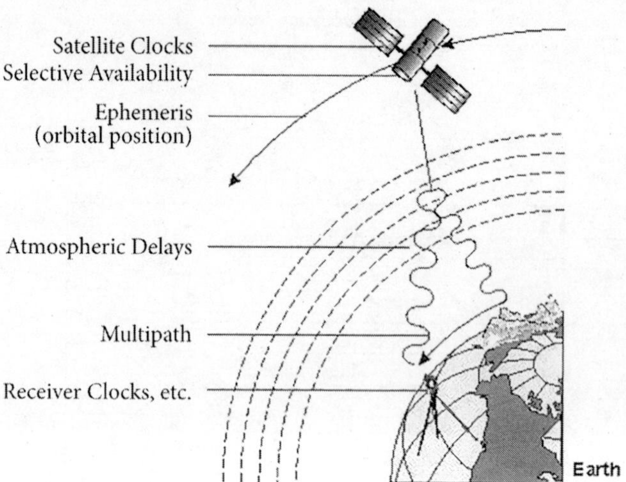

Fig. 3.29: Various sources of errors

Source: https://www.ocean-yachting.com/gps-error-sources

1. *Receiver clock errors*- A receiver's built-in clock is not so accurate as the atomic clocks on-board GPS satellites, therefore, the data may have slight timing errors.
2. *Orbital errors*- These are also known as *ephemeris errors* that are the inaccuracies in the reported locations of satellites.
3. *Number of satellites visible*- In general, lesser the number of satellites tracked, lesser is the accuracy of GNSS observations.
4. *Interferences*- Magnetic terrain, magnetic ores, electronic interferences, electric poles, etc., can cause positional errors in measurements.
5. *Undercover*- Typical GNSS units do not work indoors, underwater or underground.
6. *Multi-path error*- It occurs when the GNSS signals are reflected-off the objects falling in the path, such as tall buildings or dense forest or large mountain, before it reaches the GNSS receiver (Figure 3.30). This is called multi-path error as the signal is reaching the antenna in single line path as well as delayed path. This deviates the signal from its original path and increases the travel time of the signal, thereby causing few meters of errors.

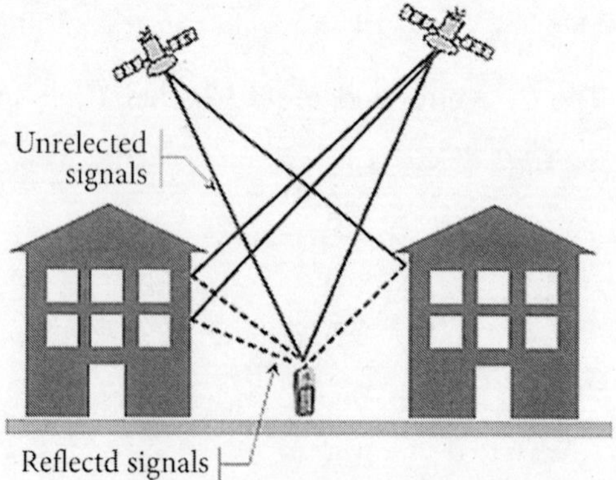

Fig. 3.30: Depiction of multipath error

7. *Ionospheric and tropospheric delays*- The satellite signals slow down as they pass through the atmosphere. The GNSS system uses a built-in model that calculates an average amount of delay to partially correct this type of error. The atmosphere consists of several layers, and temperature varies according to these layers. Although variations do occur, temperature usually declines with increasing altitude in the troposphere which ranges from 0-10 km altitude. Nearly atmospheric water vapour or moisture is found in the troposphere. The ionosphere is a layer of ionized air in the atmosphere extending from almost 80- 600 km above the Earth surface. Due to travel of signals through atmosphere, ionosphere and troposphere delays take place which is one of the reasons for producing the erroneous results. To determine an accurate position from range data, all these propagation effects and time offsets must be accounted for.
8. *Satellite geometry*- This refers to the relative position of the satellites at any given time, which may not be desirable positions, and give errors in observations. In an ideal satellite geometry, the satellites are located at wide angles relative to each other. The dilution of precision (DOP) is an indicator to the geometrical strength of the satellites being tracked at the time of measurement. Satellite geometry can affect the accuracy of GNSS positioning (Garg, 2021). Table 3.9 presents the preferred DOP value to achieve the best results. Another source of error is Geometric Dilution of Precision (GDOP) which is a measure of quality of the satellite configuration, and refers to where the satellites are in relation to one another with respect to receiver's position on the Earth (Figure 3.31). This is the spatial relationship between the GPS receiver and the satellites being tracked. The GDOP refers to three position coordinates plus clock offset in the solution. In general, the fewer the satellites available and the closer they are clustered, the more errors the readings might have. The GNSS chooses satellites that are well above the horizon, minimizing the atmospheric thickness and interference.

Table 3.9: DOP values and their ratings (Langley, 2010)

DOP value	Ratings
1	Ideal
2-4	Excellent
4-6	Good
6-8	Moderate
8-20	Fair
2-50	Poor

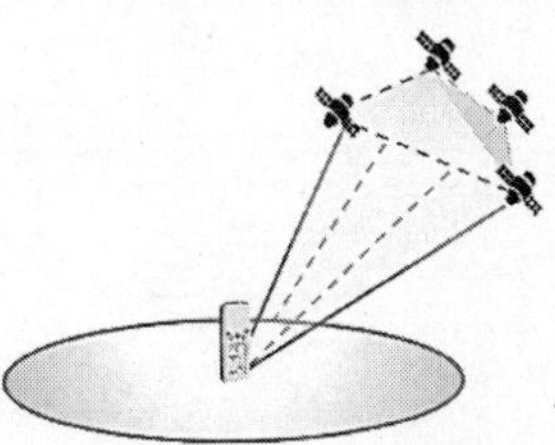

(a) Satellites clustered together Satellite geometry enclosing less volume
BAD DOP (HIGH DOP VALUE)

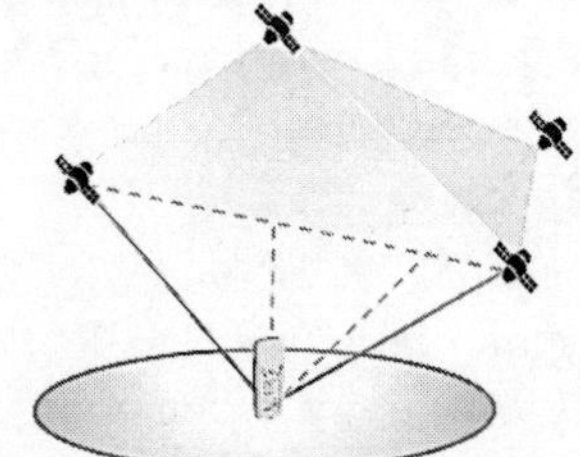

(b) Satellites far apart in the sky Satellite geometry enclosing more volume
GOOD DOP (LOW DOP VALUE)

Fig. 3.31: Geometric dilution of precision (Rathore, 2017)

The GDOP has two components: one is related to the receiver's position (PDOP) and the other is related to the time determination (TDOP). While PDOP is related to the satellite geometry (Figure 3.32), TDOP is strictly dependent on the time biases in the receiver and all the satellites. The PDOP value is further found to have two components: horizontal (HDOP) and vertical (VDOP), again related to PDOP orthogonally. The PDOP values between 2 and 4 is considered as excellent, between 4 to 6 is good, between 6 to 8 is fair, between 8 to 10 is poor, and 10 to 12 is marginal, and values above 12 is not recommended to use. The PDOP mask or filter can be used in receiver as a tolerance setting for acceptability of signal quality.

Overall estimate of accuracy = PDOP * GDOP (3.4)

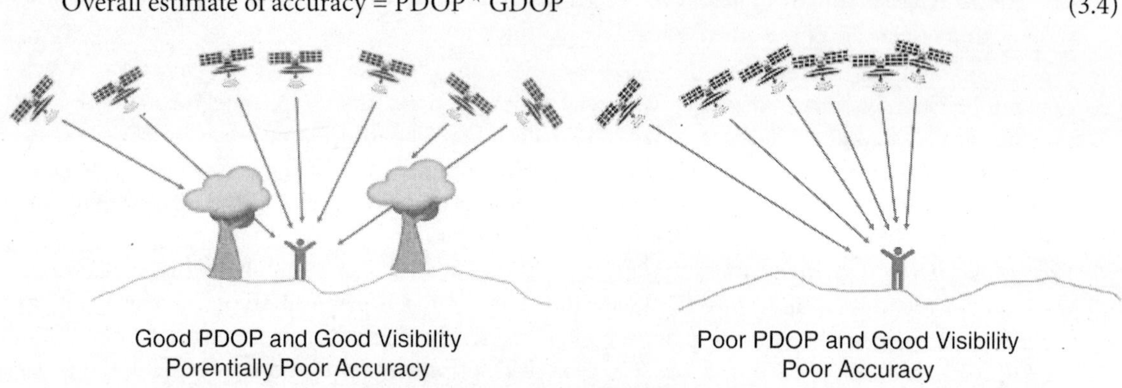

Fig. 3.32: Position dilution of precision

Source: https://www.agsgis.com/What-is-PDOP-And-Why-its-Obsolete_b_43.html)

3.4.12 Applications of GNSS technology

The GNSS has found many applications in land transportation, civil aviation, maritime, surveying and mapping, construction, mining, agriculture, earth sciences, electric power systems, telecommunications, and outdoor recreational activities (Garg, 2021). The GNSS technology has been applied in navigation, military weapons, security, location-based services, monitoring dam, bridges, dikes, landslides, and subsidence, positioning infrastructure, precision farming, forest and resource management, and many more. The GNSS networks provide valuable data for numerical weather prediction models.

The GNSS provides unique capabilities to military, civil and commercial users around the world. The US government has made the GPS signals freely accessible to anyone with a GNSS receiver. The GNSS today has become an efficient tool in the field of science, commerce, surveillance and tracking, with an accuracy of less than 1 meter. Now-a- days, GBSS are used extensively mainly by navy, airforce, army, coast guard, rescue & relief personals, civilian carriers, commercial transport, hikers, and trekkers (El-Rabbany, 2006). World-wide, there are a large number user of GPS/GNSS-enabled gadgets, like mobile phones, wrist watches etc., for a variety of applications (Xu, 2010).

Surveying and mapping: This is the most important applications of GNSS. Its use dramatically increases the productivity of mapping, resulting in collection of more accurate and reliable data in a faster way. The use of GNSS systems has revolutionized the practice of surveying. Instruments and computer software to process the results have been developed specifically for various surveying applications (Garg, 2019). Surveyors use GNSS to save time over standard survey methods. They use absolute locations to make maps and draw the property boundaries. They use GNSS receivers to survey locations that are located remotely and a direct line of sight can't be established due to obstructions, like a mountain or forest or lake. The GNSS helps to calculate the distances and heights of different places on the Earth surface. The GNSS tool can be used to establish the Ground Control Points (GCPs) to prepare the topographic mapping of real world. Today, GNSS is a vital part of surveying and mapping activities around the world.

Navigation: The GNSS is used as an in-vehicle navigation aid, and is also used by hikers and hunters. With the help of GNSS, roads or paths available, traffic congestion and alternative routes, roads or paths that might be taken to get to the destination, can be derived. The location of food store, banks, hotels, fuel station, airports or other places of interests, the shortest route between the two locations, the different options to drive on highway, can easily be obtained using GNSS. The GNSS is used for enroute navigation and for airport or harbour approaches. Some aircrafts have been equipped with two or more GNSS antenna, and the signals from all antennas are sent to one receiver to compute the attitude of the aircraft. The aircraft also has an inertial navigation system (INS- a navigation system to sense changes in a vehicle's attitude and velocity) which along with GNSS can greatly improve the overall accurate and precise navigation.

Robotics: Just like GNSS provides direction to human, the same case applies to robots. The GNSS enhances navigation of mobile robots and makes them applicable to diverse fields such as outdoor industrial work, mapping, and agriculture. Self-navigation, autonomous robots are using the capabilities of GNSS. With the help of GNSS, the cars and trucks can work without a driver.

Road traffic congestion: A navigation device has a GPRS receiver for receiving the real-time information about the average speed on a highway, indicating the traffic congestion. The device can compute the alternate route to avoid the congestion, based on historically speed data on secondary roads weighed by the current average speed in the congestion area. Highway departments survey existing roads and planned routes with the help of GNSS.

Fleet tracking: The use of GNSS technology is helpful to identify, locate and track fleet of vehicles in real-time. Automobile vehicles, such as car, truck, ship or aircraft can have a receiver that constantly keeps a track of position. Some automobile companies connect receivers into radio systems that use satellite communications to allow one master control center to track all the vehicles in the fleet anywhere available. The GNSS tracking systems are used to route and monitor delivery vans and emergency vehicles. Many emergency vehicles have also been fitted with receivers so that the vehicle closest to an emergency could be located without losing time.

The GNSS tracker is one of the important application of GNSS technology, which is a combination of software and GNSS technology. It is a hardware device, like GNSS receiver, which can log the location data, then send the data to the server. The working principle of GNSS tracker is shown in Figure 3.33.

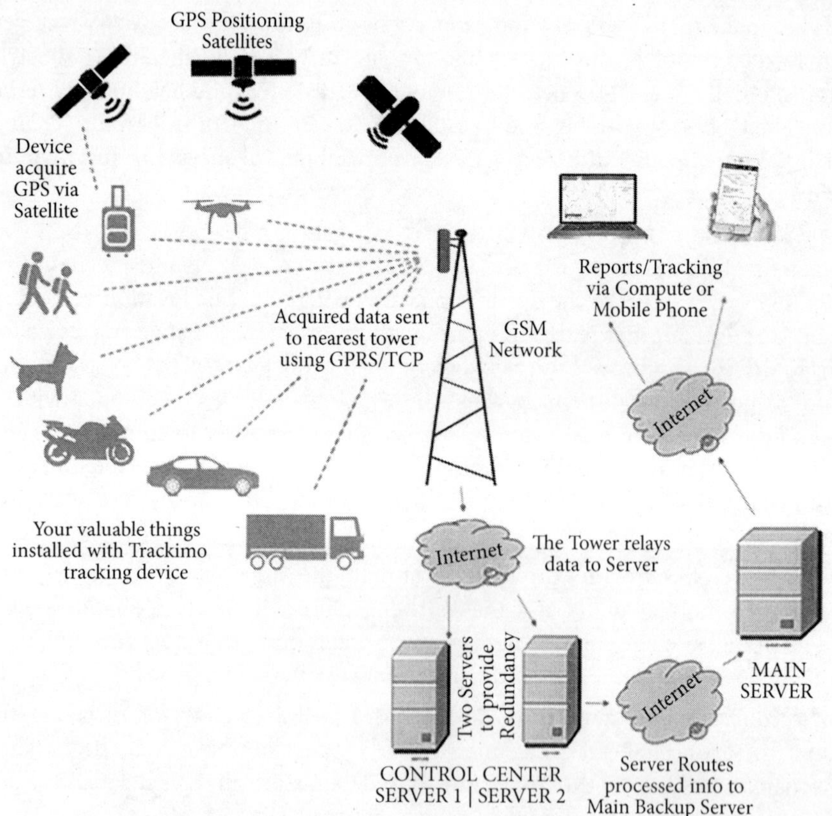

Fig. 3.33: A typical architecture of a GNSS tracker system

Source: https://www.gpstracker-factory.com/gps-tracking-device-system-works/

Tectonics: The GNSS enables direct measurement of fault movement during the earthquake, crustal motion and deformation to estimate the seismic strain built-up for creating seismic hazard maps. At the high-precision end, it is possible to measure the movements of tectonic plates, or establish major political boundaries. Receivers are able to measure the locations with a repeatability of only a few millimeters over baselines that extend several hundred kilometers. When an earthquake occurs, observations can determine which parts of the area actually moved, and in which directions and how far.

Agriculture: The GNSS-based applications in precision farming are being used for farm planning, field-based crop mapping, soil sampling, tractor guidance, crop scouting, and yield mapping. In precision agriculture farming, GNSS is used to accurately guide farm machinery engaged in ploughing, planting, fertilizing, and harvesting. Most smart phones feature GNSS map for navigation applications. The GNSS allows farmers to work during low visibility field conditions, such as rain, dust, fog, and darkness. Farmers have a specific season for planting, weeding, and harvesting, and due to the repeat in the seasons, they put the GNSS receiver on their tractors and other farming equipment which allows them to map their plantations and ensure that they return to precisely the same time when planting or weeding in the next season. This strategy is effective especially in foggy and less visibility. Its high accuracy makes it very suitable for use in mapping soil sample locations, and locating the soils suitable for a particular farming.

The combination of GNSS and GIS has given rise to the site specific farming an approach to precision agriculture. The GNSS-based applications in precision farming are used for: Farm planning, Field mapping, Soil sampling Tractor guidance, Tractor scouting, Yield mapping. Productivity can be increased with the help of precision agriculture, and the geographic information regarding the Plant-Animal-Soil requirements beforehand and then applying the relevant treatment.

Forest: Deforestation and disappearing wildlife habitats are a big problem in the modern world. Manufacturing industries use state-of-the-art GNSS technologies to produce more wood products. The rate with which large forests are being cut down for various uses is alarming than the rate at which they are being replanted, as trees take many years to grow to their full potential. The GNSS technology can make tree planting more efficient.

Defense: Army uses GNSS as a modern device for defensive purpose, like trending and rescue work. The US Department of Defense was the first to develop the GNSS system, and since then the system has been adopted by numerous military forces around the world. Today, there has been a diverse use of the app, and it can be used to map the location of vehicles and other machinery, such as missiles during a war. Military GNSS has been integrated into fighters, bombers, tankers, helicopters, ships, submarines, tanks, jeeps, and soldiers' equipment. In addition to basic navigation activities, military applications of GNSS include target designation, close air support, "smart" weapons, and rendezvous. The GBSS is very important to determine the location of terrorist attacks, and the surgical strike. This is a technique used purposely to protect the soldiers and also manage the resources.

Geo-fencing: A geo-fence is a virtual border that is set up within a GNSS unit, and the operator is informed through text or email whenever a GNSS tracker crosses the chosen region. The GNSS technology has been utilized in geo-fencing which alerts you whenever a person or object enters or exits from a chosen area.

Anti-collision device: It serves as a anti-collision device for railways and airways. It gives a warning to the driver if two GNSS-enable trains are running on the same track. To find and rescue any crashes of vehicles, trains, ship and airplanes, GNSS plays a very important role. The GNSS is used in aviation throughout the world to increase the safety and efficiency of flights. Space-based position and navigation enables determination of 3-D positions for all phases of flight from departure, enroute, and arrival, to airport surface navigation.

Mining: The development of modern positioning systems and techniques, particularly RTK GNSS, has dramatically improved various mining operations. In open-pit mines, for example, the use of RTK GNSS has significantly improved several mining operations, such as drilling, shoveling, vehicle tracking, and surveying. Satellite navigation has proven a significant increasing in productivity and improved on-site safety in mining activities, e.g., mineral and aggregate extraction with especial incidence in iron ore and coal extraction and transport tasks. In open pit mining, GNSS is very useful for tasks, such as machine guidance, grading, dozing, blasting, collision avoidance, selective mining, stock pile operations, construction of accurate benches, ramps and pads, bulk Earthworks, and rehabilitation works.

Disaster relief: The GNSS provides the location of areas where disaster relief measures could be provided. The GNSS serves as a useful tool in disaster management and rescue operations, since it provides real time situations, and the time is a critical component during the disaster. In order to save lives and reduce the loss of property, it is important to know the relevant information timely, as well as the precise location of landmarks, streets, buildings, emergency service resources, and disaster relief sites reduces the effect. The GNSS has proven to be very effective at the time of Tsunami, Katrina and Rita that have created havoc in the past in parts of the world. The rescue team with the integration of GNSS, GIS and remote sensing data planned the rescue operations by correctly locating the site and other relevant information.

Fishing: Maps of the main concentrations of fish areas, fishing ports and beach landing points, markets, processing, freezing and trans-shipment points, coastal landforms can be located with the help of GNSS. The GPS fishfinder units are significant for displaying the location of fishes. It has inbuilt features, such as maps of coastlines, rivers, and lakes as well as the resolution of nautical miles.

Astronomy: Both positional and clock synchronization data is used in astronomy and celestial mechanics calculations. It is also used in amateur astronomy using small telescope to professional's observations, for example, finding extra planets.

Unit Summary

This unit describes the modern surveying equipment used for field data collection. Firstly, EDM has been discussed along with its principle and method of observations. Secondly, Total Station is decribed. The principle used and applications of Total Stations are discuused. Various methods of observations are summarised. Lastly, the components of GPS/GNSS and working method are described. The GPS/GNSS can be used in large number of applications where location is required. These modern equipment not only require less manpower but also save lot of time and funds required to complete the survey work. The other advantage is that the data is avaiabel in digital form which can be analysed and converted into maps using the capabilities of associated software.

Exercises for Practice

a. Short questions:

3.1 How does EDM work?

3.2 What are the different types of EDM? What is the basic difference?

3.3 Explain various components of a Total Station.

3.4 What are different measurements you take with Total Station, and what parameters you compute?

3.5 What do you understand by the terms- Ellipsoid, Geoid and Mean sea level. Draw a diagram to show the three surface.

3.6 What are the advantages and disadvantages of GNSS?

3.7 Define L1, L2 and L5 frequency in GNSS.

3.8 What is Initialization in GNSS Surveying? What do you understand by the terms- Base Station and Differential Correction.

3.9 Define the terms- Ephemeris, Epoch, Dual-frequency Receiver, Pseudorandom Noise or Number (PRN), Selective Availability, Carrier Phase in GNSS, Coarse or Acquisition (C/A) Code.

3.10 What is the basic difference between Static method and Kinetic method of GNSS data collection?

3.11 Define the following acronym as applied to GPS and GNSS
GDOP, PDOP, HDOP and VDOP

3.12 Define the following terms-

- i. Atmosphere delays
- ii. Receiver clock errors
- iii. Multipath errors

3.13 Discuss the term SBAS.

b. Long questions:

3.14 What is the use of reflecting prism in a EDM/Total Station Survey? Show the travel of electromagnetic wave from instrument to prism and back for Time measurements, and Phase measurement.

3.15 Draw a neat sketch and show various components of a Total Station.

3.16 Discuss various steps involved in setting up a Total Station in the field.

3.17 Discuss the modalities of working with (i) normal Total Station, (ii) Reflectorless Total Station, (iii) Laser-based Total Station, and (iv) Smart Station.

3.18 Write various sources of errors in Total Station observations? How do you remove them?

3.19 Discuss the main segments of GNSS.16. Discuss the methodology of DGNSS data collection

3.20 Discuss the sources of error that affect the quality of GNSS observations. How GNSS errors can be corrected?

c. Unsolved numerical question

3.21 A distance is measured along a slope with an EDM which when corrected for meteorological conditions and instrument constants, is 714.652 m. The EDM is 1.750 m above the ground, and the reflector is located 1.922 m above ground. A theodolite is used to measure a vertical angle of $+4°25'15''$ to a target placed 1.646 m above ground. Determine the horizontal length of the line.

(Ans: 712.512 m)

References and Suggested Reading

1. Basak, N.N., (2017), Surveying and Levelling, Tata McGraw Hill.
2. Beshr, A. and Abo Elnaga, I. (2011), Investigating the Accuracy of Digital Levels and Reflectorless Total Stations for Purposes of geodetic Engineering. Alexandria Engineering Journal, 50, 399-405, http://dx.doi.org/10.1016/j.aej.2011.12.004
3. Dhunta, P. S., (2001), Principles of GPS, Deep Prakshan: New Delhi.
4. Duggal, S.K., (2017), Surveying, Vol. I & II, TataMc-Graw Hill.
5. El-Rabbany, Ahmed (2006) Introduction to GPS: The Global Positioning System, 2nd Edition, Artch House, Boston.
6. Garg, P.K., (2019), Principles and Theory of Geoinformatics, Khanna Book Publishers Pvt. Ltd, Delhi
7. Garg, P K., (2021), Digital Surveying Methods, New Age International Pvt Ltd., Delhi
8. Gopi, Satheesh; Sathikumar, R., and Madhu, N. (2017), Advanced Surveying, Pearson India.
9. Langley, R. (2005), The Mathematics of GPS, Political Science, The Mathematics of GPS
10. Langley, Richard B., (2010), Innovation: Better Weather Prediction using GPS, July 1, GPS World.
11. Lackner, Stefan and Lienhart, (2016), Werner Impact of Prism Type and Prism Orientation on the Accuracy of Automated Total Station Measurements, Joint International Symposium on Deformation Monitoring (JISDM, 30 March-1 April, 2016, Institute of Engineering Geodesy and Measurement Systems (5200).
12. Mishra, Gopal, (2014), Total Station in Surveying– Operation, Uses and Advantages, The Constructor, https://theconstructor.org/surveying/total-station-operation-uses-advantage/6605/.
13. Punmia, B.C., Jain, Ashok K., Jain, and Arun K., (2016), Surveying, Vol. I & II, Laxmi Publications.

14. Rajput, Krunal (2020), What is EDM in Surveying, Types, Errors, https://civiljungle.com/edm-in-surveying/.
15. Rathore, Chinmaya S., (2017), Planning a GPS Survey Part 2 – Dilution of Precision Errors, Graticule, http://graticules.blogspot.com/2017/02/planning-gps-survey-part-2-dilution-of.html
16. Reda, Amezene and Bedada, Bekele (2012), Accuracy analysis and Calibration of Total Station based on the Reflectorless Distance Measurement, Master of Science Thesis in Geodesy, School of Architecture and the Built Environment, Royal Institute of Technology (KTH), Stockholm, Sweden, December 2012.
17. Seeber, G, (2003), Satellite Geodesy, II edition, de Gruyter: Berlin.
18. Sickle, Jan Van (2015), GPS for Land Surveyors, CEC Press, London.
19. Sprient (2011), Sprinet e-book: How to choose a GNSS receiver, https://assets.ctfassets.net/wcxs9ap8i19s/27RvX9UZHaWeSguS2XpIhi/d76fedbbcb1b504460bf3e57f2a3203b/How-to-choose-a-GNSS-receiver.pdf
20. Staiger, Rudolf, (2007), Recommended Procedures for Routine Checks of Electro-Optical Distance Meters, FIG No. 9– A Technical Monograph, International Federation of Surveyors (FIG).
21. Subramanian, R., (2012), Surveying and Levelling, Oxford University Press.
22. Tulloch J. E. (2012), EDM Calibration Handbook, Department of Sustainability and Environment, Surveyor-General Victoria, Melbourne Victoria.
23. Tusat, Ekrem and Turgut, Bayram (2004), Investigation Ambiguity Resolution in GPS and the Effects of Ionospheric, Modeling on Base Vector ComponentsTS29 – Positioning and Measurement Technologies and Practices III–Applications and Processing, TS29.7 Investigation Ambiguity Resolution in GPS and the Effects of Ionospheric Modeling on Base Vector Components, FIG Working Week 2004, Athens, Greece, May 22-27, 2004
24. Ekrem TUSAT and Bayram TURGUT, Turkey
25. Xia, Z., Luo Y., Zheng Y. and Pan H. (2006), Error analysis of a reflector in total station angle measurement. Third International Symposium on Precision Mechanical Measurements, doi:10.1117/12.716307, Vol. 6280 62802X-1.
26. Xu, Guochang (2010), GPS: Theory, Algorithms and Applications, Springer
27. Yeazel, J. (2006). WAAS and its relation to enabled hand-held GPS receivers. Retrieved October 12, 2005, from http://gpsinformation.net/exe/waas.html
28. https://www.aga-museum.nl/aga-geodimeter-3-8/
29. https://owaysonline.com/global-positioning-system-on-ships/
30. https://www.directindustry.com/prod/south-surveying-mapping-instrument-co-ltd/product-160571-1650628.html
31. https://www.onlinecivilforum.com/site/setting-total-station/
32. https://www.southalabama.edu/geography/allison/GY301/Total%20Station%20Setup%20and%20Operation.pdf.
33. https://www.nsscanada.com/product-category/leica-geosystems/survey-accessories/prisms-reflectors
34. https://surveyforbeginners.wordpress.com/2019/01/01/total-station/
35. https://lasersurvey.com.au/product-category/surveying-equipment/total-stations/robotic-total-stations/

36. https://www.brainkart.com/article/Sources-of-Error-for-Total-Stations_4653/
37. https://shivkumardas.wordpress.com/agri-tech/an-introduction-to-gps-gis-and-its-uses-in-agriculture/
38. https://www.septentrio.com/en/learn-more/about-GNSS/why-multi-frequency-and-multi-constellation-matters
39. https://www.tokopedia.com/aprilianisurvey/gps-geodetik-hi-target-v90-plus-rtk
40. https://www.gps.gov/systems/gps/control/
41. https://www.smartnetna.com/hiw03_nrtk.cfm
42. http://emitech-infosystems.com/our-services/gps-dgps/
43. https://www.ocean-yachting.com/gps-error-sources
44. https://ifreebsd.ru.com/product_tag/59755354_.html
45. https://gssc.esa.int/navipedia/index.php/MSAS_General_Introduction
46. https://www.icao.int/APAC/APAC-RSO/GBASSBAS%20Implementation%20 Workshop/1-6_MSAS%20System%20Development_Rev2%20(S%20Saito).pdf
47. https://www.euspa.europa.eu/european-space/egnos/what-egnos
48. https://egnos-user-support.essp-sas.eu/new_egnos_ops/egnos-system/about-egnos
49. https://gagan.aai.aero/gagan/
50. https://www.asimindia.in/what-is-gagan-waas/
51. https://www.unoosa.org/documents/pdf/icg/activities/2008/expert/2-5b.pdf
52. https://www.oc.nps.edu/oc2902w/gps/gpsacc.html
53. https://www.sparkfun.com/GPS_Guide
54. https://www.agsgis.com/What-is-PDOP-And-Why-its-Obsolete_b_43.html
55. https://www.gpstracker-factory.com/gps-tracking-device-system-works/

04 PHOTOGRAMMETRY SURVEYING

UNIT SPECIFIES

Through this unit, we have discussed the following aspects:

- Types of photographs and their utility
- Various technical terms used in photogrammetry
- Determination of scale, and relief displacement
- Determination of height of various points through Parallax Bar observations
- Determination of various parameters from tilted photographs
- Digital photogrammetry and its utility
- Requirement of aerial-triangulation
- Exercises for computation of 3-D coordinates using photogrammetric measurements

In this unit, the utility and advantages of aerial photogrammetry for data collection are discussed. The characteristics of each types of photographs are given. The relationships to determine the scale and relief displacement would help in computing various essential parameters and removal of relief displacement errors. Questions of short and long answer types following lower and higher order of Bloom's taxonomy, assignments through a number of numerical problems, a list of references and suggested readings are given in the unit so that the students can go through them for practice.

Rationale

This unit on Photogrammetric surveying would help the students to acquire knowledge about the types of photographs and their utility in data collection and mapping. It explains the characteristics of vertical aerial photographs and tilted photographs. Relationships are derived for the determination of scale of photograph as well coordinates of points. The concept of stereovision and its utility in generating 3-D model as well determination of height of various points are covered to develop a better understanding of subject matter. Some related problems are given which can help further for getting a clear idea of the concern topics on photogrammetry. Some explanation of tilted photograph will further enhance the knowledge about their uses. Digital photographs can be analysed through the use of photogrammetric software, so a brief description of some photogrammetric software is also presented. The photographs are often used in large numbers, therefore the concept of mosaic and aerial triangulation is also given.

Pre-Requisite

Mathematics: geometry and trigonometry

Unit Outcomes

List of outcomes of this unit is as follows:

U4-O1: Describe various types of photographs and their uses

U4-O2: Describe the essential components and characteristics of vertical aerial photographs

U4-O3: Realize the role of parallax in stereo vertical aerial photographs for computation of height of various points

U4-O4: Explain the role of tilted aerial photographs and digital aerial photographs

U4-O5: Apply the parameters to generate 3-D coordinates of points

Unit-4 Outcomes	Expected Mapping with Programme Outcomes *(1- Weak correlation; 2- Medium correlation; 3- Strong correlation)*					
	CO-1	CO-2	CO-3	CO-4	CO-5	CO-6
U4-01	2	1	2	3	1	2
U4-02	2	2	3	1	3	-
U4-03	3	2	2	3	1	2
U4-04	3	3	2.	2	1	-
U4-05	2	2	3	1	-	2

4.1 INTRODUCTION

The name "*photogrammetry*" is derived from the three Greek words *phot* which means light, *gramma* which means letter or something drawn, and *metrein*, the noun of measure. The photogrammetry is defined as *the art, science and technology of obtaining reliable information about physical objects on the earth's surface and its environment, through processes of recording, measuring, and interpreting photographic images.* It derives information from the photographs, and transforms this data into meaningful results and thematic maps.

Photogrammetry discipline, like other disciplines, is constantly changing and developing. Presently, it is dependent on developments in data products, computer software and semi-auto or automatic processing methods. The computers and software are greatly impacting the developments in photogrammetry, shifting from analog to analytical and digital photogrammetric methods. Images used in photogrammetry are captured from a special (metric) camera, or from digital sensors, mounted usually on aerial platforms or drones. These images can also be recorded from a device mounted on a tripod kept on the ground, called *terrestrial images.* Photogrammetry requires certain skills, techniques and judgments to be made for taking reliable measurements on aerial photographs (both hardcopy and digital images).

The acquisition of photographs consists of project planning, image acquisition, image processing, and derivation of results. The end product from the photogrammetric process can be the coordinates of individual points, a graphical representation of the ground surface (topographic map, thematic map, etc.), or a rectified image with map-like characteristics (ortho-photos) or a digital elevation model (DEM) or digital maps. The most common use of aerial photographs is the creation of base maps for various applications, selection of ground control points (GCPs), and ground truth verification. Large-scale mapping using photogrammetry provides a cost-effective and faster approach for applications, such as urban planning, forestry, agriculture, natural resources planning, terrain analysis, road alignment, infrastructural development, etc. Aerial photographs have also been successfully used to create 3-D model of various ground features for the purpose of planning, visualisation and measurement. The 3-D mapping requires viewing the area from two different camera angles, thereby recreating the similar conditions for 3D model in software environment, as it existed at the time of photography.

Modern photogrammetry also includes airborne and terrestrial based LiDAR (light detection and ranging) data. The LiDAR data acquired from both approaches are being used for large number of applications. With good lighting and atmospheric conditions, good quality and better contrast photographs can be acquired for various mapping purposes, but if vegetation or high-rise structures are present in the area, LiDAR data may be an excellent source of input data for detailed and accurate interpretation below the vegetation cover. More recently, aerial photographs have also been acquired from a drone for creation of large scale maps.

4.2 HISTORICAL DEVELOPMENTS

The conception of using photographs for purposes of measurement appears to have been originated with the experiments of Aime Laussedat of the Corps of the French Army, who in 1851 produced the first measuring camera image for its application to topography. The first known aerial photograph was taken in 1858 by a French photographer and balloonist, Gaspar Felix Tournachon, known as *Nadar*. It was a view of the French village of Petit-Becetre taken from a tethered hot-air balloon, 80 m above the ground (Figure 4.1). In 1858, Laussedat, Meydenbauer in Germany carried out the first experiment in making the critical measurements of architectural details on the basis of two photographs of the building.

Fig. 4.1: (left) Nadar "elevating photography to the condition of art", caricature by Honoré Daunier. Published in Le Boulevard 25th May, 1862. (center) Nadar's earliest surviving aerial image, taken above Paris in 1866. (right) Boston from a tethered balloon, 13th October, 1860, by James Wallace Black.

Later, besides hot air balloons, kites, pigeons and rockets have been used which carried cameras into space for taking air-photographs. In 1882, the English meteorologist, E. D. Archibald was among the first to take photographs from kites, and then in 1889, Arthur Batut took aerial photographs from a kite, in Labruguiere, France. The first aerial photograph from a rocket mounted camera was taken by the Swedish inventor, Alfred Nobel in 1897. He is best known now-a-days for the Nobel prize. In 1903, Julius Neubranner designed a tiny breast-mounted camera for carrier pigeons.

The first successful aerial photography from an airplane piloted by Wilbur Wright was taken in 1909 by L.P. Bonvillain who took motion pictures of the military field at Centocelli, near Rome. During World War I (1914-1918). The battle maps used by both sides were created from these aerial photographs. After the war, the aerial photographs were used for civilian purposes. Sherman Fairchild took a series of overlapping photographs and made an aerial map of Manhattan Island. Along with his successful development of aerial camera, Fairchild also designed and built airplanes with high-wings and stable platform to take aerial photographs. His cameras were carried on Apollo-15, 16 and 17 to map the moon surface.

Since year 1950s, the development of the photogrammetry went through four major phases, as briefly explained below as well as shown in Figure 4.2.

a. **Graphical photogrammetry**: This was the first phase which gave rise to new inventions and applications, followed by many experiments and investigations.

b. **Analogue photogrammetry**: This was the second phase of the era of analogue photogrammetry (1900 to 1960), which led to the invention of analogue rectification and stereo-plotting instruments. During this phase, airplanes and cameras were extensively used in the World War II (1939-1945), and established aerial survey techniques for military operations. It further gave rise to new developments, such as application of radio control to photo flight navigation, and new wide-angle lenses & devices to acquire true vertical photographs. Photogrammetry thus became a popular surveying and mapping method, and several optical or mechanical and electronic components for modeling and processing as well as to reconstruct the 3D geometry from stereo-photographs. Major output during this phase was topographic maps.

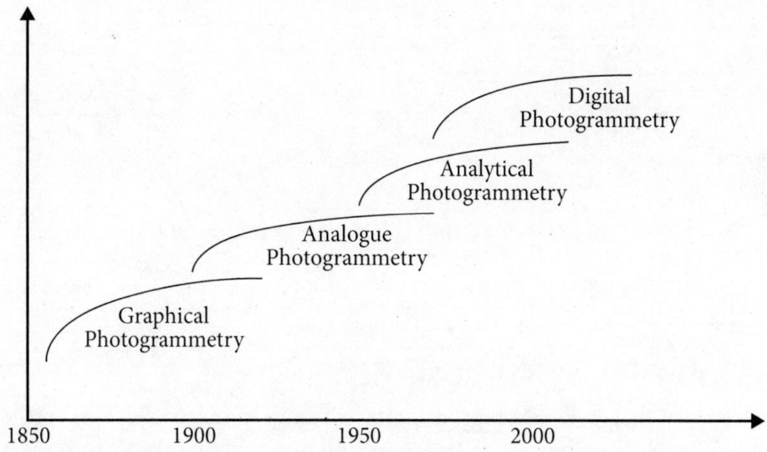

Fig. 4.2: Major phases of development of photogrammetry

c. **Analytical photogrammetry:** In 1950s, third phase began with the use of computers; and thus analytical photogrammetry was developed for accurate photogrammetric measurements. Aerial triangulation was improved greatly, and the analytical plotters developed in 1970s. Computer replaced the expensive optical or mechanical instruments. The resulting devices were digital/ analogue hybrids which used mathematical modeling assisted with digital processing. In this mode, the image coordinates were obtained using mono-comparators or stereo-comparators, and further processing was done through computer. Images in hard copy form (e.g., on film) were used in both the above methods, but the output was in digital form.

d. **Digital photogrammetry:** The fourth development phase is the digital photogrammetry phase. Digital photogrammetry, also called *softcopy photogrammetry*, requires digital images that are processed through a computer. With the availability of images with high resolution digital scanners and digital cameras, digital images are processed using photogrammetric software employing the same analytical principles for modeling. Hence, digital methods use strengths of both analytical and image processing methods for analysis and stereo-plotting works (Grave, 1996). The output from this is in digital form: Digital Elevation Model (DEM), digital maps, digital orthophotos, etc. Digital photography is used for a diverse set of commercial, industrial, agricultural, governmental and private applications. The main advantage to work in digital environment is huge saving in time and cost to derive the useful results, and using the digital output as input into other analysis systems, such as GIS (Hassani and Carswell, 1992).

Table 4.1 shows a comparison of photogrammetric techniques used.

Table 4.1: A comparison of various photogrammetric techniques

Characteristics	Analog	Analytic	Digital
Image	Film	Film	Pixels
Plotter	Analog	Analytical	Computer
Model Construc.	Mechanical	mechanic/computer	Computer
Stereo Viewing	Optical	Optical	Varies
Output	Mech./CAD	Mech./CAD	CAD
Aerotriangulation	Very limited	On/Off Line	Semi-automatic
Orthophoto	Very limited	Unavailable	Automatic
Limitations	Focal length Film Format	Film format	None
Accuracy	Average up to $\pm 15\ \mu m$	Very high up to $\pm 3\ \mu m$	Same as scanning accuracy
Cost	Very high	Very high	Reasonable to high

4.3 TYPES OF AERIAL PHOTOGRAPHS

There are two main types of images in photogrammetry. Photographs are taken with either **aerial** camera mounted in an aircraft or with terrestrial camera mounted on a tripod on the ground. Aerial camera in aircraft is usually pointing vertically down towards the ground, while in terrestrial photogrammetry it is pointing near horizontal.

a. Aerial photographs

Aerial photographs which are normally used for mapping and photo-interpretation can be classified into two main categories *viz.*, vertical and tilted photographs.

i. Vertical aerial photographs

An aerial photograph taken with the camera axis held in a vertical or nearly vertical position is classified as vertical photograph (Figure 4.3a). A tilt of camera axis up to ±3° from vertical line is acceptable. When the geometry of a vertical photograph is considered, the photograph is assumed to be taken with the optical axis truly vertical.

ii. Tilted photographs

Sometimes due to unavoidable conditions (e.g., strong winds), the camera axis is unintentionally tilted more than 3° from the vertical. The resulting photograph is called tilted photograph (Wolf, 1980). The tilted photographs may further be classified in two categories *viz.*, low oblique and high oblique photographs.

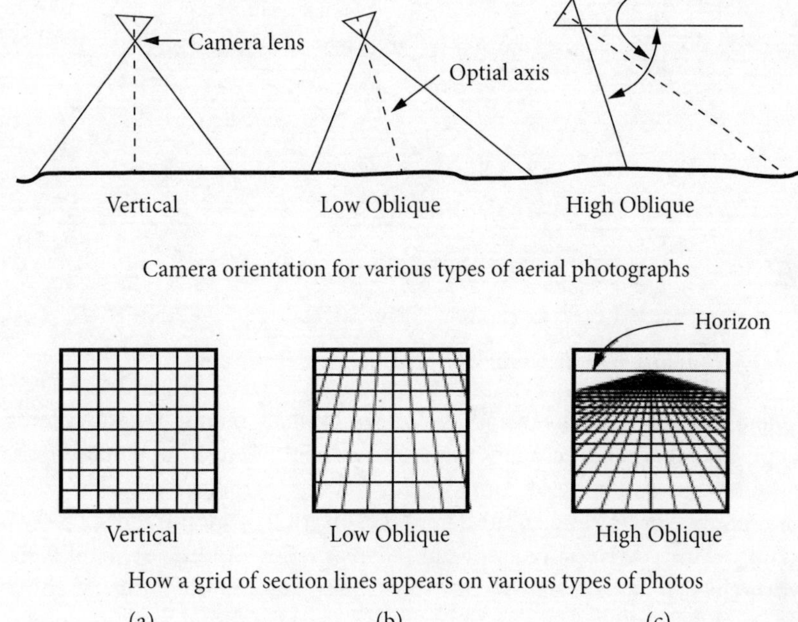

Fig. 4.3: Vertical, low oblique and high oblique photographs *(ASPRS, 1980)*

***Low oblique photographs*:** An aerial photograph taken with an intentional tilt of camera axis 15° to 30° from the vertical axis is called as low oblique photograph (Figure 4.3b). This kind of photographs is often used in reconnaissance surveys of the area as measurements can't be made directly on such photographs.

***High oblique photographs*:** A photograph in which the apparent horizon appears is termed as high oblique photograph. Apparent horizon is the line in which the Earth appears to meet the sky as visible from the aerial exposure station. The high oblique photographs are obtained when the camera axis is intentionally inclined about 60° from the vertical axis (Figure 4.3c). Such photographs are useful for mapping the international boundary and extracting the details of the territory on other side as well as military applications.

b. Terrestrial or Close-range photographs

Terrestrial or close-range photographs are taken when the camera, either hand-held or tripod mounted, is located on or near the ground. For a detailed mapping of the feature/object, the camera is set up close to the object and photographs are taken. The output may be photographs or non-topographic products, like 3D models, measurements, or point clouds. Figure 4.4 shows the close range or terrestrial photographs of a 3D building taken from three locations of camera, to determine the precise coordinates of corners or to recreate a 3D model.

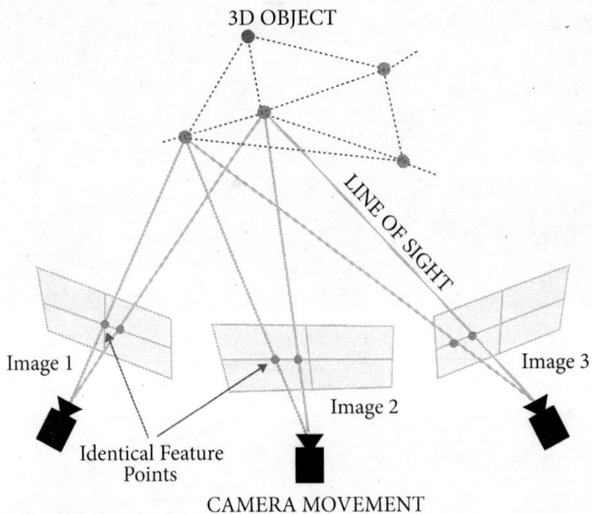

Fig. 4.4: Terrestrial or close-range photographs

Close-range photogrammetry has been applied for precise deformation measurements in industrial and engineering applications. There are many uses of **terrestrial** or close-range photogrammetry, such as 3D modeling of objects, monitoring of dam structures, buildings, structures, towers, or movie sets, architectural restoration, preserving the cultural heritage, location of monuments, medical imaging for forensic sciences and reconstructive surgery, facial reconstruction studies, structural stability studies of bridges and hydro-electric dams, earth-works, stock-piles, automobile industry, shipping industry, antenna calibration, study of traffic accidents and crime scenes by police departments, etc.

4.4 APPLICATIONS OF PHOTOGRAMMETRY

Photogrammetry has been used in several areas. Some applications are given below.

Mapping: The biggest and largest application of photogrammetry is in the field of mapping various features and preparation of various thematic maps, including topographic maps. The 3D maps can easily be created with stereo-photographs. Photographs have also been used as ground reference data for the analysis of remote sensing images.

Geology: They are used for mapping structural geology, faults, folds, lineaments, analysis of thermal patterns on Earth's surface, geomorphological investigations.

Forestry: Forest cover mapping can be carried out using the aerial photographs. Timber inventories, biomass estimation and forest types mapping have also been undertaken by these photographs.

Agriculture: Mapping soil type, soil conservation, crop planting, crop types, crop disease, crop-acreage estimation have become easy with the aerial photographs. Landuse mapping has been the most popular applications of photogrammetry so far.

Design and construction: Site planning and route alignment studies can be undertaken using photogrammetry. Photographs have been used in design and construction of dams, bridges, transmission lines, railway lines, roads, etc. They are very much useful in planning the growth of cities, new highway locations, detailed design of construction, planning of civic amenities, etc.

Cadastre: Aerial photographs have been successfully used for the determination of land boundaries for assessment of area and associated taxes. Large scale cadastral maps are prepared for re-appropriation of land.

Exploration: They are used for various exploratory jobs, such as oil or mineral exploration.

Military intelligence: The photographs are being used for reconnaissance survey, study of terrain conditions and topography, deployment of forces, planning manoeuvres, planning of operation, etc.

Medicine and surgery: Photogrammetry is used in stereoscopic measurements of human body, x-ray photogrammetry in location of foreign material in body and location and examinations of fractures and grooves, bio-stereometrics, etc.

Miscellaneous: Crime detection, traffic studies, oceanography, meteorological observations, architectural and archaeological surveys, planning new infrastructure, etc.

4.5 ADVANTAGES AND DISADVANTAGES OF PHOTOGRAMMETRY

The advantages of photogrammetry over the traditional ground surveying are numerous. The photographs provide an economical and efficient way of mapping a large area, no site access issues, provide a permanent record of features at that instant of time; this is especially useful in rapidly changing sites, such as mines, quarries and landfills. However, there are some disadvantages also (Garg, 2019). A summary of the advantages and disadvantages is given below:

Advantages:

1. It provides a permanent pictorial record of the area that existed at the time of aerial photography.
2. It covers a large area; hence mapping is economical than the traditional survey methods.
3. It is cost-effective, providing high level of accuracy.
4. It provides a bird's eye view of the area, and thus helps in easy identification of both topographic and cultural features.
5. It is particularly suitable for areas that are unsafe, hazardous, and difficult access. Photogrammetry is an ideal surveying method for toxic areas where the safety of ground surveying staff is important.
6. It can effectively be used in the office for detail mapping of several features, thus minimizes the field work.
7. The sequential photographs of the same area may be used for monitoring the area.
8. If area has to be updated after some time, the latest photographs can be used to update new or missing information, and therefore there is no need to map the entire area again.
9. The use of digital photographs ensures total flexibility of scale of mapping.
10. The coordinates of every point in the mapped area can be determined with no extra cost.
11. It requires less manual effort for mapping the area.

Disadvantages:

1. Processing and analysis of aerial photographs require experienced technical manpower.
2. Photographic coverage requires advance flight planning as well as specialised equipment and aircraft.

3. Atmospheric conditions (winds, clouds, haze etc.) may affect the flight plan as well as quality of aerial photographs.
4. Seasonal conditions, i.e., snow cover will affect the photographs and obstruct the features.
5. The ground information hidden by high-rise buildings or dense tree canopies, and roads hidden by trees on both sides, cannot be mapped accurately.
6. Accuracy of contours and cross sections will depend on the accuracy of 3D models generation from stereo-photographs which is a factor of scale of photographs and number of ground control points (GCPs) used for creating the 3D model.
7. Aerial photography is expensive for developing countries, like India, particularly if repetitive coverage of an area is required for large scale mapping and monitoring purposes.
8. For sensitive areas or restricted areas, aerial photography may not be allowed.

4.6 COMPARISON OF AERIAL PHOTOGRAPH WITH MAP

The basic geometry of a map and an aerial photograph is different, as listed in Table 4.2. Aerial photographs cannot be directly used as maps, as vertical aerial photographs do not have a uniform scale throughout and their projection system is also different. In addition, relief displacement and distortions are present on aerial photographs. Figure 4.5 shows the B&W aerial photographs and corresponding topographic map of the area. The photograph needs to be transformed from perspective projection to the orthometric view before it can be used as map. Such transformation will yield ortho-photo which can replace a map. Details of transformation of aerial photographs into ortho-photos are given in Moffit and Mikhail, (1980).

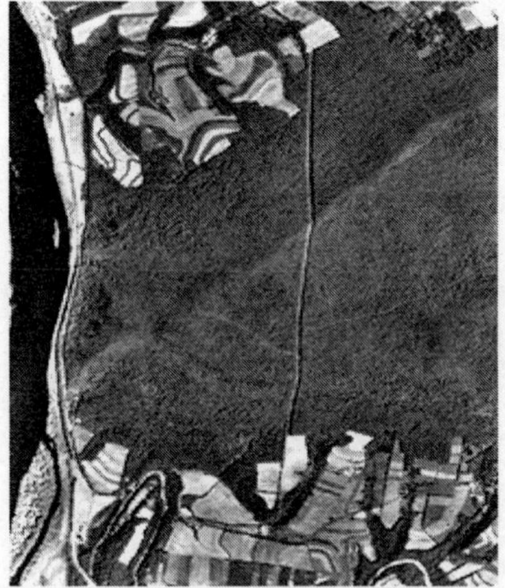
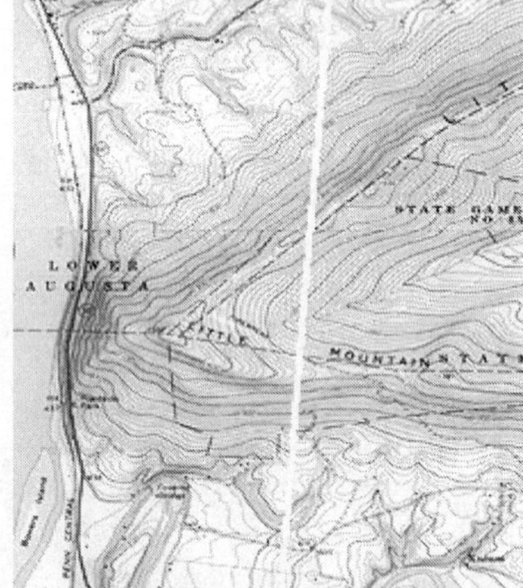

Fig. 4.5: Aerial photograph and corresponding topographic map

Table 4.2: Difference between an aerial photograph and a map

S.No.	Aerial photograph	Map
1	It is based on a central or perspective projection.	It is based on an orthogonal projection.
2	It is a real image of the ground.	It is an abstract representation of the ground. Legend is needed to read it.

S.No.	Aerial photograph	Map
3	The objects and features are represented in their true shapes and sizes.	The features are generally represented symbolically with different symbols and not with their actual sizes.
4	The scale of the photograph is not uniform throughout.	The scale of a map is uniform throughout.
5	An aerial photograph has geometric distortion that increases towards the edges of the photograph.	A map is a geometrically correct representation of the Earth surface projected.
6	Relief displacement error is present	No such error is present
7	In absence of a legend, it is comparatively difficult to read a photograph and identify the objects.	Objects/features are shown in Legend with different symbols and colours, which make it easy to read and identify them.
8	Aerial photography can be carried out for inaccessible and hazardous areas.	The mapping of inaccessible and hazardous areas is very difficult.

4.7 FLIGHT PLANNING

Acquisition of aerial photographs requires details of flight planning. Now-a-days, many digital cameras are used to capture aerial digital images. The photographic format of each photo is 23 cm x 23 cm. Acquisition of photographs requires thorough planning prior to flying to capture photographs of the project site. The project area is marked on the topographic map to study the elevation range as well as the cultural and natural features present in the area. In India, aerial photos are taken in the air by the authorised agency of Govt. of India, *viz.* Survey of India (SoI), Dehradun, and National Remote Sensing Centre (NRSC) Hyderabad. Users make a request to these agencies to obtain aerial photographs either from the archive or flying afresh through the area, on payment basis, subject to clearance from the Ministry.

It is required to know the extent of area, and decide on the scale of photographs. Also the focal length of the camera lens, size and shape of the area to be photographed, amount of end lap and side lap, relief displacement, tilt angle of photographs, drift angle and ground speed of aircraft are determined. Knowing the scale and focal length and average elevation (above *msl*) of the terrain, flying height of the aircraft can be determined. The scale and flying height are inter-related to each other (Moffit and Mikhail, 1980); higher the flying height, smaller the scale and larger the area covered. The number of flight lines, their locations, spacing between them (air-base), and their orientation are computed in advance which would depend on the characteristics of the area to be photographed. Other specifications to be known include area covered by one photograph, number of photographs taken in each flight line, total number of photographs, weather and meteorological conditions etc. Clouds in the photographs are undesirable, therefore, a clear sunny day with least wind is considered to be ideal for taking air-photos.

The vertical photographs are usually taken by a series of parallel flight lines (Figure 4.6); these lines are normally taken along the longer dimension of the area (l_1). After finishing capturing of photographs along first flight line, the aircraft takes a semi-circular turn (outside the project area) to enter into second flight line of area and continues in the similar manner till the entire area is covered, as shown in Figure. The flight line directions are generally planned in East-West or North-South direction. The land limitations, such as mountain range, lake, and sea-side may also affect planning of the flight line directions, and under such situations, flight lines are planed parallel to these details.

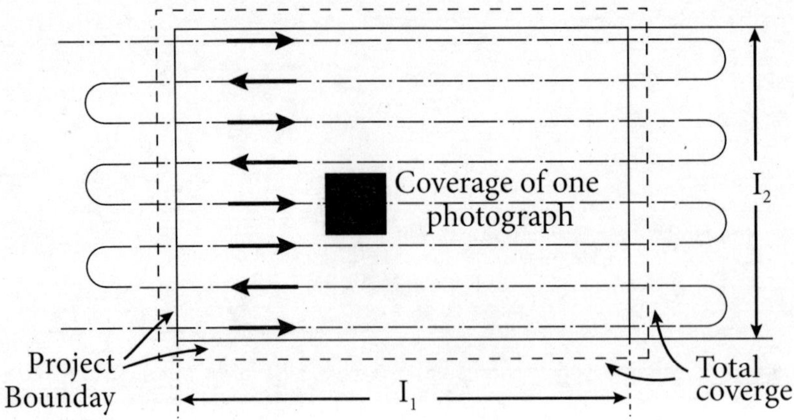

Fig. 4.6: Route in flight planning

The aircraft speed is controlled in such a way that the required forward or end overlap is maintained for all the photographs. Overlapping photographs of the ground are taken as the aircraft moves along a defined flight path (Figure 4.7). Flying at a slower speed will allow for a better turning radius. Best speed is one where lift and drag are equal, making it most efficient. The airbase or the distance between two successive aircraft locations that satisfies the amount of required end overlap is computed (Moffit and Mekhail, 1980). Once the airbase is determined and the aircraft speed is decided upon, the time between two successive exposures can be computed.

Strong winds can affect the direction of aircraft. ***Drift*** is the lateral shift of the aircraft from the flight lines that may be caused by the blowing wind or the pilot error. So, a good practice is to capture the photographs on a sunny day with no wind. Taking aerial photography in urban territories is better at midday, when shadows are shortest. ***Crab*** is the angle formed between flight line and the edges of the photographs in the direction of flight. It occurs when the aircraft is not oriented with the flight line. It reduces the effective width of photographic coverage (Kraus, 1994).

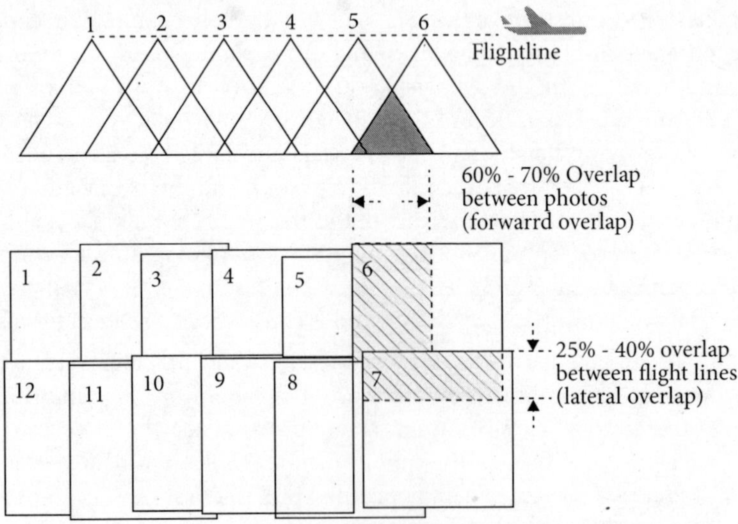

Fig. 4.7: Photography during a flight planning (Source: ASPRS, 1980)

The following are the considerations in flight planning:

1. Mark the flight plan on a map, and compute the length (L) and width (W) of the area.
2. Plan and establish the ground control points.
3. Compute the scale of photograph using the relationship:

   $S_{avg} = f / (H - h_{avg})$ (4.1)
4. Compute the coverage on the ground by one photo (*A*).
5. Compute the flight lines spacing:

   (D) = Photo coverage (*A*) * (100 – amount of side lap) /100 (4.2)
6. Calculate the number of flight lines:

   $N_1 = (W / D) + 1$ (4.3)
7. Always round-off the number of flight lines, i.e., 8.4 becomes 9, as we don't want to lose any small are uncovered.
8. Select the first flight line along North-South or East-West boundary of the area.
9. Compute the distance between two consecutive photos (or airbase B):

   B = Image coverage (*A*) * ((100 – amount of end lap) /100) (4.4)
10. The camera exposes the next photo when aircraft has moved a distance equal to airbase. If the aircraft speed is (*v*), the time (*t*) taken between two consecutive photos:

    t = Air-base (B) / Aircraft speed (*v*) (4.5)
11. Number of images per flight line (N_2) = $(L / B) + 1$.
12. Always round up the number of photos, i.e., 25.3 becomes 26.
13. Add two images at the beginning of flight line and two at the end of flight (to ensure continuous stereo coverage), i.e., additional 4 images for each flight line are taken. So, number of images per flight line:

    $= (L / B) + 1 + 4$ (4.6)
14. Total number of images for the project = $N_1 \times N_2$.
15. Estimate the cost of the project.

4.8 TECHNICAL TERMS IN AERIAL PHOTOGRAMMETRY

Exposure station: Location of an aircraft in the air at the time of taking a photograph is called exposure station (e.g., 1,2, 3, etc., in Figure 4.7).

Air-base or Camera base: The distance between two successive exposure stations along a flight line is called the air-base or camera base (e.g., 1-2, 2-3 etc., in Figure 4.7).

Perspective centre: The point of origin or termination of bundles of perspective light rays is called the perspective centre (Figure 4.8).

Perspective (Central) projection: All the projecting rays by a camera lens pass through the perspective centre to form the perspective projection (Figure 4.8). An aerial photograph is based on a perspective (central) projection. Due to relief variations of the ground objects, an aerial photograph differs geometrically from the map of corresponding area.

Flight line: The flying path an aircraft takes when taking the photographs is called flight line. It represents x-axis for photographic measurements (e.g., line 1-6 in Figure 4.7), while y-axis is perpendicular to it passing through the principal point of the photograph.

Flight strip: Each flight line during photography of an area will cover some area on the ground in the form of a corridor, called a flight strip. Figure 4.7 shows two flight strips.

Strip of photographs: The number of photographs covered during a flight strip is called strip of photographs (e.g., 6 photographs in one strip in Figure 4.7).

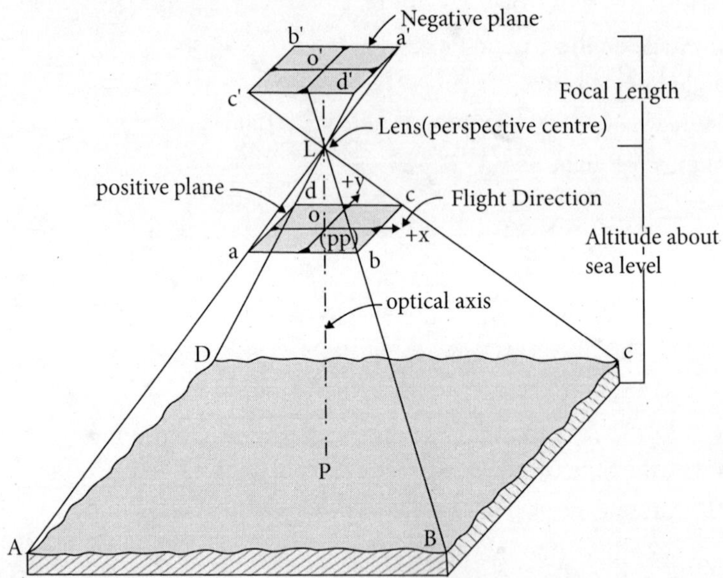

Fig. 4.8: Geometry of an aerial photograph (Garg, 2019)

Total number of photographs: The total number of photographs are obtained by multiplying the number of photographs in each strip by the number of flight lines) (e.g., 6x2= 12 photographs in Figure 4.7).

Flying height: The height at which the aircraft flies (altitude of aircraft above mean sea level) when the photographs were taken is known as the flying height (Figure 4.8). It has a direct relationship with the photographic scale; higher the flying height, smaller the scale. Flying height is always known prior to flying to the area.

Plumb line: It is a vertical line from the exposure station, indicating the direction of gravity. In coincides with the optical axis of the camera lens in case of a truly vertical photograph (Figure 4.8).

Camera axis: The camera axis represents the optical axis. It is a line passing through center of camera lens perpendicular both to camera plate (Negative) and photographic plane (Positive) (Figure 4.8).

Focal length: Distance from the optical centre of the lens to the focal plane, when the camera is focussed at infinity, is called the focal length of camera lens (Figure 4.8). Its value is known, as it is supplied by the manufacturer of camera.

Ground nadir: The point on the ground vertically beneath the perspective centres of the camera lens, denoted by letter P (Figure 4.8).

Fiducial marks: Four index marks, which are shown at the edge or corners of the photographs, are called the fiducial marks (Figure 4.9). These are used for determining the principal point of the photograph.

Fiducial axes: Straight lines joining the opposite fiducial marks on a photograph are called fiducial axes (Figure 4.9).

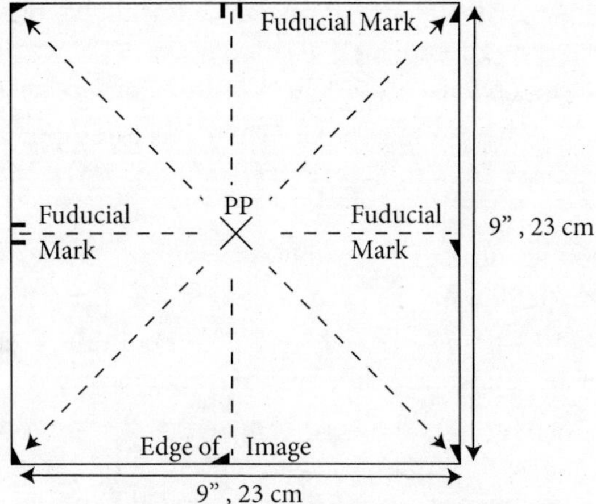

Fig. 4.9: Fiducial marks on the photograph

Principal point (PP): The geometric centre of an aerial photograph, located at the intersection of lines drawn between the opposite fiducial marks. is called the principal point (Figure 4.9).

Conjugate principal point (CPP): The principal point (PP) from one photo transferred onto adjacent photo along the flight line is called the CPP.

Photograph centre: The point on the photograph that is located at the intersection of the fiducial axes is called the photographic centre. In a perfectly adjusted camera, the photograph centre and the principal point are identical (Figure 4.9).

Photographic nadir: The point on the photograph which corresponds to the ground nadir, denoted by letter *n*. The point on the photograph at which a vertical line (plumb line) from the perspective centre to the ground nadir intersects the photograph, is called photographic nadir.

Overlap: The common region (expressed as a percentage) between two photographs is called overlap. The overlap between two successive aerial photographs in the same flight line is called *longitudinal overlap* or *forward overlap* or *end lap*, and the overlap between photographs in adjacent parallel flight lines is called the *lateral overlap* or *side lap* (Figure 4.7). The amount of end lap is kept a minimum of 60%, which is useful to generate 3D view of the common area between the photographs. The lateral overlap or sidelap is kept between 25-40%, which is used to create a mosaic of the area.

Superlap: The common overlap between three successive photographs is called superlap. It means that a photograph with 70% overlap will have 40% superlap region.

Mosaic: The process of seamlessly joining a series of overlapping air photos together to form one large image, is called a mosaic. It is created to view and analyse the large area.

Stereo-pair: Two successive photographs taken during a flight line with sufficient overlap is called a stereo-pair.

Stereo-model: In two successive photographs, the overlapping part can be utilized for stereo measurements. The stereo-pair is used to create a stereo model that can be seen in 3D using a stereoscope device.

Parallax: The apparent displacement of the position of an object with respect to a reference point, caused by a shift in the point of observation, is known as parallax.

Floating mark: It is a mark (dot or cross), associated with parallax bar or stereometers or stereoscopic plotting machines, seen as occupying a position in the 3-D space formed by the stereoscopic fusion of a stereo-pair, and used as a reference mark in examining or measuring the stereo-model.

Scale: The ratio of a distance on a photograph or map to its corresponding distance on the ground is called scale. Scale may be expressed as a ratio (1:25,000), a fraction (1/25,000), or equivalence (1 cm = 250m).

Photo-interpretation: An act of identifying the objects from images and judging their relative significance is called photo-interpretation.

Control points: The reference points precisely located on both the ground and the photo whose three-dimensional coordinates are known, are called as control points.

Orthogonal projection: Maps are based on orthogonal (parallel) projections where the projecting rays are perpendicular to the line on the ground. The advantage of this projection is that the distances, angles or areas on the plane are independent of the elevation differences of the objects, and these measurements compare well with the actual ground.

Ortho-photos: An ortho-photo is an aerial photograph that is rectified by combining photogrammetric principles with Digital Elevation Model (DEM) data. An aerial photograph does not have a constant scale throughout the photograph, whereas the scale is uniform throughout in an ortho-photo, and hence it can be used as an alternate to a map.

Angle of tilt: The angle at the perspective center between the photograph perpendicular and the plumb line is called angle of tilt. It is present in case of tilted photographs (Figure 4.10).

Isocentre: The point on the tilted photograph where the bisector of the angle of tilt (t) strikes the photograph (located in the principal plane as well as on the principal line), is called isocentre (Figure 4.10). It is denoted by letter *i*. The isocenter is a unique point common to the plane of the tilted photograph, its principal plane, and the plane of the assumed truly vertical photograph taken from the same camera station and having an equal principal distance (i.e., the isocenter is located at the intersection of three planes).

Principal line: The line on the tilted photograph which passes through the principal point and the nadir point (and the "isocentre"), is called the principle line (Figure 4.10).

Principal plane: The vertical plane through the perspective centre containing the photograph perpendicular and the nadir point (and the "isocentre") is called the principal plane.

Azimuth: The horizontal angle measured clockwise about the ground nadir from a reference plane (usually the north meridian) to the principal plane. The azimuth of a photograph is the ground-survey direction of tilt, while swing is the direction of tilt with respect to the photograph axes.

Swing: The angle about the principal point of a tilted photograph, measured clockwise from the positive y-axis to the principal line at the nadir point, is called swing. It is denoted by letter *s*. Swing also refers to rotation of the photograph (or photo-coordinate system) around the photograph perpendicular (or photograph z-axis). In air navigation, swing represents aircraft rotation about the aircraft's vertical axis and is referred to as 'yaw' or 'crab'.

Tilt displacement: It is the displacement of images on a tilted photograph is radially outward (inward) with respect to the isocenter if the images are on the low (high) side of the isometric parallel. The 'low' side of a tilted photograph is the side closer to ground.

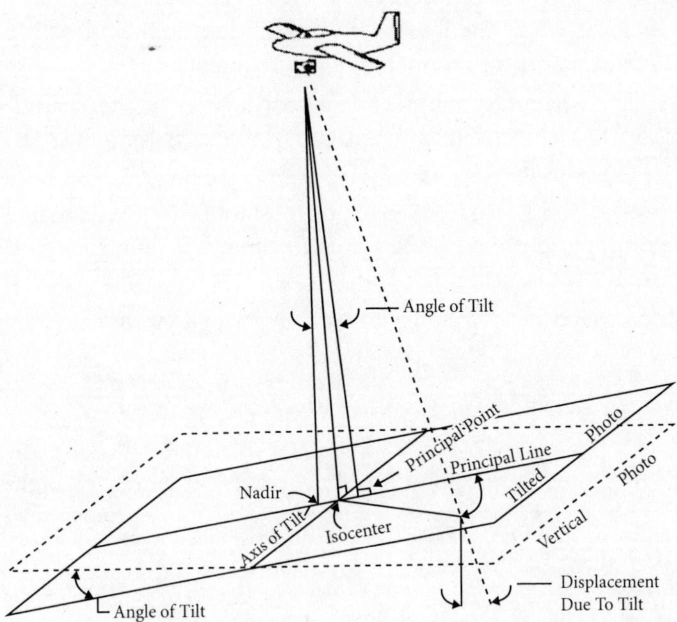

Fig. 4.10: Representation of a titled photo

4.9 SCALE OF A VERTICAL PHOTOGRAPH

The scale is a ratio of the distance between two objects on the photograph to the distance between the same points on the ground. In Figure 4.11, if *A* and B are the ground points and a and b are their corresponding images on the photograph, O is the exposure station, *f* is the focal length, and *H* is the flying height above ground, the scale *S* of the vertical photograph is computed as:

$$\text{Scale} \quad S = \frac{\text{Map distance}}{\text{Ground distance}} = \frac{ab}{AB}$$

Considering the isosceles triangles Oab and OAB, we can write:

$$S = \frac{ab}{AB} = \frac{f}{H} \tag{4.7}$$

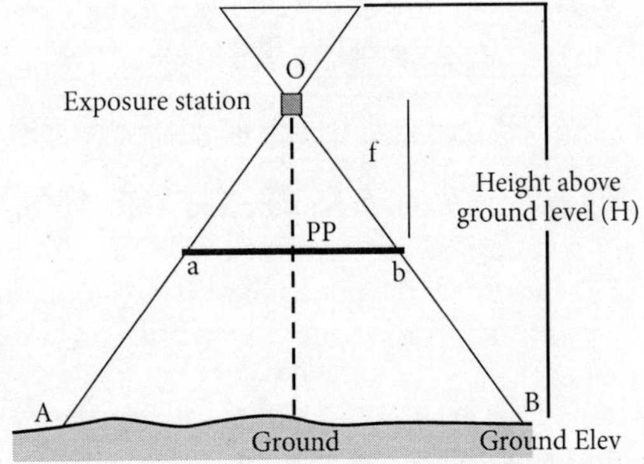

Fig. 4.11: Scale of a vertical photograph in a flat terrain

So, the scale is the ratio of focal length and flying height of the aircraft. It is directly proportional to focal length of camera lens and inversely proportional to the flying height.

This relationship is valid when the ground is assumed to be flat. But the ground is always undulating with some amount of relief. In case of undulating ground, the scale is computed as:

In Figure 4.12, four point on the ground A, B, C and D, having different elevations with mean sea level are shown. The photograph is taken from an exposure station L with a camera having *f* focal length and from a flying height H above mean sea level. Let us consider two points *A* and *B* having elevation h_a and h_b, respectively, above the mean sea level.

The scale of the photograph at point *A* may be written (using two similar triangles Lao and LAO_A) as:

$$S = \frac{ao}{AO_A} = \frac{Lo}{LO_A} = \frac{f}{H - h_A} \tag{4.8}$$

Similarly, the scale of photograph at point B will be (using two similar triangles Lbo and LBO_B):

$$S = \frac{bo}{BO_B} = \frac{Lo}{LO_B} = \frac{f}{H - h_B} \tag{4.9}$$

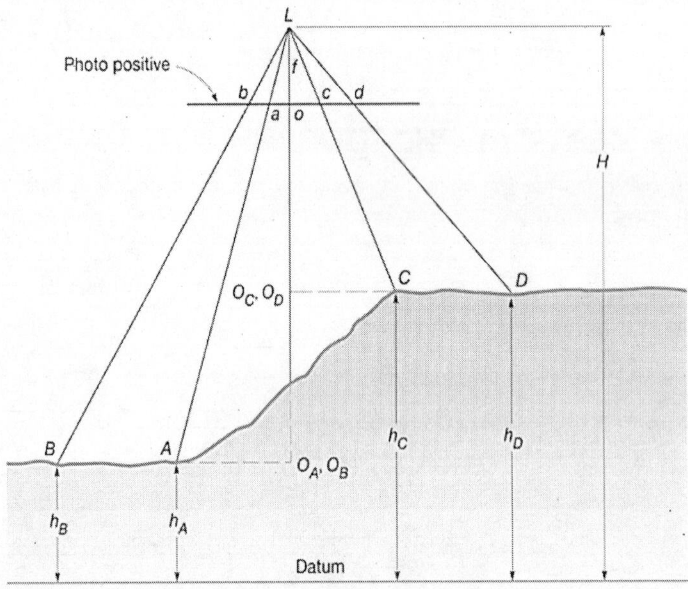

Fig. 4.12: Scale of a vertical photograph in an undulating terrain (*Schenk, 2005*)

Scale at ground points *C* and *D* can also be computed in a similar manner. Hence, it is observed from the above relationship that the scale of a vertical photograph varies at different points depending on their elevations, as *f* and *H* will be same for all the points in a given photograph. If all the points within a photograph are situated at the same elevation, the scale will be constant and uniform throughout. But in general, there is a height variation in the terrain, so scale of the photograph will vary from ground point to point. If the average elevation of the area is h_{av}, a more generalised relationship to compute the average scale S_{av} of photograph having points at different elevations may be written as:

$$S_{av} - \frac{f}{H - h_{av}} \tag{4.10}$$

A more general expression for the scale may be written as-

$$S = \frac{f}{H - h} \tag{4.11}$$

So, photographic scale is directly proportional to focal length of the camera lens. When a camera with larger focal length is used, a larger scale is obtained, and vice versa. It also varies inversely with the flying height; scale decreases as the flying height increases and vice versa.

4.10 RELIEF DISPLACEMENT OF A VERTICAL PHOTOGRAPH

The relief displacement is present on the vertical photograph due to height of various objects. It is the displacement of the objects on a vertical aerial photograph from their true plan positions. Aerial photograph is based on the central or perspective projection, and all the objects on aerial photographs appear to be displaced from their true positions. However, the objects near the principal point will be free from any relief displacement. Due to varying heights of different ground points, these points appear to be displaced on the photograph from their true geographic location. The higher objects, such as tall buildings, trees, towers, hills etc., are displaced away (outward) from the principal point, while the valleys, depressions etc., are displaced (inward) towards the principal point. The relief displacement is measured from the principal point of the photograph. It is always radial from the principal point, and therefore it is also known as the *radial displacement*.

Let us establish a relationship for relief displacement. In Figure 4.13, let AA_0 be a tower of height h above the ground. A photograph of the tower is taken from exposure station O. The a and a_0, respectively are the corresponding images of top and bottom of the tower on the photograph. The aa_0 distance (d) on the photograph is called the relief displacement as it is the horizontal displacement of tower on the photograph due to its height. The distance (d) can be computed as-

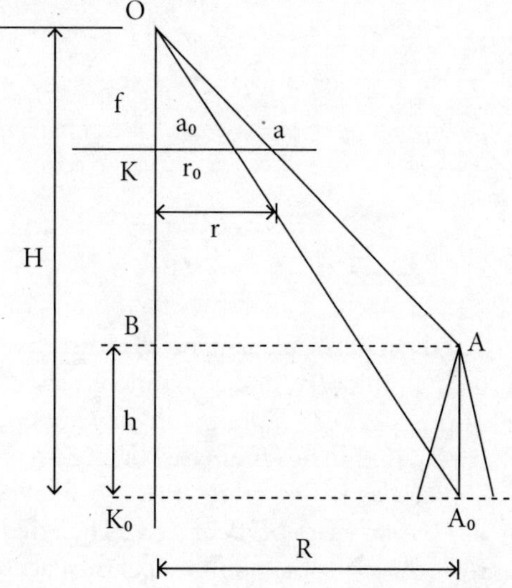

Fig. 4.13: Relief displacement of a tower on a vertical photograph (Garg, 2019)

Let f = focal length of camera lens

$R = K_0A_0$

$r = Ka$

$r_0 = Ka_0$

$$d = a\,a_0 = r - r_0 \tag{4.12}$$

From similar triangles OKa_0 and OK_0A_0, we can write as-

$$\frac{r_0}{R} = \frac{f}{H}$$

$$r_0 = \frac{fR}{H} \tag{4.13}$$

Similarly, from similar triangles OKa and OBA, we can write-

$$\frac{r}{R} = \frac{f}{H-h}$$

$$r = \frac{fR}{H-h} \tag{4.14}$$

So relief displacement from equations 4.12. 4.13 and 4.14 would be-

$$d = r - r_0 = \frac{fR}{H-h} - \frac{fR}{H}$$

$$d = \frac{fR}{H(H-h)}(H-H+h)$$

$$d = \frac{fRh}{H(H-h)}$$

As per equation 4.13, $R = \left(\frac{H-h}{f}\right) r$

$$\text{So } d = \frac{fh}{H(H-h)} \times \frac{(H-h)}{f}.r$$

$$d = \frac{rh}{H} \tag{4.15}$$

It is seen from equation 4.15 that the relief displacement (*d*) is directly proportional to radial distance of displaced image point from principal point (*r*), and as (*r*) increases (*d*) also increases. It is therefore advisable that the objects situated at the edges of a photograph may not be selected for any measurement as it may have larger error which is required to be eliminated before use. Relief displacement is directly proportional to the height of the object, as (*h*) increases (*d*) also increases. Whereas, (*d*) is inversely proportional to the flying height, as (*H*) increases (*d*) decreases. The relief displacement is maximum at the edges of photograph and is zero at the principal point. Relief displacement in the photograph is also one of the main reasons, a photograph is not directly used as a map for measurement purpose. From equations 4.13 and 4.14, we can write as-

$$\frac{r}{r_0} = \frac{H}{H-h}$$

$$\frac{r}{H} = \frac{r_0}{H-h}$$

So relief displacement can be written in terms of r_0

$$d = \frac{rh}{H} = \frac{r_0}{H-h}h \qquad (4.16)$$

If the relief displacement is known, the height of an object can be computed as:

$$h = \frac{dH}{r} \qquad (4.17)$$

Equation 4.17 has been used to approximately compute the height of the object from a single vertical aerial photograph.

4.11 STEREOSCOPY

Stereoscopy is a technique for creating a 3D model of depth perception by means of a stereo-pair. Human eyes are the best example of stereoscopy that will see two images of any object, which are only different in view angle, and orientation. The human-being has a good stereoscopic vision to view 3D with both eyes using the basic principle of stereoscopy that left eye will see left image and right eye will focus on the right image. It will allow interpreter to judge the distance and depth in the overlap region (Figure 4.14a). Human eyes fixed on same object provide two points of observations which are essentially required for creating parallax.

The perception of depth starts with the acquisition of visual information through the human eyes, and merging these images together in the human brain. When two images of the same object reach the brain, it tries to fuse both the images into a single 3D image, and starts interpreting the information. For example, the brain makes use of a number of clues (e.g., colour, shape, size, pattern, texture, orientation, shadow, location of objects etc.), to determine the relative distance and depth between various objects present in the scenes, as well as carry out the interpretation.

In the similar manner, a 3D model giving depth perception may be created from stereo-pair of photographs when viewed through a stereoscope. In 3D model, the topographic features appear much taller than in reality and slope appears much steeper due to the exaggeration of vertical heights with respect to the horizontal distances. The vertical exaggeration may be useful in photo-interpretation for identifying many features, such as slopes, low ridge and valley, topography, small depressions and elevations. Factors affecting the vertical exaggeration are photographic and stereoscopic factors.

The vertical exaggeration is fundamentally related to the ratio of air base (B) with the flying height of aircraft (*B/H*). It is directly proportional to air base, and can be correlated with focal length (*f*), distance (D) from the eye at which the stereo-model is perceived (approximately 45 cm), and eyebase (*b*). The vertical exaggeration is inversely proportional to flying height (*H*), and eye base (*b*) (approximately 6 cm). Vertical exaggeration ratio (*R*) is determined by using the formula:

$$R = \frac{BD}{bH} \qquad (4.18)$$

In stereoscopy, when a human being's two eyes (binocular vision) are focused on a certain object, the optical axes of the eyes' converge at that point forms a parallactic angle (α). The near the object, the greater the parallactic angle, and vice versa (Figure 4.14b). The brain has learned to associate distance with the corresponding parallactic angles, and provides the visual and mental impression which object is closer and which one is farther. This is the basis of depth perception. If the two objects are exactly the same distance from the viewer, then the parallactic angles will be the same, and the viewer would perceive them at as being the same distance away. The maximum distance at which distinct stereoscopic depth perception is possible is approximately 1000 m for the average adult.

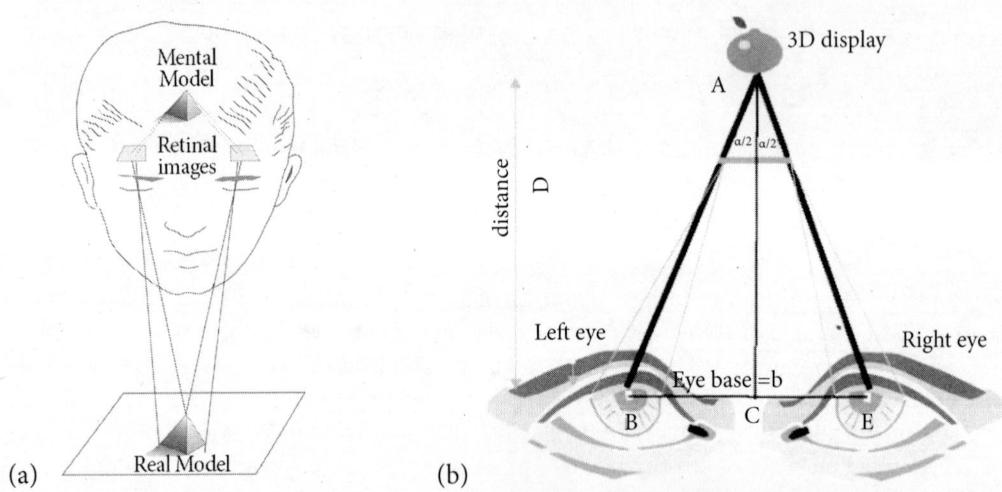

Fig. 4.14: Stereoscopy (a) Human vision creating 3D (Quirós, 2018), and (b) parallactic angle (Nam et al., 2012)

4.11.1 Stereoscopic model

A stereoscopic model can be created with a stereo-pair or stereoscopic images or a stereogram. When two successive photographs taken from two different exposure stations cover part of the common scene which is called a stereo-pair. The 3D model created for the common area to both photographs is called a *stereoscopic model* or *stereo-model*. Stereoscopic model thus provides the ability to resolve parallax differences between far and near objects. The 3D view can be created from a stereo-pair through stereoscopic process, using simple equipment, such as stereoscopes or sophisticated equipment, such as stereo-comparator, or digital photogrammetric systems.

Some people with their normal vision in both eyes and having experience in photogrammetry can develop a stereoscopic vision without the use of a stereoscope. Seeing stereoscopically with the eyes without a stereoscope can be practiced with a famous "sausage-link" exercise, as shown in Figure 4.15a. Hold your hands at eye level about 10 inches in front of you, and select the homogeneous background with a different color than your fingers. Point your index fingers against each other, leaving about 1 inch distance between them. Now look through your fingers, into the distance behind them. An extra piece of finger "like a sausage" should appear (Figure 4.15b). If you try to look at the sausage, it will disappear, it is only present if you look at something more distant than your fingers. Move your fingers closer or farther apart to adjust the length of the "sausage." Close one eye or the other and the illusion will disappear. If you focus your attention back to your finger tips the illusion will also disappear.

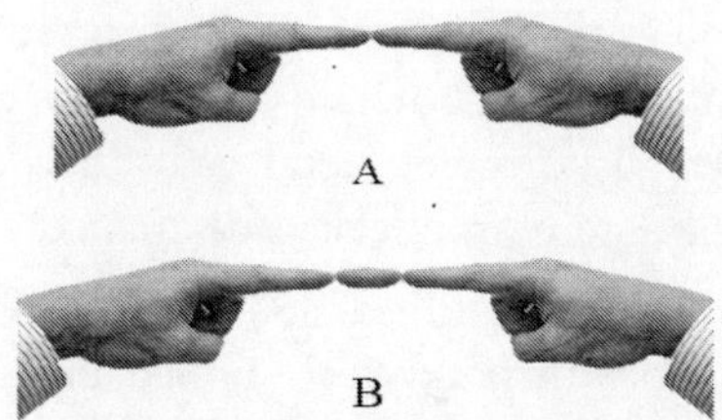

Fig. 4.15: Stereoscopic exercise with index fingers (Handerson, 2010)

4.11.2 Requirements of a stereoscopic vision

To obtain a 3D model from the two dimensional photos, stereo-pair of an area must fulfil the following essential conditions:

1. Both the photographs must cover the same common area with minimum 60% overlap.
2. Time of exposure and contrast of both the photographs must be nearly same.
3. Scale of two photographs should be the same.
4. The brightness of both the photographs should be similar.
5. Parallax must be present in the stereo-pair (i.e., the photographs must have been taken from two different exposure stations).
6. Base to height ratio must have an appropriate value; the normal value is up to 2.0 but not less than 0.25. If this ratio is too small, say 0.02, the stereoscopic view will not provide a better depth impression. Base to height ratio increases when the overlap decreases. Short focal length wide angle lens cameras give better base height ratio which is important in natural resource inventory and mapping.

4.11.3 Stereoscopes

Stereoscopic vision is the basic pre-requisite for photogrammetry and photo-interpretation in 3D environment. Many applications need information extraction with stereo-images rather than mono-images. This requires two views of a single object from two slightly different positions of camera. The photographs are taken from two different positions with overlap in order to reproduce the objects in a manner as they are individually seen by the eyes.

Stereoscope is an optical device for 3D viewing of landscapes or objects. A stereoscope helps viewing a stereo-pair; left image with the left-eye and right image with the right-eye, to create a 3D model. To do so, the photographs are to be properly oriented below the stereoscope in the same manner as they were taken at the time of photography. When each eye views the respective image, these two images help creating a 3D view in overlap region. The fusion of common area of these two images in the brain will allow the judgement of depth or distance.

There are two type of stereoscopes used for three dimensional studies of aerial photographs; (i) Lens stereoscopes, and (ii) Lens and mirror stereoscopes.

I. Lens stereoscopes

Lens stereoscope is also called as *pocket stereoscope,* as it can be kept in pocket owing to its small size. Being light-weight, it is easy to transport in the field, if required. It consists of two plano-convex lenses with magnifying capability, which are mounted on a metallic frame (Figure 4.16a). The distance between these two lenses is adjustable as per the comfort of users' eyes. The eye base average distance is approximately 65 mm for a human-being. The height of pocket stereoscope is normally 10 cm, but its legs can be folded when not in use. Distance between legs of the stereoscope and the focal length of lenses are normally so designed that the stereo-model can be created.

Figure 4.16b shows the line diagram of rays from stereo-photographs to human eyes through lens of a pocket stereoscope. Lens stereoscopes are handy, economical and light-weight, and thus convenient for studying the small format aerial photographs. They have disadvantages, such as limited magnification (2x–4x), and limited field of view to view the larger size photos in 3D at one single glance.

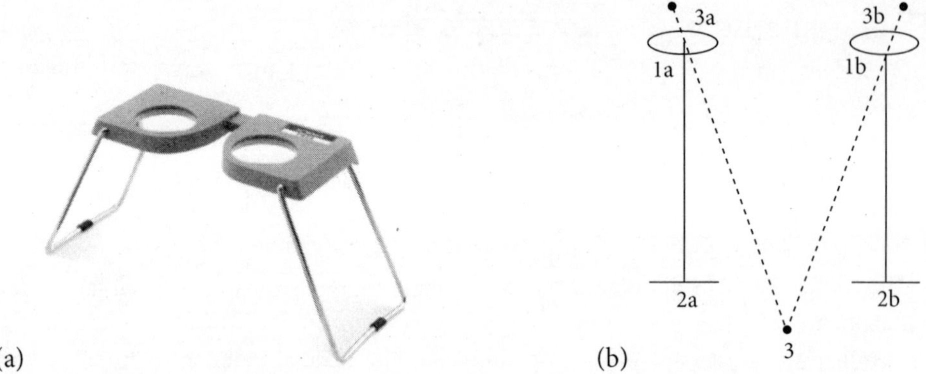

Fig. 4.16: (a) Lens stereoscope, and (b) line diagram of rays from lens stereoscope (Garg, 2019)

II. Lens and mirror stereoscopes

Mirror stereoscope is also called as *Reflecting stereoscope* for viewing of stereo-photographs in 3D. It consists of an arrangement of prisms and mirrors allowing almost entire overlap area to be viewed at a glance in the field of view (Figure 4.17a). Mirrors are fixed to both the legs at 45° angle, along with two right angled prisms near the eyepiece lens. The function of mirrors is to reflect the rays 90° towards the prisms. The rays will strike to hypotenuse of the prisms and deflect 90° towards the eyepiece lenses. The binoculars provide 3x to 8x magnification. One custom made screw situated at one foot allows to level the remaining three legs for providing a stable base to the stereoscope. Figure 4.17b shows the line diagram of rays from photographs to eyes.

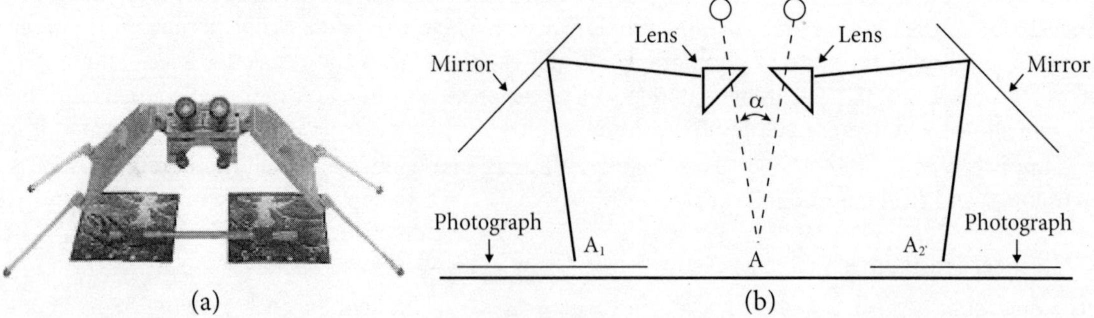

Fig. 4.17: (a) Mirror stereoscope, and (b) line diagram rays of mirror stereoscope (Garg, 2019)

The main advantage of a mirror stereoscope is that the observer can see the entire overlap area from both the images in a magnified form, utilising the property of optics, even-though these photographs may not necessarily be located just below the lens. They are however heavier to carry in the field. The mirror stereoscopes are most widely used in stereo-photogrammetry and 3D measurements, in combination with an instrument called, *parallax bar*. The measurements of points in the stereo-pair are done using parallax bar to determine the elevations of those points.

4.12 DETERMINATION OF HEIGHT FROM VERTICAL AERIAL PHOTOGRAPHS

One of the important application of stereo-photographs is the determination of elevations of various points in the overlap region. For this purpose, a parallax bar, also known as a *stereo-meter*, is used for taking the measurements. The pre-requisite is that the elevation of at least one control point is known in the common area (like ground levelling method) before the computation of elevations of other unknown points. The details of the control points are sometimes supplied along with the aerial photographs, else some permanent point is selected and its elevation is determined either from the contour map or levelling or GPS observations. The procedure for height determination from stereo-pair is described below.

4.12.1 Orienting a stereo-pair of photographs

The first step in creating a stereo-model is that the stereo-pair must be properly oriented under the stereoscope. The process of orientation is called *base lining*, which is performed as below:

1. On both the photographs, their respective principal point and conjugate principal point are marked, as shown in Figure 4.18a. Principal point and conjugate principal point are joined by a straight line and the line extended on each photo. This line represents the flight line.
2. Under a mirror stereoscope, two photographs are to be kept apart in the direction of flight line on a flat surface with overlap region inwards.
3. The stereo-pair is aligned in such a way that the line drawn on both the photos lie in a straight line, as shown in Figure 4.18a.
4. Stereo-pair is seen through the stereoscope so that left lens is over the left photograph and the right lens is over the right photograph. The line joining the centre of the lens should almost be matching with the direction of flight line (Figure 4.18b).
5. The distance between the photographs may be adjusted inward or outward till the two images are fused in the brain and a 3D model of the overlap region is created.
6. In the beginning, it might appear a bit difficult to see the two image fusing together to create a stereo-vision, but with little more practice and concentration, it will appear to be easy.
7. Once the perfect 3D model is created and the lines drawn on the photographs fall in a line, the photographs are said to be properly oriented (or base lining is completed).
8. 8. Select the features/points on both the photographs in the overlap region whose heights are to be determined. The visual interpretation elements (size, shape, shadow, tone, texture and surrounding objects) will help identifying the objects; but now with the addition of relief, a more natural view of terrain may be seen.
9. Use parallax bar for taking the measurements of these points. The difference between two parallax bar readings will provide parallax differences between the two points.

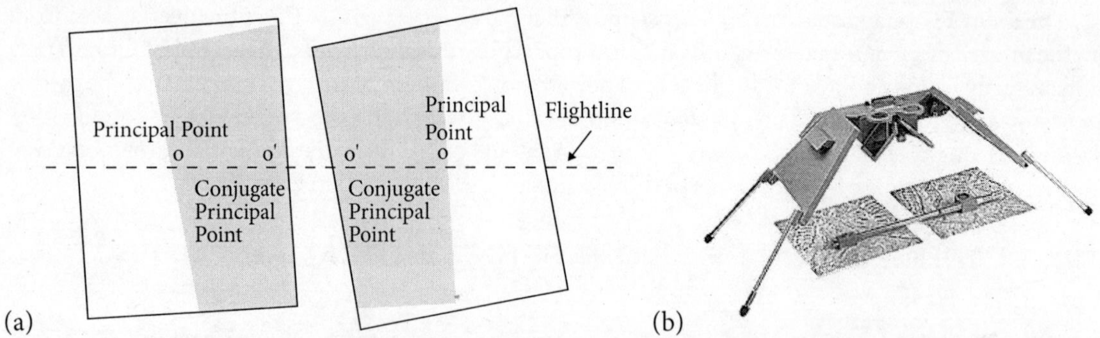

Fig. 4.18: (a) Base lining of a stereo-pair, (Aber et al., 2010) and (b) Creating a stereovision (JARS, 1996)

4.12.2 Measurements by Parallax Bar

The change in position of an image from one photo to the next due to aircraft's motion is called *stereoscopic parallax, x-parallax,* or *simply parallax*. The parallax is directly related to the elevation of the point, and is greater for high points than for low points. The parallax bar is a device used to measure the difference of parallax between any two points on the stereo-photographs, more precisely.

The parallax bar consists of a graduated metallic rod (in mm) attached with two bars, as shown in Figure 4.19. Left bar is fixed with the rod and right bar is movable with micrometer drum attached on right end of the graduated rod. A pair of glass graticules, one at each bar can be inserted into the grooves provided in the bar. Each graticule is etched with three small identical index marks (cross, dot or circle), called *floating marks*. The left hand bar is generally clamped at any required distance so that the entire overlap area is covered by the separation of left and right floating marks.

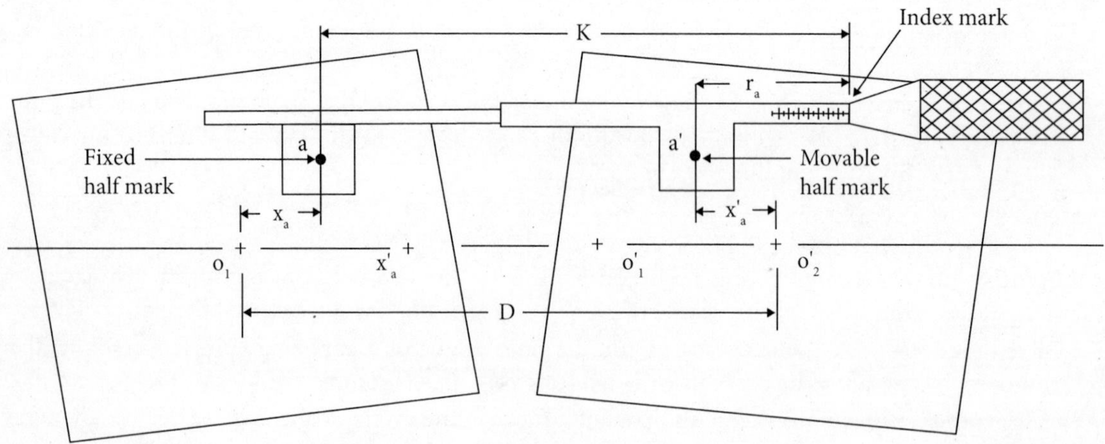

Fig. 4.19: Parallax bar measurements (Garg, 2019)

The points/objects on stereo-pair are selected whose parallax bar readings are to be taken for the computation of heights. The oriented stereo-pair is now viewed in stereo mode, and the floating marks of graticules are kept at a point/object (left mark on left image and right mark on right image). To merge the two images of the floating marks, the separation of two floating marks can be changed by the rotation of micrometer screw having a least count as 0.01mm. It means one complete rotation of micrometer drum which has 100 graduations, will change the reading on main graduated rod by 1 mm. The floating marks on clear glass will either move away or come closer. While further rotating micrometer screw, at a given instant, these marks start to fuse together into a single mark. If micrometer screw is further rotated, these marks will appear as one mark which will now rise up or go down vertically (and appears to be floating in the air). When the two floating marks appear to be a single mark, resting exactly on the terrain point, main scale on graduated rod and micrometer screw readings both are read and added together. The combined reading is called *parallax bar reading* at that point (in this figure, it is K + micrometer reading). Several parallax bar readings at that point may be taken and average value used in the computation. Similarly, the parallax bar readings of other points/objects are taken.

4.13.3 Measurement of absolute parallax

The absolute parallax of a point on a stereo-pair is determined as the algebraic difference between two distances which are measured from the corresponding principal points, parallel to the direction of flight (air base). It is also called *x-parallax*. Figure 4.20 shows a stereo-pair where locations of two points *a* and *b* are marked with their x-coordinates. Let the distance on left photo be x_a and x_b and on right photo be x'_a and x'_b.

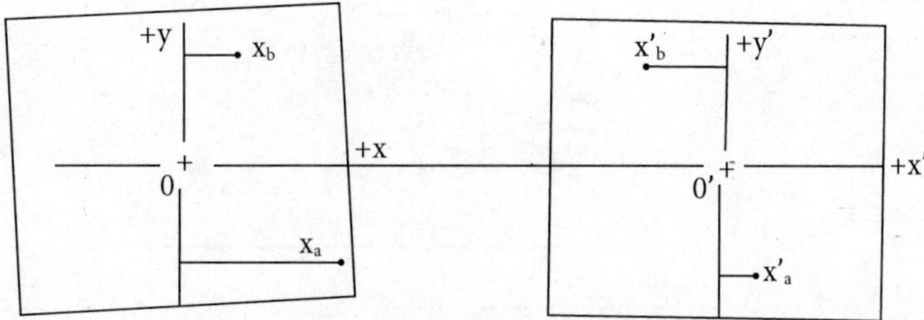

Fig. 4.20: Measurements on stereo-pair

Since, the absolute parallax of a point is determined as the algebraic difference between two distances of a point, it is computed as given below.

$$p_a = x_a - x'_a$$

$$p_b = x_b - (-x'_b) = x_b + x'_b \tag{4.19}$$

The distances on the left and right photos are measured with the help of parallax scale, which is shown in Figure 4.21. This scale has a better least count that the normal scale (ruler) available for manual measurements. It has a main scale and a vernier scale and the addition of both the readings is the final distance. The main scale is graduated in mm while the vernier scale has 11 fine graduations in 10 mm space., so one graduation reads as 1.1mm. The distance is measured in such a way that the main scale and vernier scale both are read and added.

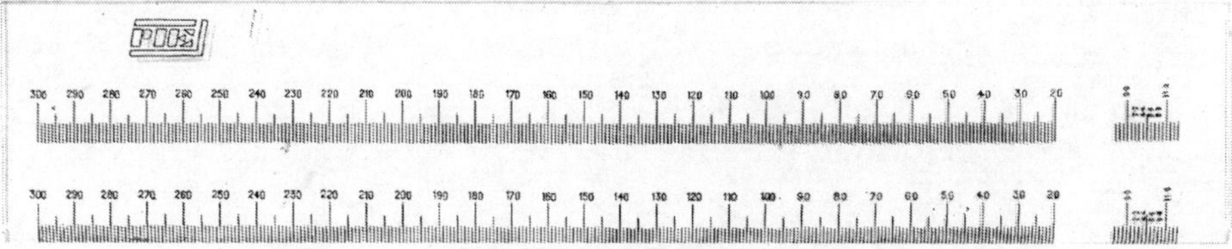

Fig. 4.21: Parallax scale

4.12.4 Height determination

Let A and B are two points whose images are a_1 and b_1 on left photo and a_2 and b_2 on right photo (Figure 4.22). The stereo images are taken from L_1 and L_2 with B as air base distance and H as the flying height. Let h_a and h_b be the heights of these point with respect to a datum. From L_2 draw two lines parallel to $L_1a'_1$ and $L_1b'_1$ to cut the photographic plane at a'_1 and b'_1. If the average parallax bar readings at two points a and b are P_A and P_B, respectively, the difference in absolute parallax (Δp) of p_a and p_b can be related with the difference in parallax bar readings.

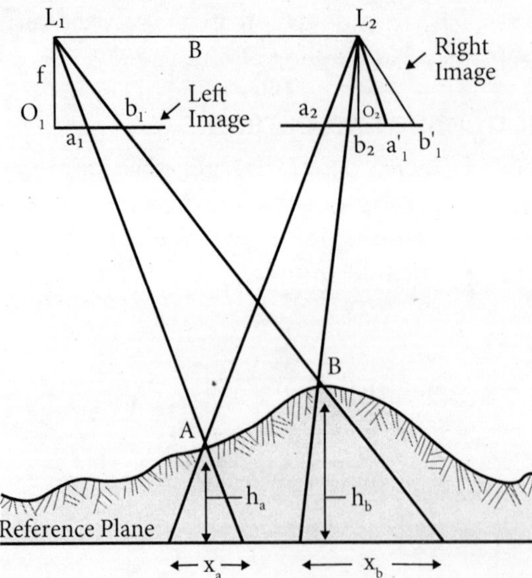

Fig. 4.22: Height observations on a stereo-pair

From Figure 4.22, the parallax p_a at point A is related as follows.

Triangles L_1L_2A and $L_2a_2a'_1$ are similar triangle, so

$$\frac{a2a'1}{B} = \frac{f}{H - h_a}$$

where B is air base, f is the focal length of camera lens, H is the flying height, h_a is the elevation of point A.

But $p_a = a2a'1$, So

$$p_a = \frac{fB}{H - ha} \tag{4.20}$$

Similarly, from similar triangles L_1L_2B and $L_2b_2b'_1$, the absolute parallax p_b is computed as -

$$p_b = \frac{fB}{H - h_b} \tag{4.21}$$

Parallax difference is-

$$\Delta p = p_b - p_a \tag{4.22}$$

$$= \frac{fB}{H - h_b} - \frac{fB}{H - h_a}$$

$$= \frac{fB(H - h_a) - fB(H - h_b)}{(H - h_a)(H - h_b)}$$

$$= \frac{fB(h_b - h_a)}{(H - ha)(H - h_b)}$$

$$= \frac{p_b(h_b - h_a)}{H - h_a}$$

$$h_b - h_a = \frac{\Delta p(H - h_a)}{p_b}$$

$$\Delta p = p_b - p_a$$

$$h_b = h_a + \frac{\Delta p(H - h_a)}{p_a + \Delta p} \quad (4.23)$$

$$\Delta p = p_b - p_a$$

$\Delta p = (x_b + x'_b) - (x_a - x'_a)$ (Refer to Figure 4.20)

Δp is the difference of parallax bar readings at *a* and *b* ($P_A \sim P_B$)

Since the elevation of a ground point *A* (h_a) is known from reference map or field observation or contour map, the elevation of unknown point *B* can be determined (say h_b) using parallax bar measurements.

Since, it is difficult to measure the coordinates of all the objects on the photos, so these coordinates can be replaced by the parallax bar readings (P_A and P_B) at points *a* and *b*.

Using the relationship in equation 4.23, the elevation of point *B* (h_b) can be computed.

In a similar way, elevation of another unknown point *C* (h_c) can be computed using the relationship given below.

$$h_c = h_a + \frac{\Delta p(H - h_a)}{p_a + \Delta p} \quad (4.24)$$

Where $\Delta p = P_A \sim P_C$

It is evident from above relationship that the elevations of various points can be computed by knowing the elevation of at least one ground control point. This method, however, determines the approximate height of the points, so care must be taken to minimise the errors. The errors in the determination of height may be due to several reasons –

1. Locating and marking the flight line on photos.
2. Orienting the stereo-pair for parallax measurement.
3. Measurements in parallax and photo-distances.
4. Shrinkage and expansion of photographs.
5. Difference in flying height of stereo-photographs.
6. Tilt present on photos, if any.
7. Error in the height of known ground control point.

4.13 TILTED PHOTOGRAPHS

Tilted photographs are those taken with the camera axis making more than 3° angle with the vertical line. These photographs are mainly used by army and military. Titled photographs can't be used directly for the measurement or mapping purpose because of inherent distortions present. Therefore, we need to understand the geometry of tilted photographs before they are used for engineering purpose.

4.13.1 Scale of a tilted photograph

The scale of a tilted photograph changes throughout the photograph. It is too small on the high side and too large on the low side (Figure 4.23). The scale also changes along the principal line (i.e., in the direction of tilt), however it does not change along any line perpendicular to the principal line. In Figure 4.23, all the standard notations have been used to represent various points in a tilted photograph. The *o* is the principal point and *n* is the nadir point on the photo. Here, *y* -axis is rotated till the new *y* -axis or the *y* '-axis coincides with the principal line having a positive direction (*no*).

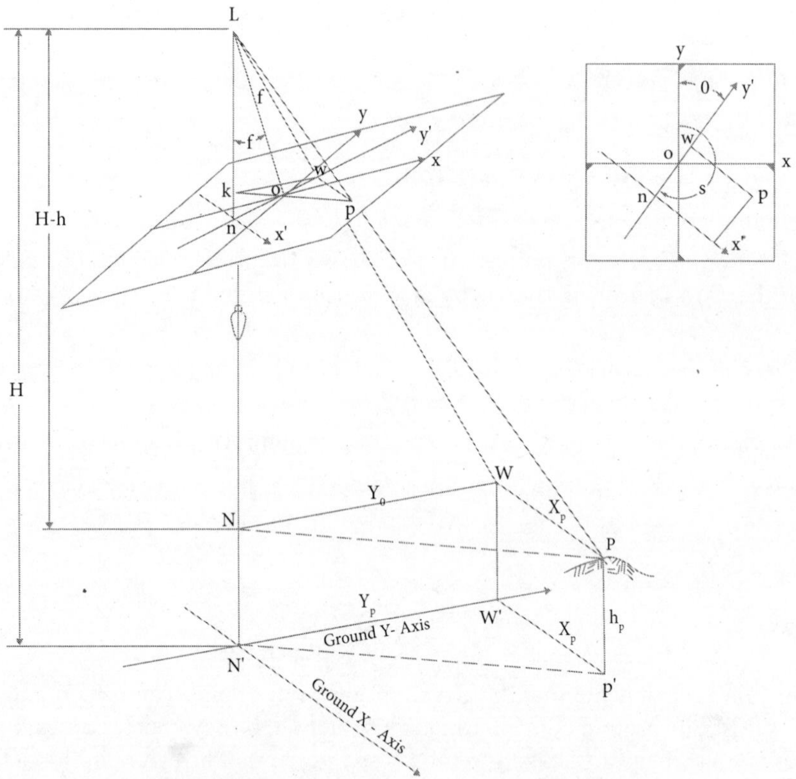

Fig. 4.23: Geometry of a tilted photograph (*Mikhail et al., 2001*)

Let this angle of rotation be θ=180°- s, where s=swing angle, *P* is the point lying at an elevation h_p above the datum, *p* is its image on the tilted photo. Its photo-coordinates are (*x, y*) and measured coordinates with respect to axes defined are (*x', y'*) by collimation marks. Since θ *is* measured clockwise, it must be negative. From *p* point a perpendicular *pw* is dropped to the principal line. Co-ordinates in rotated system can now be derived. On translating *x'* axis from *o* to *w*, the co-ordinates of point *p* on a tilted photo after the translation are-

$$\left.\begin{aligned} x' &= x\cos\theta + y\sin\theta \\ y' &= -x\sin\theta + y\cos\theta + on \\ &= -x\sin\theta + y\cos\theta + f\tan t \end{aligned}\right\} \quad (4.25)$$

$$on = f\tan t$$

A perpendicular line is dropped from *w* as *wk* on *Ln*. In fact, this denotes a horizontal line; *Ln* being a vertical line). Since *wp* is perpendicular to the principal line, it is also a horizontal line. Therefore, the plane made by *kwp* points is a horizontal plane. Also, *PP'= WW'= NN'=* h_p (from construction on the ground), and therefore, the plane formed by *NWP* points is also horizontal.

From similar Δ^s *Lkp and LNP, we can write-*

$$\frac{kp}{Np} = \frac{Lk}{LN} = \frac{Lk}{H-h}$$

But *Lk = Ln–kn = f* sec *t–y'* sin *t.*

Therefore,

$$\frac{kp}{NP} = \frac{f\sec t - y'\sin t}{H-h} \quad (4.26)$$

But *kp / NP* = scale for a point lying in a plane *kwp*. Since *p* lies on this plane, so scale of tilted photo (S_t) at *p* can be written as-

$$S_t = \frac{f\sec t - y'\sin t}{H-h} \quad (4.27)$$

Where, S_t is the scale of a tilted photograph at a point whose elevation is *h, f* is focal length, *t* is the tilt angle, *H* is the flying height above datum, and *y'* is the *y*- co-ordinate of the point with respect to a set of axes whose origin is at the nadir point and whose *y'* axis coincides with the principal line. This equation shows that scale is a function of tilt only, and scale variation occurs on a tilted photo even when the terrain is flat.

4.13.2 Tilt displacement

On a tilted photo, relief displacement is considered to be radial from the nadir point (*n*). Compared to an equivalent relief displacement on vertical photo, the relied displacement on a tilted photo will be (i) less on the half of the photograph upward from the axis of the tilt, (ii) greater on downward half of the photo, and (iii) identical for points lying on the axis of the tilt. The amount of relief displacement depends upon: (i) flying height, (ii) distance from nadir point to image, (iii) elevation of ground point, and (iv) position of point with respect to principal line and to the axis of the tilt. In tilted photos, the radial distance should be measured from the nadir point and not from the principal point.

The tilted photo and its equivalent vertical photo are identical along the isometric parallel, where they intersect. At any other position on the tilted photo, the image of a point will be displaced either outward or inward with respect to its equivalent position on the vertical photo. The characteristics of a tilted photograph are-

1. Displacement due to tilt is zero on a truly vertical photo, but increases proportionally as tilt angle increases.
2. Tilt displacement on slightly tilted photos is usually less in magnitude than displacement from elevation differences, but tilt is much more difficult to detect, calculate, and correct.
3. Images on "up side" of a tilted photo are displaced toward photo center
4. Images on "down side" of titled photo displace away from photo center

In Figure 4.24

t is the angle of tilt

p is the principal point

i is the isocenter

n is the nadir point

o is the perspective center (lens)

j is the focal length of photograph

$\alpha = \angle pOa$ = angle at the perspective center *O* measured counter-clockwise from the photograph perpendicular to any point a on the upper side of the photograph

$\beta = \angle pOb$ = angle at the perspective center *O* measured clockwise from the photograph perpendicular to any point b on the lower side of the photograph

a, b are the images of ground points in the principal plane of the photograph

a', b' are the corresponding points on the equivalent vertical photograph

+r is the distance *ia* = distance *ic*

-r is the distance *ib* = distance *ie*

d is the displacement for an image, parallel to the principal line and due to tilt, on the upper side of a photograph

d' is the displacement for an image, parallel to the principal line and due to tilt, on the lower side of a photograph

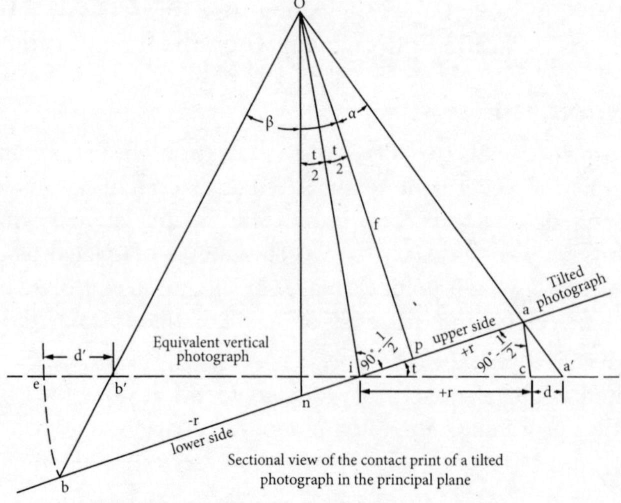

Fig. 4.24: Tilt displacement (*Kraus, 1994*)

So

$tan\ (t/2) = (pi)/f$

$(pi) = f\ tan\ (t/2)$

$tan\ (t) = (pn)/f$

$(pn) = f\ tan\ (t)$

The displacement is computed as:

$d = ia' - ic$ (4.28)

$= ia' - (ia) = ia' - (ip + pa)$

$d' = ie - ib'$

$= ib - ib' = (pb - pi) - ib'$

$$d = f\left[\tan(\alpha + t) - \tan\alpha - 2\tan\frac{t}{2}\right]$$

$$d' = f\left[\tan\beta - \tan(\beta - t) - 2\tan\frac{t}{2}\right] \quad (4.29)$$

In somewhat simpler equations, the image displacement can also be computed as:

$$d = \frac{(ia)^2}{\dfrac{f}{\sin t} - ia} \quad (4.30)$$

$$d' = \frac{(ib)^2}{\dfrac{f}{\sin t} - ib} \quad (4.31)$$

Considering the isocenter, *i*, as the center of rotation, an arc is drawn from point *a* on the tilted photograph so as to intersect the equivalent vertical photograph at a point *c*. The distances *ia* and *ic* are therefore equal and the isosceles triangle *iac* is thereby formed having two sides equal to +*r*.

$$\angle.Oip = 90^\circ - \frac{t}{2}$$

From the isosceles triangle *iac*,

$$\angle\ iac = 90^\circ - \frac{t}{2}$$

Therefore, lines *Oi* and *ac* are parallel (corresponding opposite interior angles) and,

$\Delta Oia'$ is therefore similar to $\Delta aca'$. Therefore -

$$\frac{d}{r + d} = \frac{ac}{Oi} \quad (4.32)$$

$$ac = 2r\ \sin\ \frac{t}{2}$$

$$Oi = \frac{f}{\cos\frac{t}{2}}$$

Substituting the values in equation 4.32, we get-

$$\frac{d}{r+d} = \frac{2r\sin\frac{t}{2}}{\frac{f}{\cos\frac{t}{2}}}$$

$$d = \frac{2r\sin\frac{t}{2}\cos\frac{t}{2}}{f}(r+d)$$

Using the property of trigonometry-

$$2\sin\frac{t}{2}\cos\frac{t}{2} = \sin t$$

Therefore, we can write as-

$$d = \frac{r\sin t(r+d)}{f} \tag{4.33}$$

$$df = r^2 \sin t + dr\sin t$$

$$d(f - r\sin t) = r^2 \sin t$$

$$d = \frac{r^2 \sin t}{f - r\sin t} \tag{4.34}$$

Dividing the numerator and the denominator by *sin t*, we get-

$$d = \frac{r^2}{\frac{f}{\sin t} - r} \tag{4.35}$$

The equation 4.35 applies to both sides of the photograph when the algebraic sign of *r* is observed. The equation 4.35 is usually expressed in an approximate form when computing the value of *d* for either side of a single lens photograph having a tilt angle of less than 3°.

$$d = \frac{r^2 \sin t}{f} \tag{4.36}$$

4.14 AERIAL TRIANGULATION

Aerial triangulation in photogrammetry is done to determine and calculate 3-D coordinates of points by using series of stereo-photographs, thereby reducing the field work. The results of aerial triangulation are the orientation elements of all photographs or stereo-models and the 3-D coordinates of points in ground coordinate system. It is used in different mapping tasks, such as DEM and orthophoto generation, 3D extraction and object reconstruction, surveying and cadastral purposes (3rd and 4th order networks).

Aerial triangulation is used extensively for many purposes, such as extending or densifying ground control through strips or blocks of photos for use in subsequent photogrammetric operation. Establishment of control points are required for compilation of topographic maps with stereo-plotters, locating the property corners for cadastral mapping, and developing the DTM. Determining ground coordinates of points at various time intervals are also useful to monitor the displacement of dams or deformations in structures, and precise industrial measurement, such as determination of the relative position of large machine parts during fabrication. The aerial triangulation can provide controls for photogrammetric purposes for both small scale and large scale maps. For small scale mapping (1:50,000 or so), the required accuracy is 1-5 m, and for large scale mapping (1:1,000–1:10,000), the required accuracy is 0.1-1 m.

There are several benefits of aerial triangulation besides having an economic advantage over land surveying, such as; (i) minimizing delays due to adverse weather condition, (ii) Access to non-accessible ground within the project area, and (iii) eliminating field surveying in difficult areas, such as marshes, high slopes, hazardous rocks, etc. The workflow in aerial triangulation is shown in Figure 4.25.

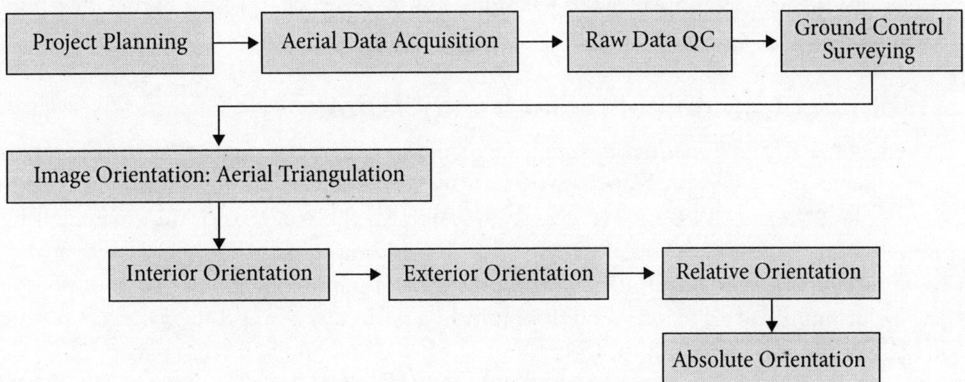

Fig. 4.25: Basic steps in aerial triangulation (Phillipson, 1996)

4.14.1 Types of aerial triangulation

There are broadly two types of aerial triangulation; (a) Radial triangulation, (ii) Block triangulation.

a. Radial Triangulation

Radial triangulation was used in 1950s in which stereo and slotted templates layouts provided photo-control for mapping purposes. A photograph is a perspective projection of the terrain on a plane, and the angles are true at the principal point, only when the optical axis of the camera has been exactly vertical at the time of exposure. In such a case, the principal point is usually taken as the radial centre (i.e., a point at which angles can be measured graphically in all directions) for radial line triangulation. The principal point is easily determined on the photograph and angles are measured at this point.

Radial triangulation is a graphical approach and based on the principle of radial line method that on a truly vertical photograph, the angles measured at the principal point in the plane of photograph to the images on the photograph are horizontal angles to the corresponding points on the ground. Here, the principal points of the neighboring photographs are transferred and the rays to various points are drawn for each photo. These planimetric bundles can be put together along a strip or in a block using two ground control points on each photo. The other points can be located by multiple intersections with 3-4 photos in a strip. Thus, the aerial photographs can be used for measuring the horizontal directions. Radial line methods can be utilized both for extension of planimetric control over large areas (radial triangulation) and for detailed plotting (radial plotting) and contouring.

Extension of planimetric control between known control points is carried out by radial triangulation using radial line principle. A minimum of 9 points per photograph, distributed as shown in middle photograph in Figure 4.26a, are provided. These points are called *minor control points* (MCPs) or *pass points*. The number of necessary ground control points can be reduced to a minimum by using the radial triangulation. Provision of ground control points constitutes a major part of survey cost, and therefore, by using radial triangulation methods, the time and cost of survey is reduced. The equipment used in the method is comparatively modest and the technique is simple in principle. The radial line methods are, therefore, preferred, particularly in developing countries. Although this simple technique provides the required accuracy, it is limited by the large physical space required for the layout of photographs.

The radial triangulation can be carried out by: (i) Graphical, (ii) Mechanical, and (iii) Analytical radial triangulation method. Graphical radial triangulation is the simplest of the above three methods, which has been explained below:

Arundel method of graphical radial triangulation

It is also known as the *Arundel method of radial triangulation*, as it was first tried in U.K. on an area near Arundel village. The basic problems in transforming the aerial photographs onto planimetric map are: (i) to obtain the control points necessary to establish the true position of principal point of each photograph, and (ii) to obtain positions of other image points with the help of which the detail plotting can be carried out. The minimum two ground control points (GCPs) per strip of photographs are necessary for bringing a strip of aerial photographs to a desired scale, and their correct position has to be provided by ground triangulation or traversing. The position of these GCPs are carefully located in the field and marked on the aerial photographs. A number of suitably located image points are chosen on either side of the principal points in order to develop a chain of radial triangulation. The principle in its simplest form is evident from Fig. 4.26b.

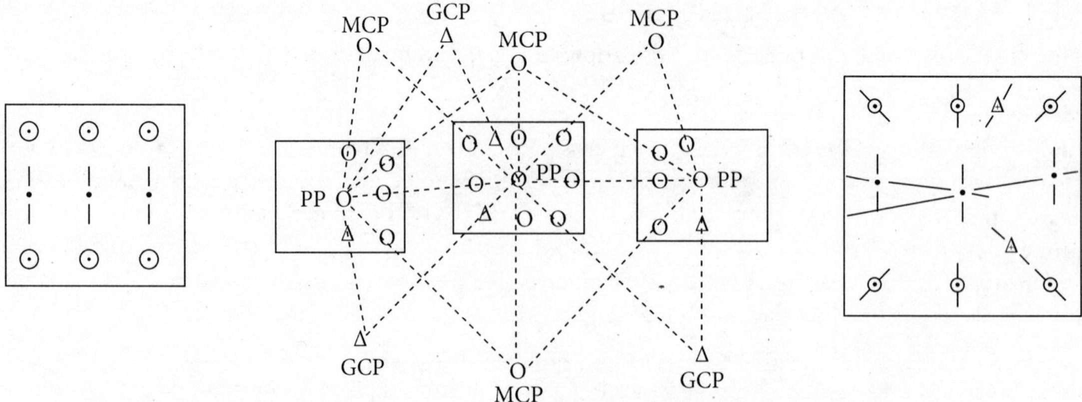

Fig. 4.26: (a) Nine points on each photo, (b) Arundel method of radial line triangulation, and (c) Rays to all points from the principal point

Graphical radial triangulation is performed with simple instruments, such as mirror stereoscope, a ruler, a pencil and a tracing sheet. The steps are given below:

1. The photographs are laid out in strips and all GCPs are identified on the photographs and their numbers marked on the photographs.
2. The principal point is obtained for each photograph. The principal points are then stereoscopically transferred to the adjoining photographs as conjugate principal points.
3. Now two points, called *minor control points* (MCPs), also called *pass points* or *wing points*, are selected on both sides of the principal point of photo, at about 2 cm from the upper and lower edge of the photograph, fulfilling the following conditions:
    a. The two points should be as nearly at the same elevation as the principal point.
    b. The points should be at a distance from the principal point, which is equal to twice the mean base of the adjoining photographs.
    c. The points should lie approximately on the bisector of the base angle on either side, and
    d. The point should serve as lateral point as well.
4. The MCPs are selected and transferred stereoscopically to adjoining photographs.
5. The lateral control points (LCPs) are selected in the centre of lateral overlaps of adjacent strips to serve as connecting points between different the strips. These LCPs are selected at least at the beginning and the end of the strip and on every third photo of the strip. These are then stereoscopically transferred to photographs of adjoining strips.
6. The radial directions from the principal point to all minor control, lateral control and ground control points appearing on the photograph are drawn, through the points. A photograph on completion of the above process looks as given in Figure 4.26c.

Due to elevation differences of terrain and variations in the flying height of aircraft, the scale of photographs generally varies considerably. The photographs of a strip are required to be brought to a common scale through graphical triangulation. The plot of the strip where all the photographs, having uniform scale, are fixed in their correct relative position is called as MCP.

It is preferable to start selecting MCPs somewhere in the middle of the strip to avoid accumulation of azimuthal errors.

Each strip is plotted on a transparent sheet to facilitate the drawing. The photographs are laid out in their correct relative directions so that the plotting is carried out in the right direction. With the first photograph in position below the tracing sheet, principal point base is transferred. The base line is extended up to the edge of the photograph on either side. The position of principal points of the first photograph and one of the two adjacent photographs is then transferred on the tracing sheet. The radial directions to all the points appearing on the photograph are also drawn. The first photograph is then removed and the second photograph is placed beneath the tracing sheet such that its base line and principal point coincides with the base line and principal point traced from the first photograph. The tracing sheet is firmly fixed and the principal point base to the next photograph and all the radials of the second photograph are traced out. Similarly, the third photograph is placed underneath the tracing sheet, and the above exercise is repeated. The photographs on the other side of the starting photograph are likewise completed. In this way, minor control plots of all the strips are prepared.

The MCPs of different strips are at different scales. To bring all of them to the same scale, a projection is made. Normally, in graphical method, scale of survey is nearly the same as the average scale of photographs. On the projected tracing sheet which is usually gridded, all known control data is plotted. Strips which contain two or more ground control points can be scaled down independently.

If there are three or more strips, the strip which has a scale equal to average photo scale, is scaled first. Other strips are brought to this common scale through lateral control points.

The scaling is then carried out by plotting the actual distance AB between the two ground control points, between which the scaling is to be done, on a straight line drawn on a separate sheet (as shown in Figure 4.27). On this line, the distance A'B' (from the minor control plot) is plotted with A and A'coinciding as shown in Figure. With a pair of bow compass, a semicircle is drawn taking B' as centre and distance BB' as radius. The scaled position of any other point C' on minor control plot is obtained by coinciding A' with A and C' falling on the line AB. Draw an arc with C' as centre and perpendicular distance from C' to the tangent as radius. The arc intersects the line AB at C, which is the scaled position of point C'.

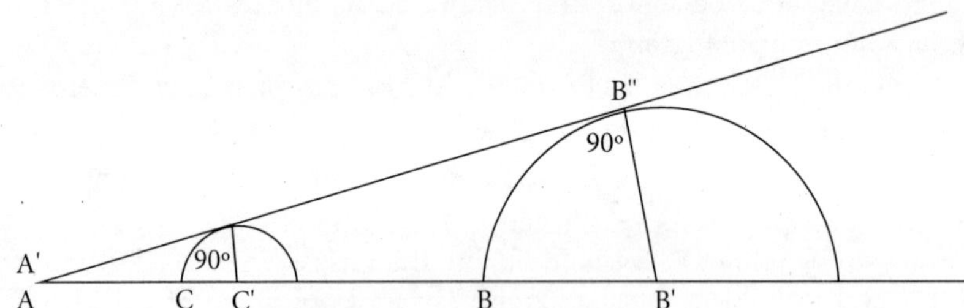

Fig. 4.27: Scaling in Arundel's method

After scaling all the points of minor control plots, these points are pricked through the tracing sheet. The positions of principal points, minor control points and lateral control points of all the strips are likewise pricked on the graph sheet and adjusted so that all ground control points fall over their plotted positions and lateral control points give positions of least error. This process brings all the minor control plots at the same scale. It may be possible to place all the strips in such a way that there are no discrepancies in the final map. However, quite often, it may be necessary to fit portions of strips together using intermediate lateral tie points. It may also be necessary to rescale some strip in two or more parts if scale errors have accumulated. If the discrepancies are large, the entire process is likely to give unsatisfactory results. Additional ground control points in strips also serve as a check. The difference in position from two adjacent strips should not normally exceed 3 mm. For better accuracy, the two points used for scaling should be farthest apart.

b. Block Triangulation

Block triangulation (bundles or independent models) provides the best internal strength as compared to the strip triangulation in radial triangulation method (Ackermann, 1975). The available tie points in consecutive strips assists in the roll angle recovery which is one of the weaknesses in the strip triangulation. In terms of computational aspect, the aerial triangulation methods are categorized as: analog, semi-analytical, analytical, and digital triangulation. The analytical and digital aerial triangulation methods tend to be more accurate than the analog or semi analytical aerial triangulation.

1. *Analog aerial triangulation*

This method uses a "first order" stereo-plotter to carry out the relative and approximate absolute orientation of the first model and cantilever extension. The strip or block adjustment is then performed using the resulting strip coordinates.

2. *Semi-analytical aerial triangulation*

Semi-analytical aerial triangulation or *independent model aerial triangulation* is partly analog and partly analytical procedure. Here, each stereo-pair of a strip is relatively oriented in the plotter. The coordinate system of each model being independent from the others; model coordinates of all control points and pass points have to be read and recorded for each stereo model. By means of pass points common to adjacent models, a 3D coordinate transformation is used to tie each successive model to the previous one to form a continuous strip. This strip may then be brought to ground coordinated system again through 3D coordinate transformation and adjusted numerically utilizing the polynomial equations. Relative orientation of each individual model is performed using a precision plotter.

3. *Analytical aerial triangulation*

Analytical aerial triangulation can be carried out using analytical relative orientation of each model first, and then connecting the adjacent models to form a continuous strip which is finally adjusted to ground controls. The input for analytical aerial triangulation is image coordinates measured by a comparator (in stereo-mode or mono-mode plus point transfer device). A bundle block adjustment is then performed by using all image coordinates measured in all photographs. An analytical plotter in comparator mode can also be used to measure the image coordinates.

4. *Digital aerial triangulation*

It uses a photogrammetric workstation which can display digital images. Digital aerial triangulation is similar to analytical aerial triangulation but here all the measurements are carried out utilizing the digital photographs. The procedure is fully automatic, but allows interactive guidance and interference. Adjustment can be carried out through bundle adjustment by fitting all photogrammetric measurements to ground control values in a single solution.

4.14.2 Orientation parameters

There are six orientation parameters, position of aerial camera and inclination of axes as (dx, dy, dz), and (ω,φ,κ). At least some control points with their known position that are visible in some photographs are required to solve these parameters. By photogrammetric methods, the coordinates can be determined only with few ground control points as contrast to ground-based triangulation methods. Using aerial resection, at least 3 control points for the calculation of 6 exterior orientation parameters of each image are required. As the process of mapping of wide areas, depending on scale, needs lots of images in different strips on a block, the number of required control points extensively increases. With these control points, an absolute orientation of the model can be carried out. In aerial triangulation method, several unknown points along with the ground control points and the coordinates of exposure stations are measured. Computation are performed to determine the coordinates of unknown points in the defined reference system. The inner orientation is performed to locate the aerial photo by using fiducial mark. Relative orientation provides a convenient means of checking most point marking and photogrammetric measurement. Pass-point and tie-points are used to connect several photos or models and strips.

4.14.3 Bundle adjustment

When working with software, the "bundle adjustment" algorithm will be necessary for triangulation. The name is derived from the 'bundle of light rays' that pass through each exposure station. Using algorithm, all photos are adjusted simultaneously to create an intersection of all light rays at each pass point and ground control points. This, in turn, it solves the unknown values of X, Y; and Z object space coordinates. The software also creates a 3D model of the area with lines, surfaces and textures.

A bundle of rays that originates from an object point and passes through the projective centre to the image points (Figure 4.28) forms the basic computational unit of aerial triangulation. Bundle block adjustment means the simultaneous least squares adjustment of all bundles from all the exposure stations, which implicitly includes the simultaneous recovery of the exterior orientation elements of all photographs and the positions of the object points (Faig, 1979).

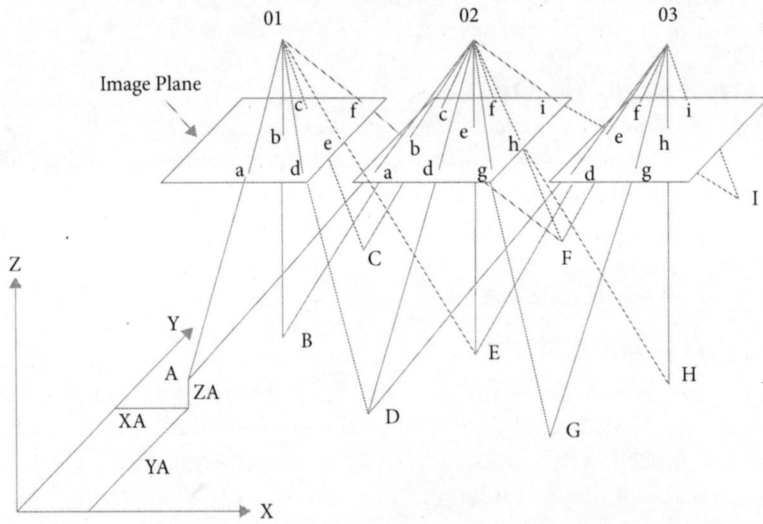

Fig. 4.28: Bundle block adjustment (Faig, 1979)

The fundamental equation of aerial triangulation is the collinearity equation which states that an object point, its homologous image point and the perspective centre, are collinear (Figure 4.29). The collinearity equations are given as:

$$F_x = x_i - x_o + c\frac{m_{11}(X_i - X_O) + m_{12}(Y_i - Y_O) + m_{13}(Z_i - Z_O)}{m_{31}(X_i - X_O) + m_{32}(Y_i - Y_O) + m_{33}(Z_i - Z_O)} = 0 \qquad 2.1$$

$$F_y = y_i - y_o + c\,k_y\frac{m_{21}(X_i - X_O) + m_{22}(Y_i - Y_O) + m_{23}(Z_i - Z_O)}{m_{31}(X_i - X_O) + m_{32}(Y_i - Y_O) + m_{33}(Z_i - Z_O)} = 0 \qquad 2.2$$

where

$$M = \begin{bmatrix} m_{11} & m_{12} & m_{13} \\ m_{21} & m_{22} & m_{23} \\ m_{31} & m_{32} & m_{33} \end{bmatrix} = \qquad 2.3$$

$$\begin{bmatrix} \cos(\Phi)\cos(\kappa) & \cos(\omega)\cos(\kappa) + \sin(\omega)\sin(\Phi)\cos(\kappa) & \sin(\omega)\sin(\kappa) - \cos(\omega)\sin(\Phi)\cos(\kappa) \\ -\cos(\Phi)\sin(\kappa) & \cos(\omega)\cos(\kappa) - \sin(\omega)\sin(\Phi)\sin(\kappa) & \sin(\omega)\cos(\kappa) + \cos(\omega)\sin(\Phi)\sin(\kappa) \\ \sin(\Phi) & -\sin(\omega)\cos(\Phi) & \cos(\omega)\cos(\Phi) \end{bmatrix}$$

Where (x_i,y_i) are the image coordinates, (x_0,y_0) are the principal point coordinates, c is the camera constant, m_{ij} is an element of the rotation matrix, (X_i,Y_i,Z_i) are the object point coordinates, (X_0,Y_0,Z_0) are the exposure station coordinates, M is the rotation matrix, k_y is the scale factor for y-axis in digital camera (this factor is 1 for film-based camera).

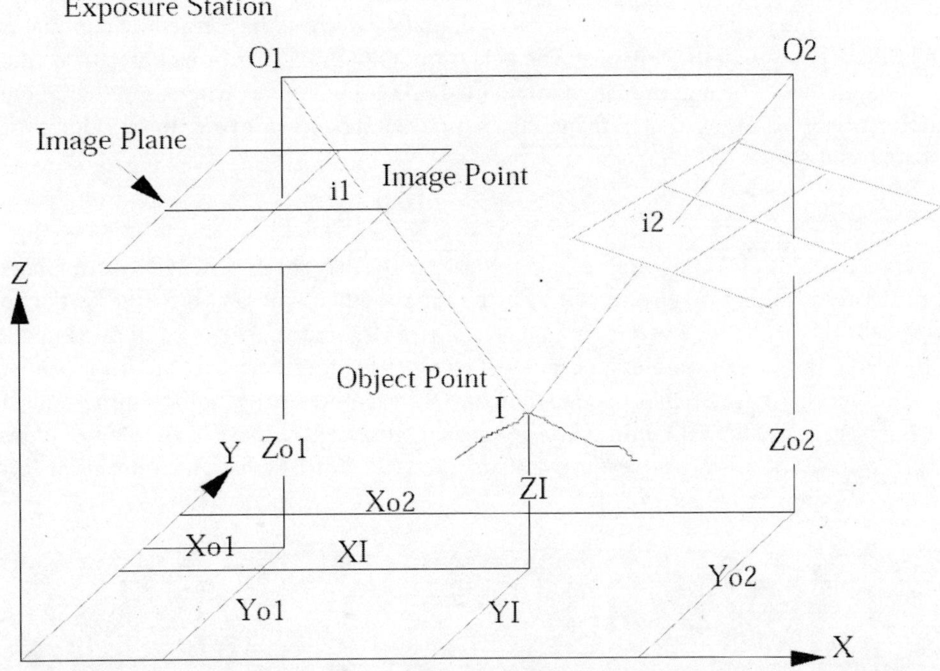

Fig. 4.29: Graphical representation of collinearity condition (Su et al., 2022)

There are 6 unknowns in the collinearity equations, namely, the exterior orientation parameters (X_0,Y_0,Z_0) The three rotation angles implicit in the rotation matrix M. The principal point coordinates (x_0,y_0) and camera constant (c) are considered to be known for the basic bundle approach. However, this might not be true. Strong imaging geometry plus a minimum of three ground control points are needed to solve the six unknowns per bundle which are then used to determine the unknown object coordinates of other measured image points.

4.15 PHOTOGRAMMETRIC MAPPING

Photogrammetric mapping means evaluating and measuring the land for the determination of topography, area, contours, and location of planimetric features, by using photogrammetric methods. The ground control points required for the photogrammetric work may be established by advanced ground-based surveying methods. Photogrammetric mapping professionals utilize their knowledge to employ the appropriate methods and technologies to image, measure, calculate, reduce, and integrate the data. Photogrammetric methods are used to obtain accurate and reliable data for mapping purposes. Many times, there is a need to analyse more than one photo as the study area is large. In such cases, several images captured by aerial photography are combined first by a process called *mosaicing*. These aerial images are analysed to create topographic maps and thematic maps for various projects.

There are two ways to carry out mapping from photogrammetric techniques; manual method and digital method. In manual method, simple equipment, like a light table, magnifying lens, stereoscopes, stereo-plotters, analytical plotters, projection table, etc., may be used. Manual method is laborious, particularly when dealing with large number of photographs. The digital method of mapping would require digital photos, a workstation and a photogrammetric software. In digital methods, skilled-manpower is required to carry out the operations on digital data. The enhancement of contrast of digital images can be done easily with the software. The analysis of photos and creation of maps are much faster than the manual method. The output from manual methods is a plot, map, whereas in digital method, it is 3D model, orthophotos, DEM, digital maps, etc., which could be used further in GIS along with other maps for spatial analysis.

4.15.1 Mosaics

The mosaic is an assembly of overlapping aerial photographs that have been matched to form a continuous photographic representation of a portion of the Earth's surface. Photo-mosaics offer the best of both high resolution images and acquiring overlapping images over a larger area. The overlap portion allows the images to be merged to form a seamless mosaic that can be used for interpretation or map preparation. The creation of a mosaic may need less overlap than for DEM capture using photogrammetry techniques. However, larger overlaps allow the mosaic to be created from central part of the photographs which is considered to have less distortion. Figure 4.30a shows six photographs which are used to create a seamless mosaic as shown in Figure 4.30b.

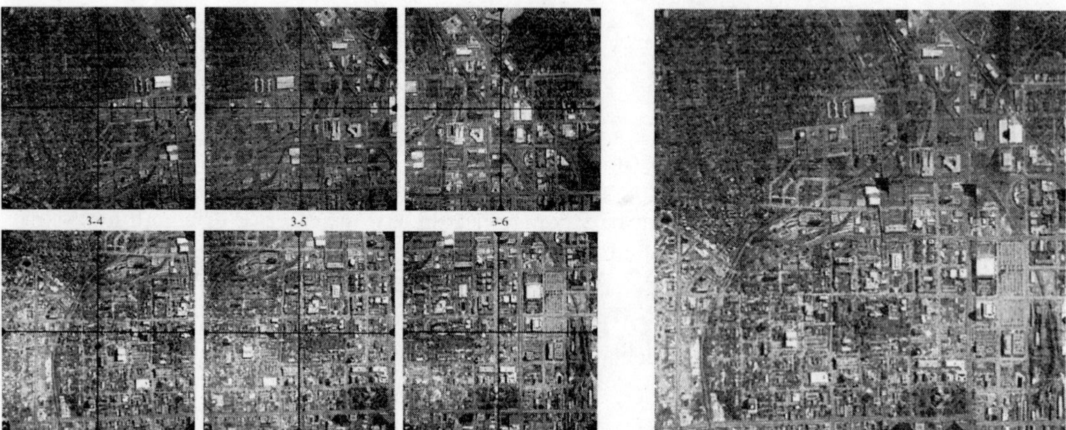

Fig. 4.30: (a) six photographs, and (b) resultant mosaic (Jenson, 2013)

The steps involved in creating a mosaic are shown in Figure 4.31. Mosaics fall into two broad categories:

1. *Uncontrolled mosaic*
   - Image details are matched in adjacent photos and photos are joined together manually.
   - No ground controls are used.
   - It is quick to prepare
   - It is not as accurate as controlled mosaics but for many purposes, such as reconnaissance surveys, they are acceptable.

2. *Controlled mosaic*
   - It is the most accurate.
   - Photo-rectification is carried out on the images.
   - Image features on adjacent photos are matched as closely as possible.
   - Ground control points and corresponding points on the images are used to create the mosaic.
   - Can be used for application where planimetric accuracy is important.

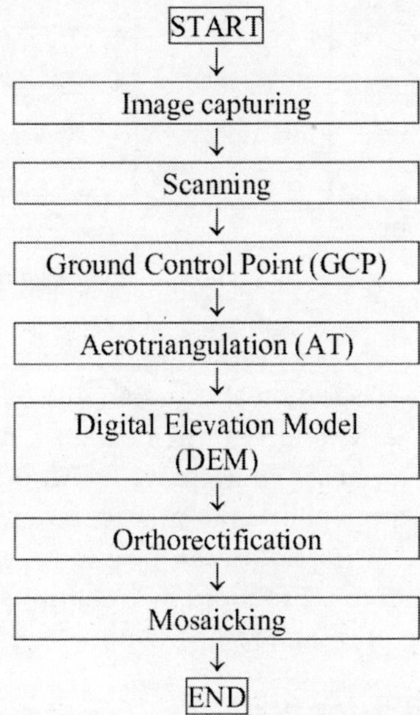

Fig. 4.31: Process involved to create a mosaic (Leh et al., 2008)

Mosaics fall into two broad categories-

i. **Photomosaics (Photomaps)**

Photomosaics are created by merging overlapping aerial photos. Additional information, such as place or road or river name can be taken form maps or ground survey. In this mosaic, there will be scale variation across the photomap, due to terrain undulations (scale of a photograph varies from point to point depending on the elevation of the point). In addition, towards the edges, there will be some relief displacement present.

ii. **Orthophotos and Orthophoto mosaic**

An aerial photograph does not have a constant scale throughout the entire image; therefore, it can't be used directly as a map. In a photograph, except at the principal point, all other points have variation in scale depending upon the undulations in the terrain. A feature, such as a tall building, will also have shape distortion because the top of the feature will have a larger scale than the bottom of it (Figure 4.32a). An orthophoto is an aerial photograph that has been rectified so that it possesses the characteristics similar to a map. It is also known as the *map substitute*.

An orthophoto is an image that shows objects in their true positions (Figure 4.32b). They are geometrically the same as conventional maps, therefore they can also be used as basic map substitutes to take direct measurements without further adjustments. The processing to create orthophotos removes the effects of relief displacement and photographic tilt. Orthophotomaps are produced using either one or multiple overlapping orthorectified aerial photos which have the benefits of both aerial photos and maps. The rectification process is performed by combining the photogrammetric principles with the DEM data (Wolf and Dewitt, 2000).

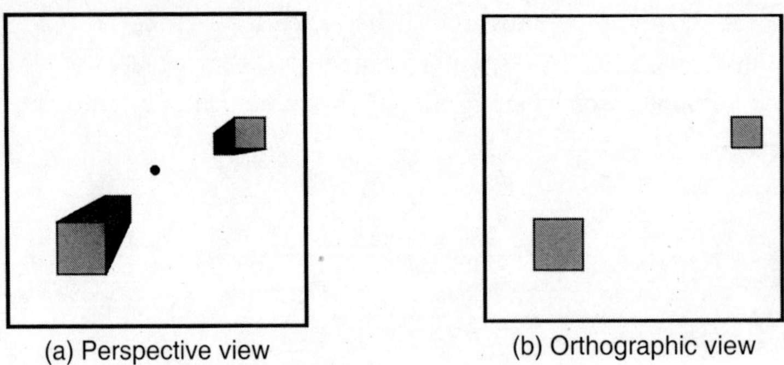

(a) Perspective view (b) Orthographic view

Fig. 4.32: Buildings on (a) aerial photograph, and (b) orthophoto

Map revision or map updation becomes easy using orthophotos. Orthophotos have been used for many years for a diverse range of applications, including GIS. The disadvantage of orthophoto is that the elevations of surface (topography) and also the height of every feature (buildings, trees etc.) above that surface should be known to create an accurate DEM, otherwise there will be some error. In past, orthophotos were produced with a photogrammetric stereo-plotter, but today with the advent of digital photogrammetric methods, orthophotos can be produced on a PC using the appropriate photogrammetric software. An orthophoto is produced by computing the scale and position distortions of each pixel of aerial photograph, re-scaling and re-positioning the pixels in a new generated image.

4.15.2 Stereo-plotting instruments

Stereoscopic plotting instruments or stereo-plotters are designed to provide accurate solutions for object point positions from their corresponding image positions in a stereo-pair. They are capable of producing accurate 3-D coordinates of x, y, and z object space coordinates after proper orientation and calibration. The modern stereoplotters can also handle oblique or terrestrial photos. The primary uses of stereo-plotters are to create topographic maps and digital files of topographic information.

An overlapping pair of aerial photos or transparencies or diapositives, are placed in two stereo-plotter projectors, as shown in Figure 4.33. This process is called *interior orientation*. With the diapositives in place, light rays are projected through them; and when rays from corresponding images on the left and right diapositives intersect below, they create a stereo-model. In creating the intersections of corresponding light rays, two projectors are oriented so that the diapositives have the exact relative angular orientation to each other in the projectors, similar to the negative in the camera at the time of exposure. The process is called *relative orientation*, which creates a true 3-D stereo-model of the overlap region. After relative orientation is completed, *absolute orientation* is performed. In this process, the stereo-model is brought to the desired scale and levelled with respect to a reference datum.

After completing the orientations, measurements from the 3-D model may be made and recorded, but it is now done in digital form. The position of any point/object is determined by using a reference mark (called the *floating mark*) in contact with the 3-D model point. At the position of the reference mark, the 3-D coordinates (x, y, and z) are obtained through either an analogue or a digital solution. Planimetric (x, y) positions and elevations (z) of various points/objects can thus be obtained.

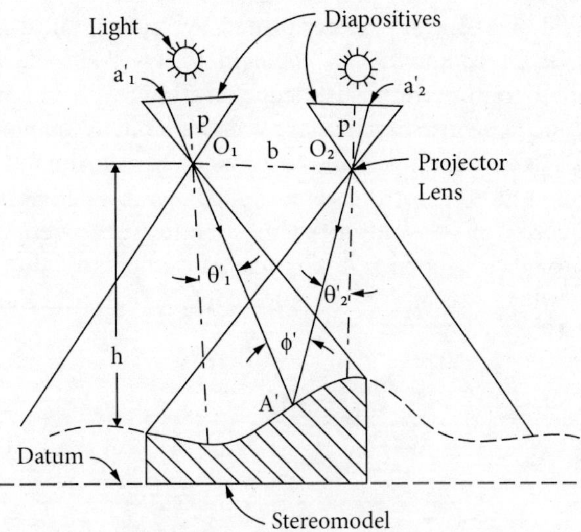

Fig. 4.33: Creating a stereo-model in stereo-plotters (ASPRS, 1980)

4.15.3 Types of stereo-plotting instruments

Several stereoscopic plotting instruments have been developed over the past; each with different features. The stereo-plotters can be classified into in to four groups: (1) *direct optical projection* instruments, (2) instruments with *mechanical or optical-mechanical projection,* (3) *analytical* stereo-plotters, and (4) *softcopy* stereo-plotters. The first-generation stereo-plotters were of direct optical projection design, creating a -3D stereo-model by projecting the transparency images through projector lenses, as illustrated in Figure 4.33. The model is formed by the intersections of light rays from corresponding images of the left and right diapositives. An operator is able to view the model directly, and make measurements on it by intercepting projected rays on a viewing screen.

Instruments based on mechanical projection or optical-mechanical projection create a 3-D model from which measurements are taken. Their method of projection, however, is a simulation of direct projection of light rays by mechanical or optical-mechanical means. An operator views the diapositives stereoscopically directly through a binocular train and carries out measurements. Figure 4.34 shows an optical-mechanical projection instrument.

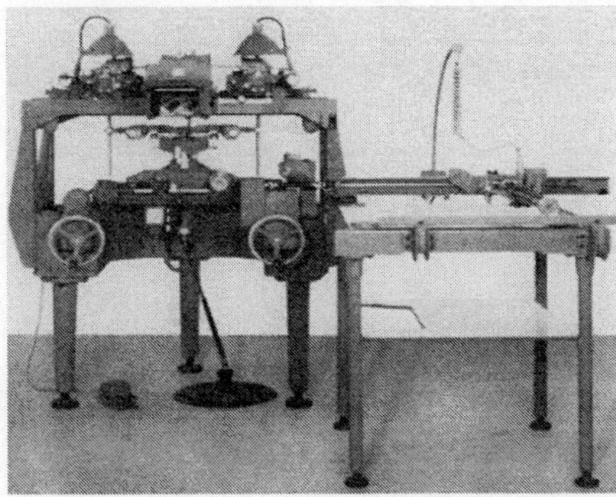

Fig. 4.34: An optical-mechanical projection instrument (Kellie and Valentine, 1987)

Photogrammetric operations can also be performed by mathematical modelling. This method is called *analytical, numerical* or *computational photogrammetry*. With development in optics and mechanics, the analogue photogrammetric instruments have improved to attain high accuracy. With the evolution of computers, analogue instruments have been replaced by analytical plotters, where single or a pair of photographs is placed which digitally records image coordinates (using mono or stereo-comparators). Analytical stereo-plotters form a stereo-model using a mathematical procedure through a computer, as shown in Figure 4.35. As with mechanical plotters, an operator views the diapositives stereoscopically directly through a binocular train. The movements of the stereoscopic images are introduced by servomotors which are under computer control. The procedure to process photos in analytical plotters is given in Figure 4.36.

Fig. 4.35: An analytical plotter (Egles and Kasser, 2001)

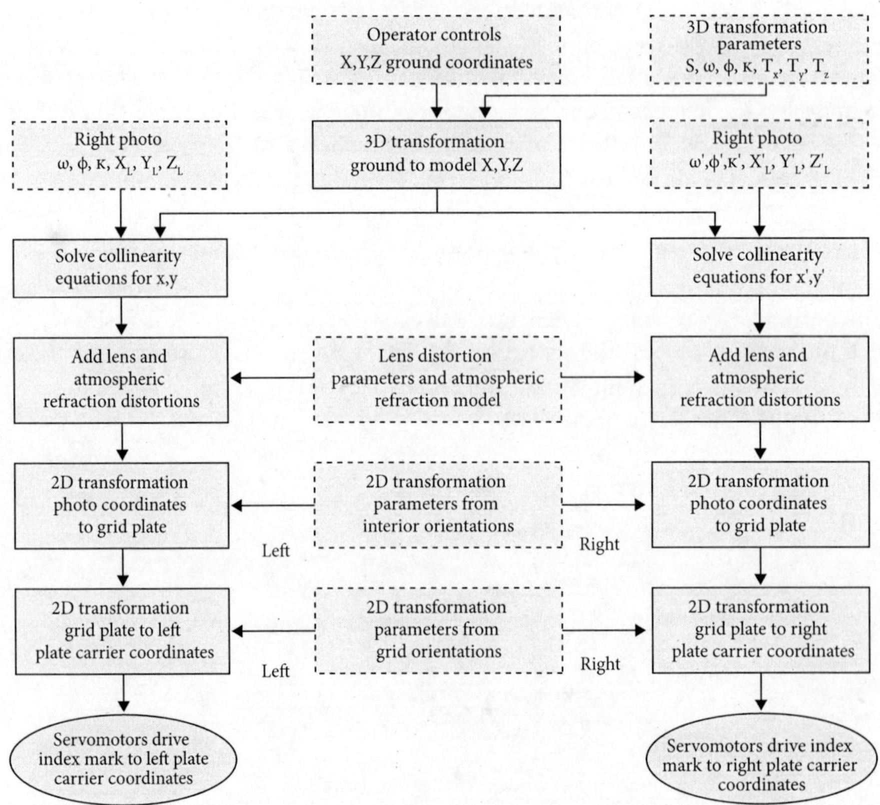

Fig. 4.36: Procedure used in analytical photogrammetry (Grave, 1996)

Softcopy stereo-plotters are the most recent developments. These plotters operate in the same manner as analytical stereo-plotters, except that instead of viewing film (hard-copy) diapositives through binocular optics, softcopy photographs are displayed on a computer screen. Special viewing system enables the operator to view the left image with the left eye and the right image with the right eye in order to see in stereo.

In digital or soft copy photogrammetry, digital images are used with wide range of data processing operations in Digital Photogrammetric Workstation (DPW). A DPW is shown in Figure 4.37. The developments in digital photogrammetry have resulted in DPW and specialised software. Some of the well-known digital photogrammetric systems are LH Systems, Z/I Imaging, and ERDAS. The DPWs are more user-friendly than the analytical plotters, and have image processing capabilities, such as enlargements, reductions, contrast enhancements, etc. The DPW is more reliable and accurate, since no calibration is required (Manugula et al., 2018). Flow diagram of the processes involved is shown in Figure 4.38. The DPWs have the potential to automate photogrammetric applications, such as aerial triangulation, DEM generation, and orthophoto production. As more photogrammetric procedures will be automated, the operation of a DPW would require less specialized skilled manpower (Egles and Kasser, 2001).

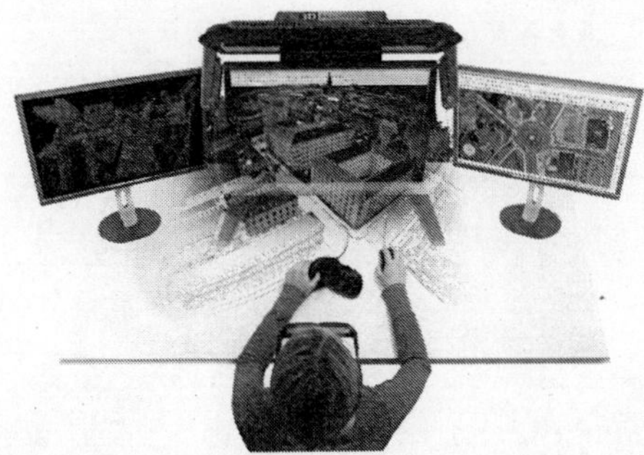

Fig. 4.37: A digital workstation supported with photogrammetric software (Eglas and Kasser, 2001)

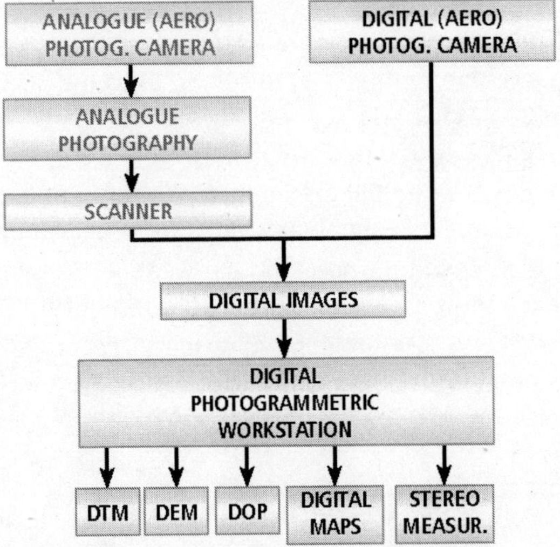

Fig. 4.38: Processing of photos in a DPW (Balenovie et al., 2011)

4.15.4 Photgrammetric software

There are several specialized software for photogrammetry, which could be used to analyze the 2D images captured by photogrammetric camera. Mapping or the surface creation process is also done with these photogrammetry software. The software can be categorized into two classes; (a) Free source, and (b) Commercial software. A comparison of some of the photogrammetry software is presented in Table 4.2.

Table 4.3: Comparison of some photogrammetric software (Formlabs, 2021)

Quality	Speed	Features	User-friendliness
3DF Zephyr	★★★★☆	★★★☆☆	★★★★★
Agisoft Metashape	★★★★☆	★★☆☆☆	★★★★☆
Autodesk ReCap	★★★★★	★★★☆☆	★★★☆☆
COLMAP	★★★★☆	★★★★☆	★★☆☆☆
iWitness	★★★★★	★★☆☆☆	★★★☆☆
Meshroom	★★★★☆	★★☆☆☆	★★☆☆☆
RealityCapture	★★★★★	★★★★★	★★★★☆
Regard3D	★★★☆☆	★★☆☆☆	★★☆☆☆
VisualSFM	★★★☆☆	★★★☆☆	★★☆☆☆

a. Free Source Software

i. **MicMac:** MicMac is useful for projects involving environmental protection, cultural heritage imaging and preservation or forestry. It can use from close-range to aerial images.

ii. **Meshroom:** Meshroom is a photogrammetric computer vision framework. This 3D reconstruction software is easy to use. It can create textured mesh automatically using a node-based workflow.

iii. **3DF Zephyr Free:** It is the free version of the software *3DF Zephyr*; a complete and efficient software for photogrammetry. It is a good software for the beginners for 3D processing. It offers 3D reconstruction tools and basic editing tools.

iv. **Visual SFM:** Visual SFM is a 3D reconstruction tool, using Structure from Motion (SfM). This GUI application is an easy photogrammetry software to use; matching and making the automatic reconstruction. It is a simple software tool with an automatic process.

v. **Colmap:** Colmap is a general-purpose SfM software. It has graphical user interface (GUI), and offers all the basic tools needed to create a 3D model using one or several photographs.

vi. **Regard3D:** Regard3D is a SfM program, allowing to generate 3D models from a series of photographs. This program is offering powerful tools with tutorials available on the website to get started.

b. Commercial Software

i. **ContextCapture:** Previously *Acute 3D*, it is used to create a finished 3D model with photographs without any human intervention. This process is easier than 3D scanning and more precise than 3D modelling. This photogrammetry solution allows working on large infrastructure projects, such as cityscapes. This program enables users to edit 3D meshes, generate cross-sections, and extract ground and break-lines.

ii. **Reality Capture:** Reality Capture is a complete photogrammetry solution, and claims to be 10 times faster than any other photogrammetry solution. This program is able to calculate meshes and textures, and allows working with a lot of different file formats.

iii. **3DF ZEPHYR:** This software allows reconstructing a 3D digital representation with images, automatically. This software has a basic and free version. The advanced versions allow to get laser-scanned objects, or to work more precisely with aerial photographs.

iv. **Autodesk ReCap Pro:** Autodesk ReCap (previously known as *Autodesk ReMake*) is a software allowing to create accurate 3D models using reality captures. It can also measure and edit point cloud data. It has a wide range of tools, for example, to clean the unwanted objects to work more specifically on a precise object.

v. **Trimble Inpho:** Trimble Inpho is a photogrammetry software dedicated to geospatial use. It can transform aerial images into precise points cloud and surface models. It is a complete software with a wide range of modules.

vi. **iWitnessPRO:** iWitnessPRO is a professional photogrammetry software, supporting both close-range photogrammetry and aerial photogrammetry. It is easy to use, supports GCPs and allows to generate orthoimages and digital surface models (DSMs).

vii. **PhotoModeler Pro 5:** PhotoModeler Pro 5 is a plugin that is available with the 3D modeling software Rhino. It is able to create realistic and accurate models from images. The PhotoModeler photogrammetry software can also extract measurements and models from images taken with ordinary cameras.

viii. **Correlator 3-D:** Correlator3D is a mapping software developed for terrain modeling and aerial photogrammetry. This software can produce dense Digital Surface Models (DSM), Digital Terrain Models (DTM), dense point clouds, ortho-mosaics and vectorised 3-D features.

Unit Summary

This unit covers various aspects of photogrammetry. Various types of photographs and associated technical term used in photogrammetry are defined. Concept of scale and relief displacement is explained. Relationships have been established to determine the scale of the photograph as well as reliedf displacement. The principle of stereoscopy and use of stereoscope are covered. Use of parallax bar to determine the elevations of various points has been discussed. Aerial triangulation plays an important role for providing horizontal as well as vertical controls from several aerial photographs; and therefore it has been described. The use of digital aerial photographs which can be employed for large number of applications, including creation of DEM, DSM etc. At the end, some photogrammetric software are also given.

Solved Examples

Example 4.1:

A vertical photograph was taken at an altitude of 1200 m above *msl.* Find out the scale of photograph for a terrain lying at an average elevation of (i) 80 m and (ii) 300 m, if the foal length of the camera is 15 cm.

Solution:

H = 1200 m, f = 15 cm, and (i) h = 80 m, (ii) h = 300 m

The scale of a photograph is given by = f / (H- h)

i. Scale = 15 / (1200–80) = 1 / 7467. It means 1 cm on the photo is equal to 7467 cm on the ground.

ii. Scale = 15 / (1200–300) = 1 / 6000

Example 4.2:

A camera of 152 mm focal length lens is used to take a vertical photograph from a flying height of 2780 m above mean sea level. If the terrain is flat having an elevation of 500 m above *msl*, determine the scale of the photograph.

Solution:

f = 152 mm, H = 2780 m, h = 500 m

S = f / (H – h)

= (152 / 1000) / (2780 – 500)

= 0.152 / 2280 = 1 / 15000

The scale of photograph is 1 : 15000

Example 4.3:

A camera of focal length 20 cm is used to take vertical photographs of a terrain having an average elevation of 1600 m above *msl*. At what height above *msl* an aircraft must fly to take the photographs at 1:10,000 scale.

Solution:

f = 20 cm, h = 1600 m, and scale = 1:10,000

S = f / (H- h)

H – h = S f or H = S f + h

H = (20/100) (10,000) + 1600 m

= 3600 m

Example 4.4:

A line AB 300 m long, lying at an elevation of 600 m above *msl*, measures 9 cm on a vertical photograph. The focal length of the camera lens is 30 cm. Determine the average scale of the photograph if the average elevation of terrain is 700 m above *msl*.

Solution:

Ground distance = 300 m, map distance = 9 cm, f = 30 cm, h = 600 m

Map distance / Ground distance = S = f / (H-h)

0.09 / 300 = 0.30 / (H–600)

H–600 = (0.30 * 300) / 0.09

H = 1000 + 600 = 1600 m

Now

$S_{av} = f / (H - h_{av})$

$S_{av} = 0.30 / (1600–700)$

$= 1 / 3000$

Example 4.5:

A line AB appears measures 12.70 cm on an aerial photograph and the corresponding line measures 2.54 cm on a map at a scale 1/50,000. The terrain has an average elevation of 200 m above *msl.* Calculate the flying height of the aircraft, above *msl,* if the photograph was taken with camera of 16 cm focal length.

Solution:

Photo distance = 12.70 cm, map distance = 2.54 cm, map scale = 1 / 50,000, h_{av} = 200 m, f = 16 cm

Photo distance / map distance = Photo scale / map scale

12.70 / 2.54 = Photo scale / (1 / 50000)

5.0 = 50000 * Photo scale

Photo scale = 1 / 10,000

Now S = f / (H–h)

1/ 10,000 = 0.16 / (H – 200)

(H–200) = 0.16 * 10000

H = 1600 + 200

H = 1800 m

Example 4.6:

A square flat area on the ground with side as 100 m and uniform height, appears as 1 cm^2 on a vertical aerial photograph. The topographic map shows that a contour of 650 m passes through the area. If focal length of camera lens is 150 mm, determine the flying height of the aircraft.

Solution:

1 cm^2 = 100 m^2

S = 1 cm = 100 m

1 cm = 100 × 100 cm

Scale = 1/10,000

h = 650 m, f = 150 mm = 0.15 m

S = f / (H – h)

1 / 10,000 = 0.15 / (H – 650)

H = (0.15 * 10000) + 650

= 1500 + 650 = 2150 m

Example 4.7:

In aerial mapping, a camera with a 100 cm focal length lens was used. Determine the height at which the airplane must fly, so that a 1 km long road on the ground fits exactly on the 20 cm photographic format of the camera.

Solution:

f = 100 cm = 1 m, size of object, d1= 1 km =1000 m, size of image, d2 = 20 cm = 0.2 m
Scale = 0.2 / 1000 = f / H

H = (1000 * 1) / 0.2

H = 5000 m Or 5 km

Example 4.8:

A rectangular area 130 kmx120 km is to be mapped from aerial photographs taken to a scale of 1/20000. The focal length of the lens of the camera to be used is 152 mm and each print is to be 230 mm square. Provision is to be made for a 60% overlap between successive exposures and a 25% lateral overlap. Find (a) the average height above ground at which the aircraft must operate; (b) the time interval between exposures in any one strip if the operating speed of the aircraft is 200 km/h; and (c) the minimum number of photographs required.

Solution:

a. Average height above ground level:

$$S_{av} = \frac{f}{H - h_{av}}, \quad \text{but } h_{av} = 0 \text{ at ground level}$$

$$\text{Hence, } H = \frac{f}{S_{av}} = \frac{152\text{mm}}{1/20000} = 152 * 20000 * 10^{-3} = 3040\text{m}$$

b. Let the flight line be parallel to the 130 km length. Since there is 60% overlap between successive exposures, the effective length of each photograph is 40% of 230 mm, i.e., 0.4*230 = 92 mm

The ground distance covered by this photo length is

92 mm *20000 *10^{-3} = 1840 m

$$\text{Number of photograph per strip} = \frac{130{,}000}{1840} = 70.65 \cong 71 \text{ photos}$$

The operating speed of the aircraft is 200 km/h. To cover a length of 130 km, the aircraft needs 130 / 200 = 0.65 hour.

Since the exposures are at regular intervals

$$\text{Time interval between exposures} = \frac{0.65}{70.65} \text{hour} = 33.12 \text{ sec}$$

c. The width of the area to be photographed is 120 km. A 25% lateral overlap results in an effective photo length of 230 mm is 0.75 * 230 =172.5 mm. The ground distance covered by this width is 172.5* 20000 *10^{-3} =3450 m.

$$\text{Number of strips} = \frac{120{,}000}{3{,}450} \cong 35 \text{ strips}$$

Minimum number of photograph required = 71*35=2485

Example 4.9:

Two points A and B of elevations of 400 m and 200 m, respectively above *msl*, appear on a vertical photograph having focal length of 20 cm and taken with flying height of 2000 m above *msl*. The photographic co-ordinates of A and B are respectively (2.75, 1.39) cm, and (-1.80, 3.72) cm. Determine the length of line AB on the ground.

Solution:

The ground co-ordinates (X and Y) are given by

$X = x (H - h) / f$

and $Y = y (H - h) / f$

where x and y are photo coordinates.

$X_A = (2.75 / 100) * (2000–400) / 0.20 = 220$ m

$Y_A = (1.39 / 100) * (2000–400) / 0.20 = 111.2$ m

$X_B = (-1.80 / 100) * (2000–400) / 0.20 = -162$ m

$Y_B = (3.72 / 100) * (2000–400) / 0.20 = 334.8$ m

$AB = \sqrt{[(X_A - X_B)^2 + (Y_A - Y_B)^2]}$

$AB = \sqrt{[\{220 - (-162)\}^2 + (111.2 - 334.8)^2]}$

$AB = 10^2 \sqrt{[(15.592)^2 + (5.0)^2]}$

AB = 442.87 m

Example 4.10:

An area of 20 km x 16 km was to be covered by vertical aerial photograph at 1: 15,000 scale with the longitudinal overlap as 60% and the side overlap as 30% scale. Determine the number of photographs (take 23 cm x 23 cm as standard format) required to cover the entire area.

Solution:

A= L1 x L2 = 20 km x 16 km = 20000 m x 16000 m, Longitudinal overlap (OL) = 60% = 0.6, Side overlap (OS) =30% = 0.3, Scale 1: 15000, Photograph size = 23 cm x 23 cm = 0.23 m x 0.23 m = (L x L)

N1 = Number of photographs per flight line = $[L_1 / \{(1 - OL) \text{ scale} * L\}] + 1$

= [20000 / {(1 – 0.6) 15000 * 0.23}] + 1

= 14.49 + 1 = 16 (taking round figure to cover entire area)

N2 = Number of flight lines = $[L_2 / \{(1 - OS) \text{ scale} * L\}] + 1$

= [16000 / {(1 – 0.3) 15000 * 0.23}] + 1

= 6.62 + 1 = 8 (Taking whole no of flight lines)

Total number of photograph required = N1 x N2

= 15 x 8 = 120 No

Example 4.11:

The distance from the principal point of a photograph to the image of the top of a tower is 6.44 cm and the elevation of the tower above the above datum is 250 m. What is the relief displacement of the tower if the scale is 1:10,000 above datum and the focal length of the camera is 20 cm?

Solution:

r = 6.44 cm, h = 250 m, scale =1:10000, f = 20 cm

d = r h / H

and S = f / H or H = f / S

So, d = r h S / f

d = (6.44 / 100) * 250 / (0.20 * 10000)

= 6.44 / 800

= 0.00805 cm

Example 4.12:

The vertical photograph of a flat area having an average elevation of 250 m above datum was taken with a camera of 20 cm focal length. A line PQ, 250 m long, measures 8.50 cm on the photograph. A tower TQ appears on the photograph, and the distance between the images of top and bottom of the tower TQ measures 0.46 cm on the photograph. If the distance of the image of the top of the tower from principal point of the photograph is 6.46 cm, determine the height of tower TQ above datum.

Solution:

Ground distance = 250 m, map distance = 8.50 cm, f = 20 cm, r = 6.46 cm, d = 0.46 cm

S = f / H = 0.20 / H

[(8.50 / 100) / 250] = 0.20 / H

So , H = 0.20 * 25000 / 0.085

H = 588.24 m

d = r h / H

(0.46 / 100) = (6.46 / 100) * h / 588.24

h = 0.46 * 588.24 / 6.46

h = 41.90 m

Example 4.13:

The tower AB 50 m high above the ground, appears in a vertical photograph, was taken at a flying height of 2500 m above *msl*. The distance of the image of the top of the tower from principal point of the photograph is 6.35 cm. Determine the displacement of the image of the top of the tower with respect to the image of its bottom. The elevation of the bottom of the tower is 1250 m above *msl*.

Solution:

r = 6.35 cm= 0.0635 m, h = 50 m

H = Flying height above the bottom of tower = 2500 – 1250 = 1250 m

The displacement of the image of the top of the tower with respect to the image of its bottom = d = r h / H

d = 0.0635 * 50 / 1250

d = 0.254 cm = 2.54 mm

A stereopair of vertical photographs is taken from a flying height of 1233 m above *msl*, with a 152.4-mm focal length camera, and air base as 390 m. With the stereo-photos properly oriented along the flight line, the coordinates of two points *a* and *b* were measured on left photo and right photo as x_a

= 53.4 mm, y_a = 50.8 mm, x_b' = –7.1 mm, y_b' = –46.7 mm , and x_a' = –38.3 mm, y_a' = 50.9 mm, x_b = 88.9 mm, y_b = –46.7 mm, respectively. Find out the elevations of ground points A and B and the horizontal length of line AB.

$P_a = x_a - x_a'$

$P_b = x_b - x_b'$

$P_a = 91.7$ mm

$P_b = 96.0$ mm

$h_a = H\,B\,f / P_a$

$h_b = H\,B\,f / P_b$

$h_a = 585$ m

$h_b = 614$ m

$X_a = B\,x_a / P_a$

$Y_a = B\,y_a / P_a$

$X_B = B\,x_b / P_b$

$X_B = B\,x_b / P_b$

$X_A = 227$ m

$Y_A = 216$ m

$X_B = 361$ m

$Y_B = -190$ m

$AB = \sqrt{[(X_{A-}X_B)^2 + (Y_A - Y_B)^2]}$

$AB = \sqrt{[(227 - 361)^2 + \{(216-(-190)\}^2]}$

$AB = 427$ m

Example 4.14:

A stereopair of vertical photographs is taken from a flying height of 1233 m above *msl.* With the stereo-photos properly oriented along the flight line, the coordinates of two points *a* and *b* were measured on left photo and right photo as x_a = 53.4 mm, y_a = 50.8 mm, x_b' = –7.1 mm, y_b' = –46.7 mm, and x_a' = –38.3 mm, y_a' = 50.9 mm, x_b = 88.9 mm, y_b = –46.7 mm, respectively. Further, coordinates of a control point C are measured as x_c = 14.3 mm and x_c' = –78.3 mm. If the elevation of point C is 591 m above *msl,* calculate the elevations of points A and B, using parallax differences.

$P_c = x_c - x_c' = 14.3 - (-78.3) = 92.6$ mm

$\Delta p_{ac} = P_a - P_c = 91.7 - 92.6 = -0.9$ mm

$h_a = h_c + [(H - h_c)\,\Delta p_{ac} / P_a]$

$h_a = 591 + [(1233-591)\,(-0.9) / 91.7]$

$h_a = 585$ above *msl*

Similarly, $h_b = h_c + [(H - h_c)\,\Delta p_{bc} / P_b]$

$\Delta p_{bc} = P_b - P_c = 96.0 - 92.6 = 3.4$ mm

$h_b = 591 + [(1233-591)\,(3.4) / 96]$

= 614 m above *msl*

Example 4.15:

On a stereopair, the parallax difference between the top and bottom images of a tree is measured as 1.3 mm. The photographs were taken at 915 m height above ground. If the average photo base is 88.2 mm, determine the height of the tree?

Solution:

$h_a = \Delta p\, H / (b + \Delta p)$

$h_a = 1.3 * 915 / (88.2 + 1.3)$

$h_a = 13$ m

Example 4.16:

In a pair of overlapping photographs (mean photo base length 89.84 mm), the mean ground level is 70 m above the datum. Two nearby points are observed and the following information is obtained.

Point	Height above datum (m)	Parallax bar reading (mm)
X	55	7.34
Y		9.46

If the flying height was 2200 m above datum and the focal length of the camera was150 mm, find the :(i) Air base (ii) The height of Y above datum,(iii) The difference in height between point X and Y.

Solution:

$f / (H\text{-}70) = 89.84 / B = 150 / (2200\text{-}70)$

B= 1275.728m

$P_x / f = B / (H\text{-}h_x)$

$7.34 / 150 = 1275.728 / (2200\text{-}55)$

$P_x = [1275.728 / (2200\text{-}55)] \times 150$

$P_x = 89.22$ mm

$P_y\text{-}P_x = \Delta P$

$(9.46\text{-}7.34) = P_y\text{-}89.22$

$P_y = 91.33$mm

$P_y = f B / (H\text{-}h_y)$

$91.33 = 150 \times 1275.728 / (2200\text{-}h_y)$

$(2200\text{-}h_y) = 150 \times 1275.728 / 91.33$

$(2200\text{-}h_y) = 2095.25$

Difference in height between A and C= 104.75- 55= 49.75m

$h_y = 104.75$m

Difference in height between X and Y= 104.75- 55= 49.75 m

Example 4.17:

On the overlap of a pair of vertical aerial photographs taken at a height of 2500 m above mean sea level with a 152 mm focal length camera are shown two points A and B. Point B is a point on a pass through a range of hills, while point A is the center of a bridge in a valley. In order to estimate the amount of rise between these points, parallax bar measurements were made as follows:

Point A: mean reading 5.90 mm

Point B: mean reading 11.43 mm

The mean level of the valley containing the principal points of the two photographs is 82.00 m above mean sea level, while a BM on the bridge near A was 74.55 m above mean sea level. If the respective photo bases are 89.1 mm and 91.4 mm, calculate the height of B above A.

Solution:

$$p = \frac{fB}{H-h} = \frac{bH_o}{H-h}$$

H_o = 2500-82.0 =2418 m above mean sea level, and elevation of A is 74.55m above mean sea level.

$$P_a = \frac{bH_o}{H-h_a} \qquad b = \frac{b_1+b_2}{2} = \frac{89.1+91.4}{2} = 90.25\text{mm}$$

$$\therefore p_a = \frac{90.25 * 2418}{2500-74.55} = 89.97\text{mm}$$

From the parallax bar readings, the difference in parallax is found as

$dp_{ab} = p_b - p_a$ =11.43 -5.90 =5.53 mm

$\therefore p_b = p_a + dp_{ab}$ =89.97+5.53= 95.50 mm

$$\Rightarrow p_b = \frac{bH_o}{H-h_b} \Rightarrow h_b = H - \frac{bH_o}{p_b} = 2500 - \frac{90.25 * 2418}{95.50} = 214.93\text{m}$$

Therefore, the height of B above A is 140.40 m

Exercises for Practices

a. Short questions

4.18 Explain in types of aerial photo based on the alignment of optical axis.

4.19 Discuss some of the principal uses of terrestrial photogrammetry. Draw the diagram to explain the Exposure station, focal length, flying height, and optical axis of aerial photograph.

4.20 Draw the diagram to explain the fiducial marks, principal point, conjugate principle point, superlap, mosaic.

4.21 What are the different methods of scale determination?

4.22 State any four advantages that an aerial photograph offers over ground based mapping.

4.23 What is the relevance of base-height ratio in aerial photogrammetry? Write the relationship to determine the (i) average scale of a photograph, (ii) relief displacement, (iii) absolute parallax of a point.

4.24 What is a relief displacement? On what factors, it depends? Why relief is not present at the principal point a truly vertical photograph.

4.25 Define stereovision. What are the essential conditions to create a stereovision?

4.26 Draw line diagrams to see the stereo-vision from a stereo-pair using (i) Lens stereoscope, and (ii) Mirror and lens stereoscope.

4.27 Draw a line diagram of Parallax Bar, and show various components.

4.28 What do you understand by Digital photogrammetry? What is the difference between an ortho-photo and normal photo?

4.29 Define interior orientation in aerial photogrammetry.

4.30 What is isocentre in a tilted photograph? What is the utility of tilted photographs?

b. Long questions

4.31 Establish a relationship to determine the scale of an aerial photograph in an undulating terrain. Why the scale of a photograph is not uniform throughout?

4.32 What do you understand by relief displacement? Derive a relationship to compute the relief displacement from a vertical photograph.

4.33 Briefly explain about image parallax and its relation with stereoscopic viewing of aerial photographs.

4.34 Discuss the salient features of (i) Lens stereoscope, and (ii) Mirror and lens stereoscope.

4.35 Describe the base lining procedure from a stereo-pair.

4.36 Discuss the use of a Parallax Bar to determine the height of points in a stereo-pair.

4.37 Explain in brief the concept of digital ortho-photo. Write down the steps involved in the generation of an ortho-photo. List some of the applications of ortho-photo.

4.38 Establish a relationship to derive the scale of a tilted photograph.

Unsolved numerical questions

4.39 A vertical photograph is taken at an altitude of 1200 m above mean sea level of a terrain lying at an elevation of 80 m. The focal length of camera is 15 cm find out the approximate scale of the photograph.

(Ans: 1:7467)

4.40 Compute the scales (maximum, minimum, and average) of a photograph, if the highest terrain, average terrain, and lowest terrain heights are 610, 460, and 310 m above mean sea level, respectively. The flying height above mean sea level is 3000 m and the camera focal length is 152.4 mm.

(Ans: Smax = 1:15,700, Smin = 1: 17700, and Savg = 1:16,700)

4.41 The length of an airport runway is 160 mm on a vertical photograph. On a map at a scale of 1:24,000, the runway length is measured as 103 mm. Determine the scale of the photograph at runway.

(Ans: 1:15,400)

4.42 A camera with a focal length of 152.35 mm and a photo size 230 mm* 230 mm is used to photograph an area from a height of 2400 m above *msl*. The average ground elevation is 420 m above *msl*, determine the average scale of the photographs. Also, determine the ground area covered by a single photograph.

(Ans: 1:13,0008, 940,100m2)

4.43 A vertical aerial photograph was taken from a flying height of 1385 m above datum with a 152.4 mm focal length camera. Two ground points A and B appear on the photograph, as *a* and *b*, and their measured photo-coordinates are $x_a = -52.35$ mm, $y_a = -48.27$ mm, $x_b = 40.64$ mm, and $y_b =$

43.88 mm. Determine the horizontal length of line AB if the elevations of points A and B are 204 and 148 m above datum, respectively.

(Ans: AB = 1036 m)

4.44 The distance measured between two points on a map is 2 cm. The corresponding distance on an aerial photograph is 10 cm. Calculate the scale of the photograph when the scale of the map is 1: 50,000.

(Ans: 1:10,00)

4.45 A road segment of length 1 km measures 6 cm on a vertical photograph. The focal length of the camera is 150 mm. If the terrain is nearly plain, determine the flying height of the aircraft.

(Ans: 2500 m)

4.46 If the length covered by each photograph is 1.5 km with 60% overlap, and length of the strip is 18 km, determine the number of photographs in the strip.

(Ans: 30)

4.47 A rectangular area 130 km x 120 km with average height as 160 m above *msl* is to be mapped from aerial photographs at a scale of 1:20000. The focal length of the camera lens to be used is 152 mm and each print size will be 23 cm square. A 60% overlap between successive photographs and a 25% lateral overlap is to be kept. Find (a) the average height above *msl* at which the aircraft must operate; (b) the time interval between two successive exposures in a strip if the operating speed of the aircraft is 200 km/h; and (c) minimum total number of photographs required to cover the entire area.

(Ans: (a)3200 m, (b) 33.12 sec, and (c) 2485)

4.48 If the height of a tower is 50 m, flying height of the aircraft above the base is 5000 m and the image of the top of the tower is 20 cm, from the principal point, what will be the relief displacement?

(Ans: 0.2 cm)

4.49 On an aerial photograph taken at an altitude of 1500 m above ground, the relief displacement of the image of a flagpole was measured as 1.6 cm. The distance from the center of the photograph to the image of the top of the flagpole is 11.0 cm. If the base of the flagpole is at an elevation of 200.0 m above msl, determine the height of the flagpole above ground.

(Ans: 18 m)

4.50 An overlapping pair of vertical photographs taken with a 152.4 mm focal length camera has an air base of 548 m. The elevation of control point A is 283 m above *msl*, and the parallax of point A is 92.4 mm, determine the flying height above *msl*.

(Ans: 1187 m)

4.51 A vertical photograph taken from 535 m above *msl* was used to determine the height of a tower. The elevation at the base of the tower is 259 m above *msl*. Determine the height of the tower, if the relief displacement of the tower is measured as 54.1 mm, and the radial distance to the top of the tower from the photo center is 121.7 mm.

(Ans: 123 m)

References and Suggested Readings

1. Aber, James S.; Marzolff, Irene and Ries, Johannes B. (2010), Small-Format Aerial Photography: Principles, Techniques and Geoscience Applications, Elsevier Science.
2. ASPRS, (1980), Manual of Photogrammetry, ASPRS publication, 4th Ed., p. 1056.
3. ASPRS, (1983), Multilingual Dictionary of Remote Sensing and Photogrammetry, ASPRS publication, p. 343.
4. Balenovic, I., Seletković, A., Pernar, R., Marjanović, H., Vuletić, D., Paladinić, E., Kolic, J., and Benko, M. (2011). Digital Photogrammetry – State of the Art and Potential for Application in Forest Management in Croatia. South-east European forestry, 2, 81-93.
5. Ebner H, Heipke C, Mayer H, and Pakzad, K., (2003) Photogrammetric Image Analysis, ISPRS Conference on "Photogrammetric Image Analysis" (PIA '03), Munich, Sept. 17-19, 2003.
6. Egels, Yves and Kasser, Michel, (2001), Digital Photogrammetry, CRC Press.
7. Formlabs (2021), Photogrammetry: Step-by-Step Guide and Software Comparison, https://formlabs.com/asia/blog/photogrammetry-guide-and-software-comparison/
8. Faig, W. (1979), Canadian Experiences in the Adjustment of Large Photogrammetric Blocks with Auxiliary Vertical Control, International Symposium on Current Trends in Aerial Triangulation, Brisbane, Queensland
9. Garg, P K (2019), Principles and Theory of Geoinformatics, Khanna Book Publishers Ltd., New Delhi.
10. Grave, Cliff, (1996) Digital Photogrammetry: An Addendum to Manual of Photogrammetry, ASPRS Publication.
11. Henderson, Scott, (2010), Sausage Finger, Cooloptical illusions, Trick, https://coolopticalillusions.com/sausage-finger-trick/
12. Hassani, Mahmoud and Carswell, James, (1992), Transition from Analogue to Digital Photogrammetry, Journal of Advances in Remote Sensing, Vol I. No. 3 –VII.
13. JARS, (1996) Stereoscopes, Japan Association of Remote Sensing, http://sar.kangwon.ac.kr/etc/rs_note/rsnote/cp7/cp7-3.htm.
14. Jenson, J R (2013), Remote Sensing of the Environment: An Earth Resource Perspective, Pearson Education India.
15. Kellie, Andrew and Valentine, Wayne Valentine, (1987), Photogrammetry, In: Brinker, R.C., Minnick, R. (eds) The Surveying Handbook. Springer, Boston, MA. https://doi.org/10.1007/978-1-4757-1188-2_18
16. Kraus, K., (1994), Photogrammetry, Verd. Dümmler Verlag, Bonn.
17. Leh, Mansoor; Bajwa, Sreekala; Chaubey, Indrajeet and Cothren, Jackson, (2008), A remote sensing based methodology of delineating stream channel morphology. 10.13031/2013.25124.
18. Manugula, S S and Bommakanti, Veeranna Bommakanti (2018) Photogrammetry, GIS & Remote Sensing, Ebooks2go Inc.
19. Mikhail, Edward M, Bethel, James S and McGlone, J Chris (2001) Introduction to Modern Photogrammetry, John Wiley & Sons.
20. Moffit, F.H. and Mikhail, E., (1980), Photogrammetry, 3rd Ed., Harper & Row Publishers, New York.
21. Nam, Kyoung; Park, Jeongyun; Kim, In and Kim, Kwanggi (2012), Application of Stereo-Imaging Technology to Medical Field. Healthcare informatics research. 18. 158-63. 10.4258/hir.2012.18.3.158.

22. Quirós Rosado, Elia. (2018). Introduction to Applied Photogrammetry and Cartography for Civil Engineering
23. Phillipson, Warren R. (1996) Manual of Photographic Interpretation, ASPRS Publication.
24. Samson, A. A., (2003), M. Tech. Thesis, Dept. of Civil Engineering, IIT Kanpur, India.
25. Schenk, T. (2005), Introduction to Photogrammetry, Department of Civil and Environmental Engineering and Geodetic Science, The Ohio State University 2070 Neil Ave., Columbus, OH 43210.
26. Sijmons, Koert, Introduction on Photogrammetry, ITC, Netherland. https://www.scribd.com/document/277243918/Introduction-on-Photogrammetry-by-Koert-Sijmons.
27. Su, Yuan-Fong; Lin, Yan-Ting; Jang, Jiun-Huei and Han, Jen-Yu. (2022), High-Resolution Flood Simulation in Urban Areas Through the Application of Remote Sensing and Crowdsourcing Technologies. Frontiers in Earth Science. 9. 1-11. 10.3389/feart.2021.756198.
28. Thomas, Robson, S.O., Kyle, Stephen and Boehm, Jan, (2014) Close Range Photogrammetry and 3D Imaging, Walter de Gruyter, GmbH, Berlin.
29. Wilkinson, Benjamin E., DeWitt, Bon A. and Wolf, Paul R.(2014), 4th Ed, Elements of Photogrammetry with Application in GIS, McGraw-Hill Education India Pvt.Ltd–New Delhi
30. Wolf, P., 1980. Elements of Photogrammetry, McGraw Hill Book Co, NY.
31. Wolf, P.R. and Dewitt, B.A. (2000) Elements of Photogrammetry with Applications in GIS. 3rd Ed., McGraw-Hill Education, New York.
32. Wróżyński, R., Pyszny, K., Sojka, M., Przybyła, C. and Murat-Błażejewska, S. (2017) Ground Volume Assessment using 'Structure from Motion' Photogrammetry with a Smartphone and a Compact Camera, Open Geosciences formerly Central European Journal of Geosciences, Vol 9, Issue1.
33. http://www.durban.gov.za/City_Services/engineering%20unit/Surveying_Land_Information/Documents/AerialPhotogrammetry.pdf
34. http://www.state.nj.us/transportation/eng/documents/survey/Chapter7.shtm#7.3 Introduction to Photogrammetry by T. Schenk, Autumn Quarter 2005
35. http://professionalaerialphotographers.com/content.aspx?page_id=22&club_id=808138&module_id=158950, History of Aerial Photography
36. file:///C:/Users/user/Downloads/8435637.pdf

05 REMOTE SENSING

Unit Specifics

Through this unit, we have discussed the following aspects:

- Remote sensing components
- Various parts of electro-magnetic spectrum (EMS)
- Interaction of EMS with the atmosphere
- Scattering process and Spectral signature of objects
- Various types of resolutions and their need
- Types of sensors and their salient features
- Types of satellite platforms and their characteristics
- Properties of digital remote sensing data
- Geometric and radiometric correction of satellite images
- Enhancement procedure to improve the quality of remote sensing images
- Image transformations and their utility
- Image classification methods
- Accuracy assessment of thematic maps

In addition to the basic principle of remote sensing data analysis, the characteristics of sensors for different types of data collection has been explained. The practical utility of optical, thermal, microwave, and hyperspectral images is presented for various applications. The unit mainly focusses on the charactrisristics and utilisation of optical remote sensing images for the preparation of various kinds of thematic maps. Questions of short and long answer types are given following lower and higher order of Bloom's taxonomy, and a list of references and suggested readings is given at the end so that the students can go through them for acquiring more knowledge.

Rationale

This unit provides details of various types of remote sensing data and their resolutions. Each of the data has a specific application required for creating the thematic maps, but this unit specifically covers the characteristiscs and analysis of optical remote sensing images. It is important to understand the interactions of EMS with the atmosphere, which is also given in this unit. Since satellite images contain radiometric and geometric errors, so students will learn to apply these corrections to satellite images. Several images have poor contrast due to unfavourable atmospheric conditions, so these image are to be enhanced using various methods as explained here. Image transformations, as explained here, also play an important role for the identification of several features from satellite images. Classification from images become an important activity, so two broad approaches of classifications are discussed here along with their advantages and disadvantages to provide a deeper understanding. At last, a thematic map without accuracy has no meaning, so accuracy assessment method has been described.

Pre-Requisites

Mathematics: geometry and trigonometry, Earth surface, Computer, Software.

Unit Outcomes

List of outcomes of this unit is as follows:

U5-O1: Describe various components of EMS and their utility in remote sensing

U5-O2: Explain the interactions of EMS with the atmosphere and their impact on satellite images.

U5-O3: Realize the role of sensors and data products available from them at various resolutions.

U5-O4: Describe various pre-processing methods to improve the geometry and quality of satellite images.

U5-O5: Apply the reflectance characteristics and other properties of objects for classification of satellite images using supervised and unsupervised classification techniques, including their accuracy assessment.

Unit-5 Outcomes	Expected Mapping with Programme Outcomes (1- Weak correlation; 2- Medium correlation; 3- Strong correlation)					
	CO-1	CO-2	CO-3	CO-4	CO-5	CO-6
U5-01	3	1	2	3	2	1
U5-02	2	2	3	1	2	2
U5-03	2	2	3	3	-	3
U5-04	3	3	2	2	3	-
U5-05	2	3	2	3	1	3

5.1 INTRODUCTION

The word "remote" means "from a distance", and "sensing" in this case means "to record." So remote sensing can be defined as the collection of information about an object, area, or phenomenon and subsequent analysis of data acquired by a device from a remote distance (Lillesand and Kiefer, 1996). More precisely, it can be defined as the "*art of science & technology of obtaining reliable information about the physical objects and environment through the process of recording, measuring & interpreting images/data obtained from remotely distant sensor systems*". For example, with eyes it is possible to gather information about the surroundings from a distance. Reading newspaper, watching TV, listening a lecture during classes, are all are the examples of remote sensing. The human eyes capture the light reflected by these objects and the brain interprets the colour, shape, pattern, etc., and translates this data into useful information. The human eye however is limited to a small part of the total electromagnetic spectrum (EMS), i.e., visible light.

Remote sensing involves an interaction between the incident radiation (EMS) and the targets/objects. It also involves the sensing of reflected /emitted energy from the objects with the help of sensors. The sensors on-board satellites detect the of radiations from the objects/targets. In remote sensing, various kinds of devices and sensors are used to record the electromagnetic radiation outside the visible range, especially the near infrared, middle infrared, thermal infrared and microwaves. Detection and discrimination of objects/features are done through recording of radiant energy reflected or emitted by objects or surface material in various wavelength regions of EMS.

The amount of radiations, emitted and reflected from the Earth's features/objects, depends on the physical and chemical properties of the objects or material. It also depends on the surface roughness, angle of incidence, intensity and wavelength of radiant energy. The variation in energy helps in identification of various objects on the Earth surface. So, the sensors play an important role in data capture and dissemination of objects/targets.

Remote sensing systems provide spatio-temporal information on Earth surface processes from local to global scale. The images from these systems help in a wide range of disciplines, such as surveying and mapping, geography, geology, computer science, zoology, agriculture, forestry, botany, meteorology, soil science, urban planning, military, oceanography and civil engineering. Information, such as land use, vegetation types, urbanisation, soils, water, geology, forest, surface elevation and snow, can be derived from remote sensing images. Temporal remote sensing images are helpful in monitoring the land use change, flood, water pollution, deforestation, forest fire, snow cover, urban sprawl, crop damage, disaster monitoring, etc.

Today, remote sensing images have become an integral part of many national level government schemes or projects. With the availability of very high resolution images from CARTOSAT-3, SPOT-7, Sentinel, IKONOS, WorldView, QuickBird, GeoEye, AMSR, TRMM, SSM/I, RADAR, SAR, etc., remote sensing applications are growing rapidly. Integration of remote sensing data with other thematic layers in a Geographic Information System (GIS) provides additional benefits and flexibility to be used in a variety of applications requiring spatial modelling. In addition, the remotely sensed data and software available on open source platforms have popularized it to be used in various disciplines. Today remote sensing has become an interdisciplinary tool, and is being applied in various disciplines.

This chapter specifically provides the details of data from optical sensors, their characteristics and methods of classification. However, a small description is also given about microwave data and images.

5.2 ADVANTAGES AND DISADVANTAGES OF REMOTE SENSING

The remote sensing technology has some advantages and disadvantages. These are summarized below:

Advantages:

1. Provides data of large areas, giving bird's eye view
2. Provides global data, including data of remote and inaccessible regions
3. Easy and rapid collection of data
4. Provides permanent record of the ground
5. All weather, and day & night data collection
6. Temporal images can be used for monitoring of the area
7. Multispectral data allows identification of features which otherwise may not be seen by human eyes.

8. Digital data can be integrated directly with other digital data in GIS.
9. Relatively economical production of various thematic maps
10. Accurate and fast determination of information
11. Rapid analysis using image processing software

Disadvantages:

1. The interpretation of imagery requires specialized skill
2. The data from high resolution sensors may be expensive
3. Field visit is a must as the analysis needs ground verification
4. Data from multiple sources is difficult to integrate and requires accurate georeferencing
5. Objects can be misclassified if sufficient reference data is not available
6. Requires complete infrastructure of hardware and software

5.3 APPLICATIONS OF REMOTE SENSING

In the present day world, here are large number of application of remote sensing. For more applications, refer to Garg (2022). Some of the applications are given below:

1. Mapping

Mapping from remote sensing image is the most important application. Mapping applications of remote sensing include: land surveying techniques accompanied by the use of a GPS, generating DEMs from remotely sensed data, baseline topographic mapping, water resources mapping, road map, damage delineation (tornadoes, flooding, volcanic, seismic, fire), mapping boundaries for tax/property evaluation, target detection, etc. This is particularly useful in remote and inaccessible areas.

2. Land Cover and Land Use

Land use applications of remote sensing include natural resource mapping, soils, water, urban, wasteland mapping. The temporal data is very useful to determine the changes in the land use and land cover which is useful in planning and management of resources and infrastructure.

3. Agriculture

Satellite images can be used as mapping tools to classify crops, examine their health and viability, and monitor the farming practices. The remote sensing data provides field-based information including crop identification, crop area determination and crop condition monitoring (health and viability). These data are employed in precision agriculture to manage and monitor the farming practices at individual field level, including crop optimization and management of technical operations. The images can help determine the location and extent of crop stress, and develop a treatment plan that optimizes the use of agricultural chemicals. Remote sensing technology can be used to prepare maps of crop types and their extent, needed for agricultural agencies. This information can be used to predict crop yield, derive crop production statistics, facilitate crop rotation records, assess soil productivity, identification of factors influencing crop stress, assessment of crop damage and monitoring the farming activity.

The spectral reflectance of vegetation depends on stage type, changes in the phenology (growth), and crop health, and thus can be measured and monitored by multi-spectral sensors. The observation of vegetation phenology requires multi-temporal images (data at frequent intervals throughout the growing season). Remote sensing can aid in identifying crops affected by conditions that are too dry or wet, affected by insect, weed or fungal infestations or weather related damage. The infrared wavelength is highly sensitive to crop vigour, crop stress and crop damage. Detecting damage and monitoring crop health requires high-resolution, multi-spectral and multi-temporal images.

4. Environmental Study

Coastlines are environmentally sensitive interfaces between the ocean and land, and respond to changes brought about by economic development and changing land use patterns. To determine the impact of human activities in coastal region, there is a need to monitor changes, such as coastal erosion, loss of natural habitat, urbanization, effluents and offshore pollution. The dynamics of the ocean and changes in the coastal region can be mapped and monitored using remote sensing techniques. Remote sensing can be used to study the deforestation, degradation of fertile lands, pollution in atmosphere, desertification, eutrophication of large water bodies and oil spillage from oil tankers. Remote sensing satellites, like MODIS, have helped to study the climate and environment. Retrievals of the Sea Surface Temperature (SST) and Land Surface Temperature (LST) from space provide information for interactions between ocean/land and atmosphere, such as evaporation processes and boundary layer dynamics. Remote sensing images are extensively used for weather forecasting as well as to warn people about the impending cyclones.

5. Forest mapping

Forest types inventories require detailed measurements of stand contents and characteristics (tree type, height, density). Using remote sensing data, various forest types can be identified and delineated. For mapping differences in forest cover (canopy texture, leaf density), multi-spectral images are required, and to get detailed species identification high resolution images are needed. Multi-temporal images datasets offer phenology information of seasonal changes of different species. Stereo-images would help in the delineation and assessment of density, tree height and species. Hyperspectral imagery can be used to generate signatures of vegetation species and certain stresses (e.g., infestations) on trees. The RADAR data is valuable for monitoring the forest in the humid tropics because its all-weather imaging capability. The LiDAR data allows capturing 3-dimensional structure of the forest. The multiple return systems of LiDAR are capable of detecting the elevation of land and objects on it. The LiDAR data helps estimate a tree height, a crown area and number of trees per unit area.

6. Geology

Geology involves the study of landforms, structures, and the subsurface to understand physical processes that create and modify the Earth's crust. Geological applications of remote sensing include: bedrock mapping, lithological mapping, structural mapping, mineral exploration, hydrocarbon exploration, environmental geology, sedimentation monitoring, and geo-hazard mapping.

7. Risk mapping

One of the essential tools for risk assessment is remote sensing data of the area. Remote sensing images can provide valuable data for mapping different types of disasters and mapping the parameters associate with risk. The high-resolution imagery has opened new possibilities in the process of risk estimation, mitigation, and management. This data is useful for both effective damage estimation as well as emergency management. The post-disaster images of event, like volcanic eruptions, tsunamis, fires, floods, hurricanes, and earthquakes have been used globally for saving lives.

8. Hydrology

Hydrology is the study of water on the Earth's surface, whether flowing above ground, frozen in ice or snow, or retained by soil. Examples of hydrological applications include: wetlands monitoring, soil moisture estimation, snow pack monitoring, measuring snow thickness, determining the snow-water equivalent, ice monitoring, flood monitoring, glacier dynamics monitoring, river/delta change detection, drainage basin mapping, watershed modelling, irrigation canal leakage detection, and irrigation scheduling.

9. Oceans and coastal monitoring

The oceans provide valuable food-biophysical resources, and are an important link in the Earth's hydrological balance. Coastlines are environmentally sensitive interfaces between the ocean and land, and they respond to changes brought about by economic development and changing land-use patterns. Ocean applications of remote sensing include; ocean pattern identification, storm forecasting, fish stock assessment, sea surface temperature, oil spill, and shipping.

10. Urban growth

Remote sensing images have been found as one of the most effective tools for monitoring and mapping the environmental changes and urban growth. The advantage of remotely sensed data is the synoptic and repetitive coverage that can help study the growth in urban areas. It helps create a base for urban environmental impact assessment, monitor urban growth, detect the urban change, land cover distribution, and land use.

5.4 COMPONENTS OF A PASSIVE REMOTE SENSING SYSTEM

A complete passive remote sensing system has various components, as discussed below, and shown in Figure 5.1:

a. Source of illumination (Sun)
b. Atmosphere through which signals travel
c. Emitted or reflected signals (from an object or phenomenon)
d. Atmosphere
e. Sensor (from a satellite platform), and
f. Ground receiving station
g. Archive centre and the data products
h. Data interpretation, analysis, and applications

First and foremost, requirement in remote sensing is that we need to illuminate the object in order to collect the images. Sun (A) is the best source of energy illuminating the objects/targets on the Earth surface. Electromagnetic radiation (EMR) from the Sun interacts with the atmosphere (B) and gets absorbed, scattered or transmitted, due to molecules, dust gas particles present in the atmosphere. The radiations reach the Earth surface and interact with the objects/targets (C) where scattering, transmittance, absorption and reflection processes will take place. These process will depend upon the physical and chemical characteristics of the objects/targets. The reflected energy from the objects again interacts with the atmosphere (D) to finally reach the sensor system (E). These analogue signals received by the sensors are converted into digital signals, and transmitted back to ground receiving station (F). From ground receiving station, signals are pre-processed and sent to data archive centre where these are stored and given to the users on demand (G). The users then carry out the interpretation and analysis of images, and apply the results to the application (H) in hand.

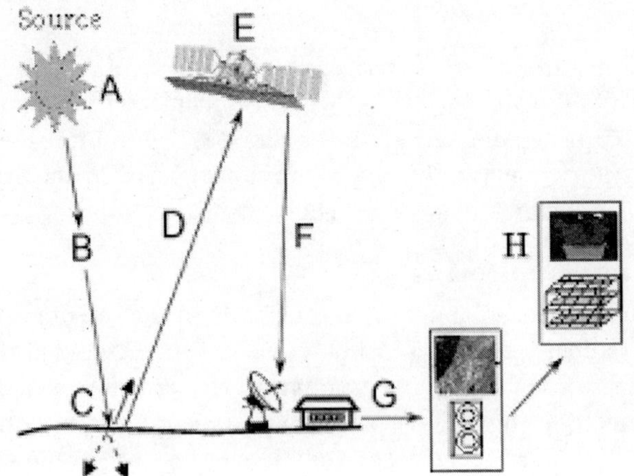

Fig. 5.1: complete passive remote sensing system (Jenson, 1986)

5.5 TECHNICAL TERMS

There are various technical terms to understand the use of remote sensing, as given below:

Electro-magnetic spectrum (EMS)

The EMS consist of a range of energy which contains several parts, such as the gamma ray, x-ray, ultraviolet, visible, infrared, microwave (radar), radio waves. Different parts of the EMS have different wavelengths and frequencies which travel with the same speed as the velocity of light (2.98×10^8 m/s).

Reflected energy

Electromagnetic energy when strikes the objects on the Earth surface can be reflected/emitted, absorbed, scattered or transmitted. The part of the incident energy that is returned back from the objects and measured by the sensor is called reflected energy.

Absorption

It is the process by which electromagnetic energy is absorbed and converted into some other forms of energy.

Transmission

It is the amount of radiations of different wavelengths that a medium (e.g., atmosphere) will transmit or allow to pass through.

Platform

A remote sensing platform is usually a satellite or an airplane, carrying different sensors.

Sensor

A sensor is an electronic device that detects EMR, and converts them into signals that can be recorded as digital numbers, in the form of bits and bytes, and displayed as an image.

Spectral band

The sensors are designed to operate in several wavelength ranges to gather the EMR reflected from or emitted by the ground features/objects. The wavelength range is called a *channel* or a spectral band or simply a *band*. The sensors can collect the data in a number of spectral bands, and may be grouped as; panchromatic (single band), multispectral (more than one band), or hyperspectral (usually over 100 bands) sensors.

Image

The picture resulting from the sensing process is called an image. A remote sensing image can be in paper format or digital format. A digital satellite image is also called a *raster image* which can be displayed on a computer monitor.

Gray scale

It is a medium to calibrate the variations in the brightness of an image that ranges from black to white with intermediate gray values.

Pixel

The word pixel is made from "*picture element*". It is smallest element in an image. As the image is composed of row and columns, one small grid is called a pixel.

Digital Number (DN)

Digital Number in remote sensing system is a variable assigned to a pixel, usually in the form of a byte. In an 8 bit image (2^8), it ranges from 0–255 into 256 grey levels. The DN value is very important as digital analysis of remote sensing images is based on the variation in these values.

Swath

The satellite moves north to east in the orbit and collects EMR reflected/emitted from the ground objects. The width of the area covered on the ground when satellite sensor scans the Earth surface while moving in an orbit is called a swath. This swath width is different for different satellites and sensors, for example, it is 185 km wide by earlier LANDSATs.

Path-Row number

Since the reflected radiations from the ground are continuously recorded by the sensor, each orbit is given a unique number for identification of the ground scene/area, called Path number. The length of the scene defines the Row number. Thus, each scene can be located with its unique Path-Row number. It varies from sensor to sensor and satellite to satellite.

Histogram

A graphical representation of DN values in a set of data is called a histogram. In a histogram, individual DN values are displayed along x-axis, and the frequency of their occurrence is displayed along y-axis.

Brightness of an image

It can be defined as the amount of energy output by a source of light relative to the source to be compared. Brightness is a relative term, and it depends on our visual perception. For example, in some cases we can easily say that the image is bright, or image is dull.

Contrast of an image

Contrast can be simply explained as the difference between the maximum and minimum DN values in an image. It is an important factor in any subjective evaluation of image quality. Contrast is the difference in visual properties that makes an object distinguishable from other objects and with the background.

Classification

It is the computational process of assigning the pixels or objects into a set of categories, or classes, having common spectral characteristics or DN values.

Thematic map

A map that displays the spatial distribution of an attribute related to a single topic, theme, or subject, is called a thematic map. Usually, a thematic map displays a single attribute, such as soil type, vegetation, rivers, geology, land use or habitation.

5.6 ELECTROMAGNETIC SPECTRUM (EMS)

The EMS consist of a broad spectrum ranging from very low frequency to very high frequency or from very long wavelength to very short wavelength. The electromagnetic radiation (EMR) travels with the velocity of light, and does not require a material medium to propagate. All EMRs travel in sinusoidal form with different wavelengths and frequencies. It can be broken down in two components; *Electric field* and *Magnetic field*, which are mutually perpendicular to each other (Figure 5.2). The distance between the two successive peaks in the same phase is called the *wavelength*, and is measured in cm, mm or µm. Whereas, the *Frequency of wavelength* is defined as the number of peaks passing through a given point at any given instant of time, and is measured in Hz, GHz.

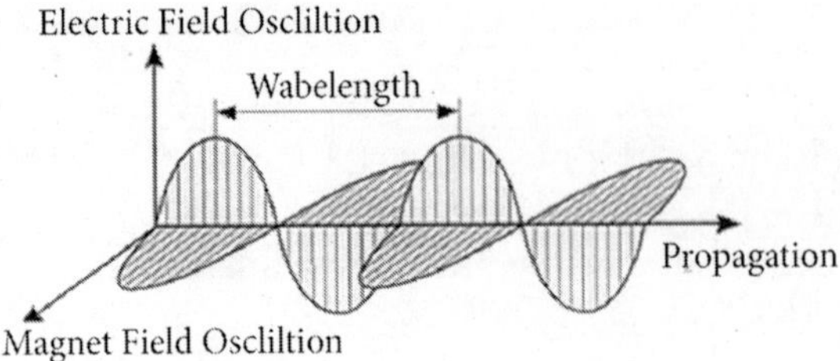

Fig. 5.2: Propagation of EMR

The frequency (ν), wavelength (*f*) and velocity (c = 2.9979x10[8] m/s) of the EMR can be related as-

$$\lambda = c / \nu \tag{5.1}$$

According to this relationship, the wavelength of the emitted radiation is inversely proportional to its frequency. Lower the wavelength, higher the frequency and vice versa.

All electromagnetic waves travel with a wide range of frequencies, wavelengths, and photon energies. The EMS ranges from the shorter wavelengths to the longer wavelengths, as shown in Figure 5.3. It has γ-ray and x-ray regions, ultraviolet (UV) region, visible region, infrared (IR) region, microwaves and radio waves. Each part of this spectrum has some specific utility. Table 5.1 presents various wavelength regions and their specific applications. The γ-ray and x-ray regions are not used in remote sensing but these regions are useful in medical science for investigation. The Ultraviolet (UV) radiations (0.3 to 0.4 µm) is mostly absorbed by Ozone at an altitude of between 20 and 40 km in the atmosphere, so, a very little part of UV reaches the Earth surface.

Visible spectrum (0.4 to 0.7 µm) is an important part of the EMS, as human eyes are sensitive to detect this region. The visible region occupies a very tiny portion of the entire EMS; the longest wavelength is red and the shortest is violet. This portion of the spectrum is associated with the concept of **colours;** Blue, green, and red being the *primary colours*. All other colours in the spectrum can be formed by combining the three blue, green, and red primary colours in various proportions. Sun energy reflected from the Earth during daytime is recorded in this part of wavelength region by many satellite sensors.

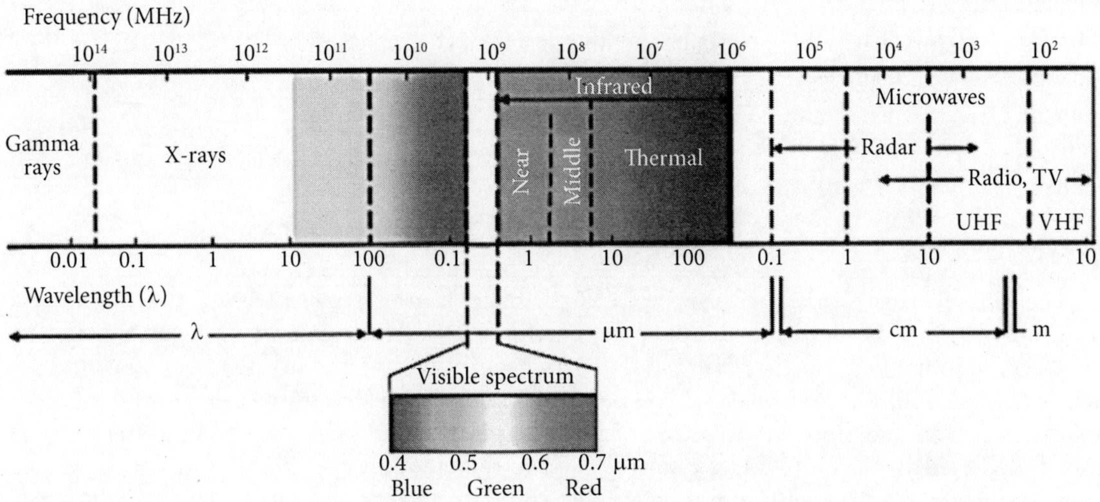

Fig. 5.3: Various part of the EMS

https://rs9956125.wordpress.com/2018/11/02/2-remote-sensing-out-of-this-world/)

Table 5.1: Principal divisions of the EMS (Gupta, 1991)

Wavelength	Description/Application
γ- rays	Gamma rays are used in medical sciences.
x-rays	x-rays are used in medical sciences.
Ultraviolet region (0.3- 0.4μm)	This region is beyond the violet portion of the visible wavelength. UV radiation is largely scattered by earth's atmosphere, and hence not used in field of remote sensing.
Visible region (0.4-0.7μm) **Violet** (0.4-0.446μm) **Blue** (0.446-0.50μm) **Green** (0.50-0.578μm) **Yellow** (0.578-0.592μm) **Orange** (0.592-0.620μm) **Red** (0.620-0.7μm)	It is most important portion of spectrum used in optical remote sensing. Human eyes can detect this light. This part of the spectrum can be associated with the colours.
Infrared region **NIR** (0.75-1.4μm) **SWIR** (1.4-3μm) **MIR** (3-8μm) **LWIR** (8-15μm) **FIR** (15-100μm)	Infrared region is a longer portion of the spectrum. Reflected IR (0.7μm-3.0μm) is used for remote sensing. Thermal IR (3μm-14μm) is the radiation emitted from Earth's surface in the form of heat.

Microwave region (1mm -30cm) **L band** (15cm-30cm) **S band** (7.5cm-15cm) **C band** (3.75cm-7.5cm) **X band** (2.5cm-3.75cm) **Ku band** (1.7cm-2.5cm) **K band** (1.13cm-1.7cm) **Ka band** (0.75cm-1.13cm)	This is a large part of wavelength used in remote sensing. The main advantage of this spectrum is its ability to penetrate through rain, fog and clouds.
Radio waves >30cm	This is the longest portion of the spectrum, mostly used for commercial broadcast and meteorology.

The infrared (IR) region covers the wavelength range from approximately 0.7 μm to 100 μm; more than 100 times as wide as the visible part. The IR can be divided into two categories based on their radiation properties. The near infrared (NIR) and shortwave infrared (SWIR), also known as the *Reflected IR*, refers to the main infrared component of the solar radiation reflected from the Earth's surface. The middle-wave infrared (MWIR) and long wave infrared (LWIR), also known as the *Thermal Infrared*. Reflected IR region (0.7 to 3.0 μm) is used in remote sensing in similar way as the visible region. The region from 0.7 to 0.9 μm is detectable with film, and is called the *photographic IR band*. The thermal IR region (3.0 to 100 μm) is quite different than the visible and reflected IR portions, as this energy is emitted from the Earth's surface in the form of heat. Principal "atmospheric windows regions" occur in the thermal region.

The **microwave region** (1 mm to 1 m) covers a large part of wavelength used in remote sensing. The shorter wavelengths have properties similar to the TIR region, while the longer wavelengths are used for radio broadcasts. Longer wavelengths can penetrate through clouds, fog and rain. This portion of the EMS is used in active remote sensing. Radar images are acquired at various wavelength bands.

5.7 BLACK BODY RADIATIONS

All objects with a temperature above absolute zero (°K, or -273°C) would emit energy in the form of electromagnetic radiation. A blackbody is a body which absorbs all the radiations falling on it, and also a perfect emitter over all wavelengths. While, a white body is a non-absorber and non-emitter, it is a perfect reflector. Ideally, we do not have a perfect black body or a white body. Natural objects behave in-between a perfect black body and a perfect white body; called the *grey body*. The emissivity of a perfect black body is 1, while for a perfect white body it is zero.

The characteristics of blackbody radiation can be described in terms of several laws:

1. Planck's Law:

The energy of an EMR can be quantized. The basic unit of energy for an electromagnetic wave is called a *photon*. The energy E of a photon is proportional to the frequency f of wavelength–

$E = hf$ (5.2)

It can also be written in terms of wavelength as-

$E = h\,(c/\lambda)$ (5.3)

Where h is the Planck's constant = $6.62606957 \times 10^{-34}$ joule·second.

The energy E radiated is inversely proportional to the wavelength of EMS, as product (h^*c) is a constant quantity. It means that lower the frequency, higher will be the energy carried by the photons and electrons. Since, γ-rays, x-rays, etc., have higher frequency due to their lower wavelength, these are able to penetrate through the skeleton/human body, and extremely useful in medical sciences.

Planck's law determines the spectral energy density of the emission at each wavelength (E_λ) at a particular absolute temperature (T). The spectral radiance can also be measured per unit wavelength (λ) instead of per unit frequency. In this case, spectral energy is given by:

$$E_\lambda = \frac{8\pi hc^2}{\lambda^5\left(e^{(hc/\lambda kT)} - 1\right)} \tag{5.4}$$

where h is the Planck's constant

c is the velocity of light = 2.9979 x 10^8 m/s

λ is the wavelength of radiation

k is the Boltzmann's constant = 1.38 X 10^{-23} J /K

T is the Absolute temperature of the radiating body in degree Kelvin (K)

2. Wien's Displacement Law:

This law states that the frequency of the peak of emission increases linearly with absolute temperature (T). As the temperature of the body *increases*, the overall radiated energy increases, and the peak of the radiation curve moves to shorter wavelengths, as shown in Figure 5.4. The maximum radiation is derived from the Planck's formula, and the product of the peak wavelength and the temperature (T) is found to be a constant.

$$\lambda_{max} = \frac{b}{T} \tag{5.5}$$

Where, b is a constant of proportionality known as Wien's displacement constant = 2.898×10^{-3}mK.

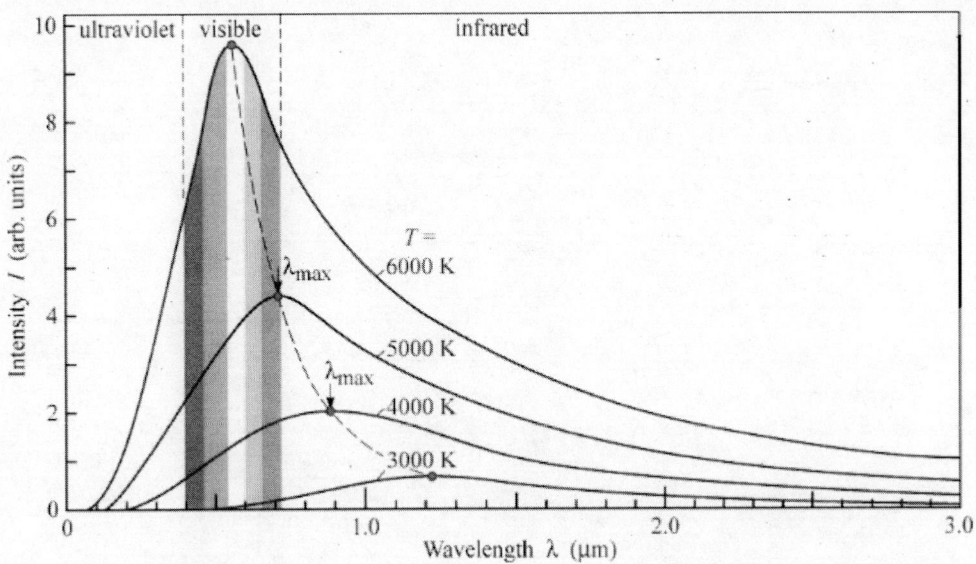

Fig. 5.4: Intensities of EMR at different wavelengths and temperatures (Castleman, 1996)

3. Stefan–Boltzmann Law:

It relates to the *total* energy emitted *(E)* by a body with its absolute temperature (T). The wavelength of the peak of blackbody radiation curve decreases in a linear fashion as the temperature is increased (Figure 5.4). *The Stefan-Boltzmann Law states that the total amount of energy per unit area emitted by an object is proportional to the fourth power of the absolute temperature,* and can be represented as:

$$E = \sigma T4 \qquad (5.6)$$

Where σ **is Stefan–Boltzmann constant** = 5.67×10^{-8} W/(m^2 K^4)

The Sun produces more spectral energy at 6,000°K than the Earth at 300°K. As the temperature increases, the total amount of radiant energy increases and the radiant energy peak shifts from higher wavelength to shorter wavelengths.

5.8 INTERACTION OF EMR WITH ATMOSPHERE

When EMR strikes a material/object on the ground, it is called *incident radiation*. This incident radiation will first interact with atmosphere and then on the surface of Earth, and then again with atmosphere before it reaches the sensor. Four types of interactions will take place with the atmosphere and surface/objects; scattering, absorption, transmission and reflection, as shown in Figure 5.5. These interactions in the atmosphere would depend on the angle of incident radiation, wavelength of radiation and atmospheric conditions. Whereas, on the surface of objects, these would depend upon the angle at which radiations strike the surface, the compositional and physical properties of the surface, and the wavelength of incident radiation.

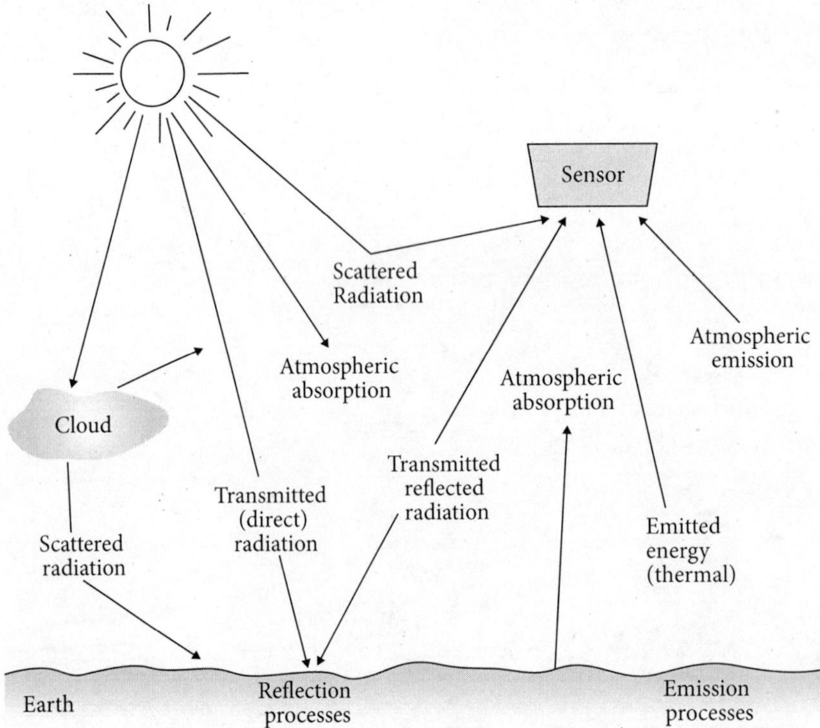

Fig. 5.5: Four types of interactions: transmission, reflection, scattering and absorption (Colwell, 1983)

5.8.1 Types of interactions

The four types of interactions are explained below.

1. Reflection

It is the process whereby incident radiation bounces-off the surface in a single predictable direction. Reflection is caused by the surfaces that are smooth relative to the wavelengths of incident radiation. As per Snell's law, the angle of reflection is always equal and opposite to the angle of incidence. The amount of reflected energy will depend upon the material of the object, wavelength region, and the atmospheric condition.

2. Scattering

It is the process by which small particles diffuse a portion of the incident radiation in all directions. Scattering occurs when incident radiation is dispersed or spread out unpredictably in different directions. Scattering occurs with the surfaces that are rough relative to the wavelengths of incident radiation. Scattering of radiation by the constituent gases and aerosols in the atmosphere causes degradation of remotely sensed images. In the real-world, scattering is much more common than the reflection. Scattering produces *blurring* of the objects in remotely sensed images, resulting in poor resolution.

Three types of scattering commonly take place in the atmosphere: Rayleigh scattering, Mie scattering and Non-selective scattering. These will depend on the wavelength of incident radiant energy, and the size of gas molecule, dust particle, and/or water vapor droplet interacting with the EMR. These are explained below;

i. Rayleigh Scattering

Rayleigh scattering mainly consists of scattering from the gases present in the atmosphere. It is primarily caused by air particles i.e., O_2 and N_2 molecules. Rayleigh scattering takes place when the dimension of the particles present in the atmosphere is much smaller than the size of wavelength λ, and can be represented as:

Rayleigh scattering æ $1 / \lambda^4$ (5.7)

Since, scattering is inversely proportional to the fourth power of the wavelength, radiations in the shorter wavelength (blue of visual part) will scatter much more strongly than the radiations in red wavelengths, as shown in Figure 5.6. Due to Rayleigh scattering, the colour of the sky therefore appears blue, and during the sunset it appears as red. During the sunset, the rays actually follow a longer path through the denser atmosphere, and the shorter wavelength radiation is strongly scattered out of the line of sight, leaving only the radiation in longer wavelengths (i.e., red and orange) to reach our eyes. Multispectral remote sensing data taken in the blue portion of spectrum is therefore has limited use due to this scattering.

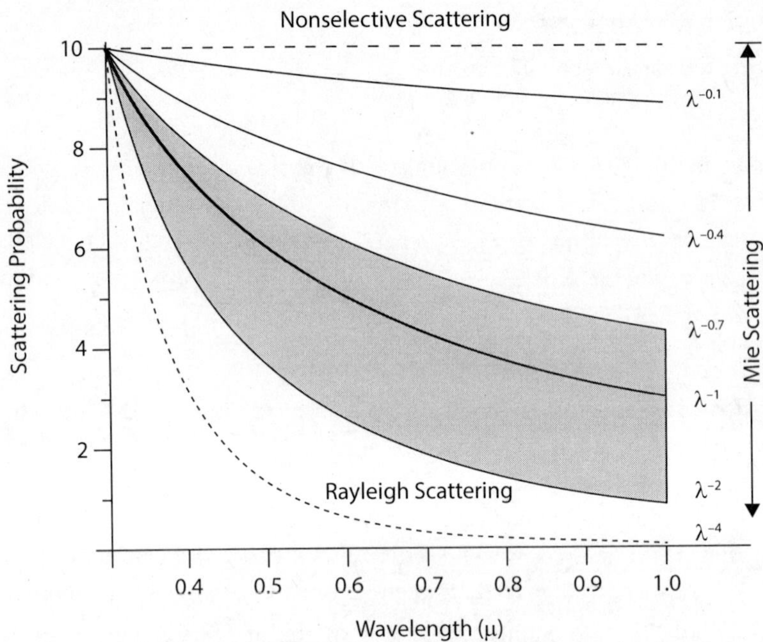

Fig. 5.6: Types of scattering (Gibson, 2000)

i. Mie Scattering

Mie scattering is caused by pollen, dust, smoke, water droplets, and other particles in the lower portion of atmosphere. This scattering is most affected in the lower 4.5 km of the atmosphere. Mie scattering (Figure 5.6) occurs when diameter of the particle present in the atmosphere is roughly of the same size as wavelength (λ). This scattering deteriorates the quality of multispectral images under heavy atmospheric haze. The greater the amount of smoke and dust particles in the atmospheric column, the more violet and blue light will be scattered away and only the longer orange and red wavelength light will reach our eyes.

i. Non-selective Scattering

Non-selective scattering is produced when the particles in the lower atmosphere are several times bigger than the diameter of the wavelength (λ). It takes places when the lower atmosphere contains suspended aerosols of diameter at least 10 times larger than the wavelengths. It is called as non-selective, as all wavelengths are scattered, not just blue, green, or red, equally, as evident from Figure 5.6. The large particles of smoke, water vapor, water droplets, ice crystals, present in the atmosphere, scatter all wavelengths of visible light almost equally, and due to this, the cloud appears as white. This scattering can severely reduce the information content of remotely sensed data that the imagery loses the contrast.

3. Absorption

It is the process by which incident radiation is taken in by a medium. A portion of the absorbed radiation is converted into internal heat energy, which may be subsequently emitted at longer thermal infrared wavelengths. Absorption of UV in the atmosphere is mainly due to electronic transitions of the atomic and molecular oxygen and nitrogen. The primary gases that are responsible for the atmospheric absorption of energy are; ozone, water vapour and carbon dioxide. Ozone in the stratosphere absorbs about 99% of the harmful solar UV radiation, and eventually protects us from diseases, like skin cancer.

There is a very little absorption of the EMR in the visible part of spectrum. The absorption in the infrared (IR) region is mainly due to water vapour (H_2O) and carbon dioxide (CO_2) molecules. The water and carbon dioxide molecules have absorption bands centred at the wavelengths from near to far infrared (0.7 to 15 μm). In the far infrared region, most of the radiations are absorbed by the atmosphere. In microwave radiation, there is no absorption, and that's why it is able to penetrate through atmosphere. Absorption reduces the solar radiance, and may alter the apparent spectral signature of the target being observed.

4. Transmission

In contrast to the absorption, transmission is **the process by which** incident radiations pass through the atmosphere and reach the Earth's surface. Visible wavelengths are able to penetrate to the Earth's surface with little atmospheric interaction, as shown in Figure 5.7. Microwaves and Radio waves have little interaction with the atmosphere, and easily penetrate to the surface.

5.8.2 Atmospheric windows

Atmospheric windows are those regions in the atmosphere where transmission of EMR from source to object and back to the sensor is maximum, and other losses are minimum. In optical remote sensing, many wavelengths within the visible and infrared portions of the electromagnetic spectrum (0.40-2.50 μm) have the ability to penetrate the Earth's atmosphere. Such wavelength regions, as shown in Figure 5.7, white regions in the curve, are considered to be very suitable in remote sensing. In these regions, sensors may be designed to capture the reflected radiations, providing images with good contrast. It is therefore important that a sensor is designed to operate in these regions where the atmospheric losses are minimum. These windows are found in the visible, near-infrared, certain bands in thermal infrared and the microwave regions, as given in Table 5.2.

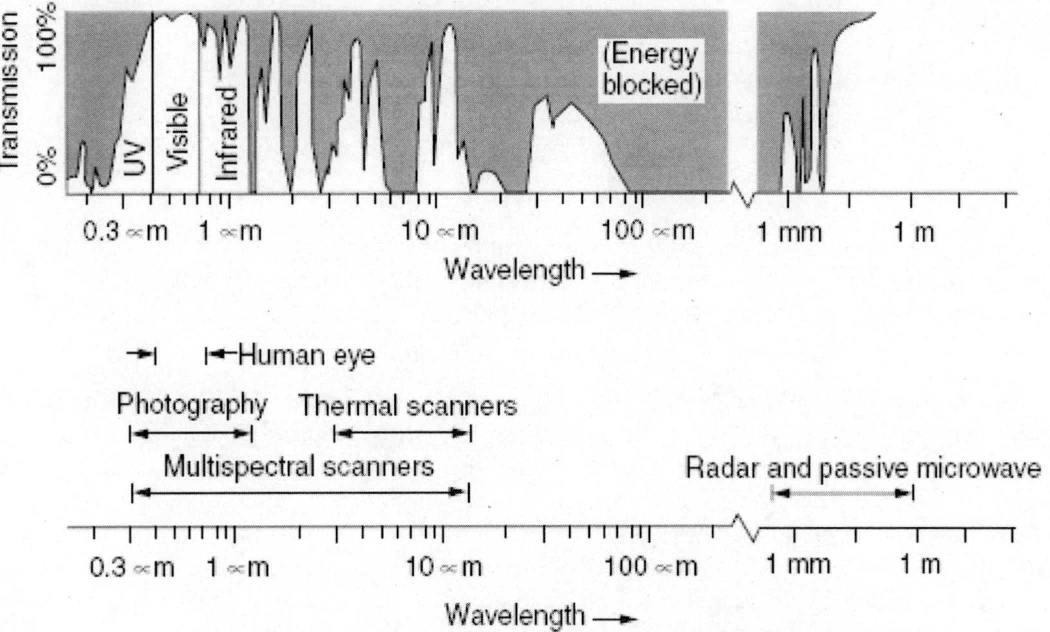

Fig. 5.7: Atmospheric transmission process (Colwell, 1983)

Table 5.2: Major atmospheric windows used in remote sensing (Garg, 2019)

Atmospheric window	Wavelength (μm)	Characteristics
Upper ultraviolet, Visible and photographic	IR 0.3-1.0 approx.	95% transmission
Reflected infrared	1.3, 1.6 and 2.2	Three narrow bands
Thermal infrared	3.0-5.0 and 8.0-14.0	Two broad bands
Microwave	>5000	Atmosphere is mostly transparent

5.9 Spectral Signature of Objects

Spectral signature is the variation of reflectance or emittance of a material with reference to wavelengths (Figure 5.8). For any given material, the amount of solar radiation that is incident on it and then reflected back will vary with the wavelength. Every object due to its own chemical composition and physical properties reflects and emits EMR over a range of wavelengths. This EMR from objects over different wavelengths help in separating them distinctly based on their reflected response for a given wavelength, due to their different spectral signatures.

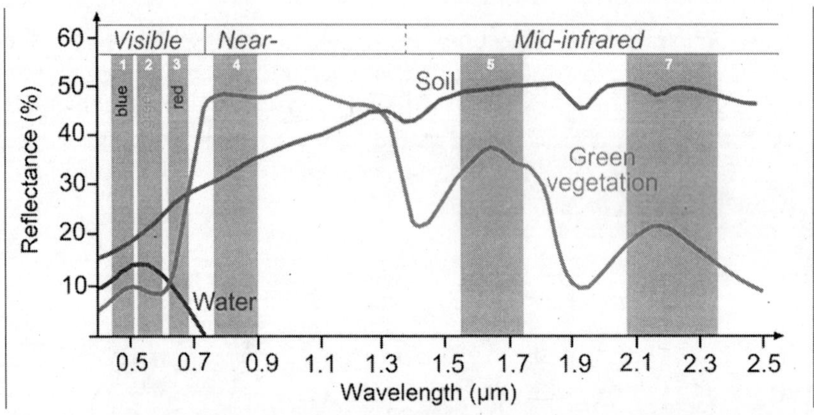

Fig. 5.8: Spectral signature curves (Chavez et al., 1994)

Figure 5.8 shows typical spectral signature curves for three features; water, soils and green vegetation, where the wavelength regions are plotted on x-axis and percentage reflectances on y-axis. The percentage reflectance can be measured and computed as-

Reflectance (%) = (Incident energy/reflected energy)*100 (5.8)

By comparing the reflectance response patterns of different objects, several useful information may be derived. For example, water and vegetation may reflect somewhat similarly in the visible wavelengths but are distinctly separable in the infrared region. The underlying principle is that the two objects may exhibit different reflectances in a certain wavelength region, and thus these two objects are easily identifiable from an image. For example, at certain wavelengths, soil reflects more energy than the green vegetation, while at other wavelengths it absorbs more (reflects less) energy. Thus, various kinds of surface materials can be distinguished from each other by their different characteristics of spectral signature. The spectral response can also vary with time and with wavelength, even for the same object. Satellite sensors normally record different reflected energy in the red, green, blue, or infrared bands of the spectrum, called *multispectral images*. The ability of sensors to detect small changes in reflectance pattern provides a basis for analysis of multispectral remote sensing data.

The spectral signature of various objects can be accurately measured in the field using an instrument, known as *Spectro-radiometer* (or *Radiometer* or *Ground Truth Radiometer*). It is a field-based instrument that measures the intensity of radiation reflected from various objects in several wavelengths of EMS (Figure 5.9). It also helps in measuring the differences in the reflectance pattern of various objects (or spectral signatures) as a function of wavelength. The reflected radiation of various objects is measured in the field through spectro-radiometer in different wavelength regions, and these values are plotted on a graph, representing spectral signature of features. The associate software can automatically generate the signatures curves for various objects so that a detailed analysis can be carried out.

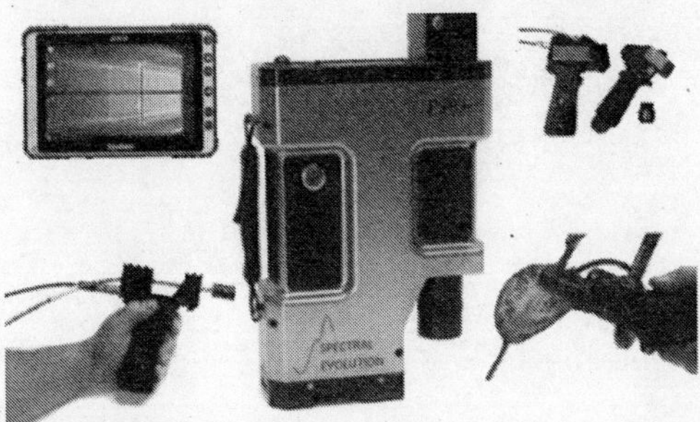

Fig. 5.9: Field spectro-radiometer (Prolite, 2021)

Other possible applications of spectro-radiometer are:

- Inputs into models for plant growth, estimating crop biomass & crop yield, estimating leaf area index and crop loss due to disease, insect infestation, etc.
- Effects of drought on plant growth and yield
- Soil moisture and fertility studies
- Irrigation scheduling studies
- Water pollution and contaminants
- Land surface reclamation studies
- Mineral mapping
- Ground truth for remote sensing image analysis.

5.10 TYPES OF ORBITS

A space-borne remote sensing platform is placed in an orbit in which it moves continuously. From geometrical characteristics point of view, orbits of the space-borne platform can be circular, elliptic, parabolic or hyperbolic. But in practice, elliptical orbits are used. There are various satellite platforms to collect remote sensing data. These satellite move in two major orbits; Geo-synchronous orbits and Sun-synchronous orbits. The satellites moving in these are orbits are called Geo-synchronous satellites and Sun-synchronous satellites, respectively (Figure 5.10).

1. Geo-synchronous Satellites

Geo-synchronous or *Geo-stationary satellite* moves in an orbit so that it covers one revolution in the same time as the Earth to rotate once about its polar axis. The satellites revolve in the same direction as that of the Earth (west to east) at an angular velocity equal to the Earth's rotation rate. In order to achieve this orbital period, geo-synchronous orbits are generally at high altitude of 36,000 km above the equator, making an inclination of 0° from the equatorial plane. Thus from any point on the equator, the satellite moving in such orbit appears to be stationary. The INSAT series of satellites launched by Indian Space Research Organization, Department of Space, Government of India, is an example of these satellites.

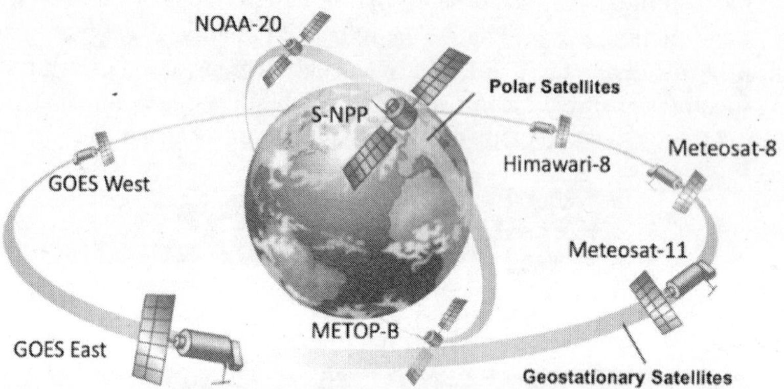

Fig. 5.10: Sun-synchronous and Geosynchronous orbits (Jenson, 2007)

The geo-synchronous satellites, like INSAT, MeteoSAT, GOES, GMS etc., are used for communication and meteorological purposes. Satellites in the geo-synchronous orbit are located at any particular longitude to get a continuous view of that particular region, thus providing frequent images of the same area in a day. Such images are very useful in weather forecasting and meteorological studies. The signals from these satellites are used for communication and television broadcast purposes. Since the images from these satellites cover a large area, the spatial resolution of such images is poor, and thus can't be used for detailed natural resource mapping.

2. Sun-synchronous Satellites

Sun-synchronous or *Polar satellites* move in low orbits (approximately 700-900 km) above the equator. The orbital period typically varies from 90-103 minutes, covering several orbits per day. These satellites make more than one revolution around the Earth in a single day. Sun synchronous orbits maintain a constant angle between the aspect of incident sun and viewing by the satellite, so that the sun-lit portion of the Earth is always below the satellite. These orbits maintain nearly 87° inclination from the equatorial plane. These satellites are passing close to poles that's why they are called Polar satellites. The National Oceanic and Atmospheric Administration (NOAA) series of satellites, like NOAA 17, NOAA 18, IRS, LANDSAT, SPOT. all are examples of polar orbiting satellites.

Due to the rotation of the Earth on its own axis, each time the satellite moves in the orbit, it observes a new area below it. The satellite's orbit period and the rotation of the Earth together are synchronized to allow complete the coverage of the Earth's surface, after a few days. It is generally between 14-22 days, and is known as *revisit period* or *repeat period* of the satellite. Revisit period is the time elapsed between two successive views of the same area by the same satellite. Images from sun-synchronous satellites have good spatial resolution, thus very useful for resource surveys and thematic mapping.

These are many polar satellites with steerable sensors, which can view off-nadir areas before and after the satellite passes over a ground in an orbit. With the off-nadir viewing capability of the satellites, revisit time can be reduced considerably than its repeat period. The revisit time is important, especially when frequent images of an area are required to be analysed for change detection, such as floods, earthquake, forest fire, etc. These satellites will cover areas at high latitudes more frequently than the equatorial zone due to the increasing overlap in adjacent orbit paths.

5.11 TYPES OF REMOTE SENSING PLATFORMS

Essentially, there are three different types of platforms (Figure 5.11) that are used to collect information on Earth's surface features, used to carry out analysis and interpretation. These platforms are:

1. **Ground based platforms**: Such platforms are operational from or near the ground kept near the object under investigation. The studies from the data collected by ground-based platforms are carried out extensively, both at laboratory and in the field. The results greatly help in the development of reflectance pattern and design of sensors for the characterization of Earth surface features, as well as detailed mapping of minute changes, such as cracks, deformations. Terrestrial cameras, handheld cameras, spectro-radiometers, laser based equipment, and GPS are the examples used for laboratory and field experiments to collect information about the Earth features.

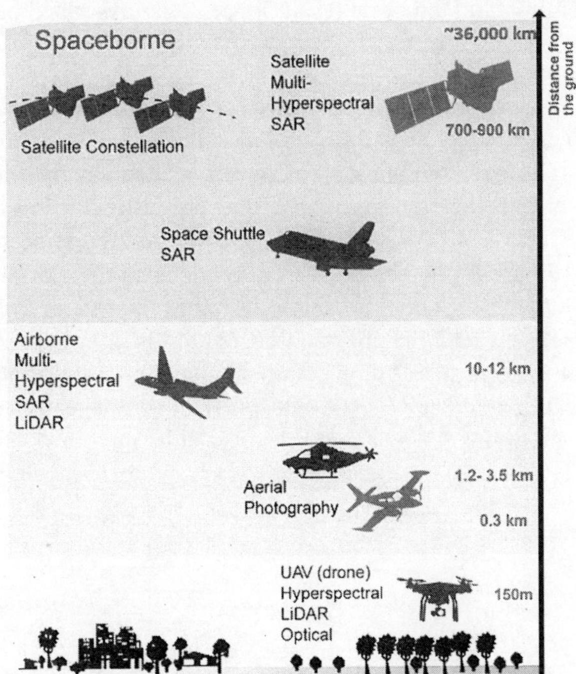

Fig. 5.11: Various platforms used in remote sensing (Lechner et al., 2020)

2. **Air-borne platforms**: These platforms are used to collect the data from a low height above the ground or to test the performance of the sensors before they are actually mounted in the space-borne platforms. These systems cover a larger area than the ground-based methods, and offer speed in data collection. Important airborne platforms include; balloons, aircrafts, drones/Unmanned Aerial Vehicles (UAVs), and laser scanners. Aircrafts and drone offer an economical method of collecting the high resolution ground data. Laser based equipment are also being mounted on aerial platforms to collect the data. The drones (UAVs) are also gaining popularity to collect large scale data about the land surface.

3. **Space-borne platforms**: These platforms operate at much greater heights, such as Landsat, SPOT and IRS remote sensing satellites, the NOAA series of meteorological satellites, the GOES and INSAT series of geostationary satellites. They carry the sensors which have been tested earlier on previous platforms, and collect the data for a large area in a short time.

5.12 DIFFERENT TYPES OF RESOLUTIONS

A satellite image can be best described in terms of its resolution. In remote sensing, the term resolution is used as the capability to identify the presence of two or more objects. Objects closer than the spatial resolution appear as a single object in the image. An image showing finer details is said to have higher resolution as compared to the image that shows coarser details. In remote sensing, four different types of information are needed, such as spatial information, spectral information, temporal information, and radiometric (intensity) information. This information from an object is gathered by using the multispectral sensors/scanners onboard various satellite platforms.

Four important types of resolutions used in remote sensing work are: spatial resolution, spectral resolution, radiometric resolution and temporal resolution, and are described below.

1. Spatial Resolution

A digital image consists of an array of pixels in rows and columns, and each pixel contains information about a small area on the land surface. Spatial resolution is the size of the smallest dimension on the Earth's surface over which an independent measurement can be made by the sensor. It is the minimum separation between the two objects that a sensor is able to record distinctly. It is usually described by the instantaneous field of view (IFOV) of the sensor. The IFOV of the ground seen from the detector of a sensor is also called a *pixel* (Figure 5.12). The IFOV is dependent on the altitude of the satellite; higher the altitude, larger is the IFOV. The spatial resolution describes the size of a pixel of a satellite image covering the Earth surface (Figure 5.12). High spatial resolution images have < 5 m, while low spatial resolution images are > 500 m pixel size. An object of smaller dimension than a pixel but with good contrast with respect to its background can also be detected. Due to this reason, several features, such as roads, canals, railway lines and drainages are detectable on satellite images.

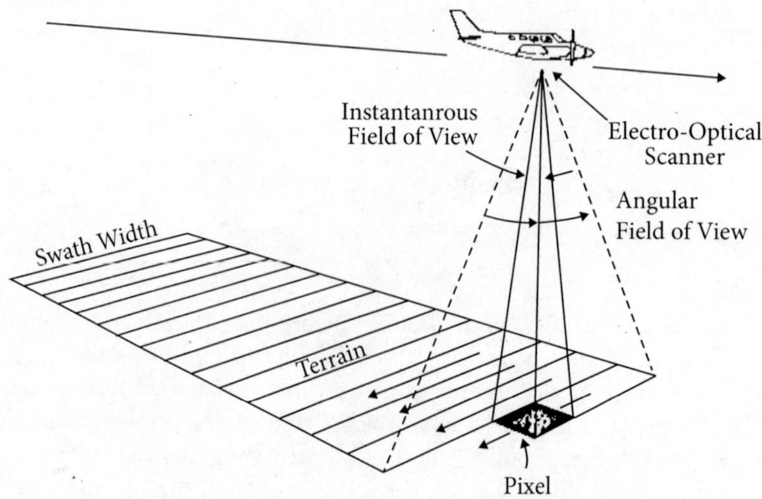

Fig. 5.12: Concept of IFOV and pixel (Garg, 2019)

Remote sensing systems with spatial resolution more than 500 m are generally considered as low resolution systems, e.g., MODIS and AVHRR sensors. The moderate resolution systems have the spatial resolution between 100–500 m, for example, IRS WiFS (188m), band 6, i.e., thermal infrared band of the Landsat TM (120m), and bands 1-7 of MODIS having resolution 250-500 m. High resolution systems have spatial resolution approximately 5-100 m, such as Landsat ETM+ (30 m), IRS LISS-III (23 m MSS and 6 m Panchromatic) and AWiFS (56-70 m), SPOT-5 (2.5-5 m Panchromatic). Very high resolution systems provide less than 5 m spatial resolution, such as GeoEye (0.45 m for Panchromatic and 1.65 m for MSS), IKONOS (0.8-1 m Panchromatic), and QuickBird (2.4-2.8 m). Figure 5.13 shows a comparison of four images taken at different spatial resolutions. Image (A) at 1 m spatial resolution is the best for identification of features, while the image (D) is the worst as at 250 m spatial resolution, one can only see the pixels with grey shades but without any information contents.

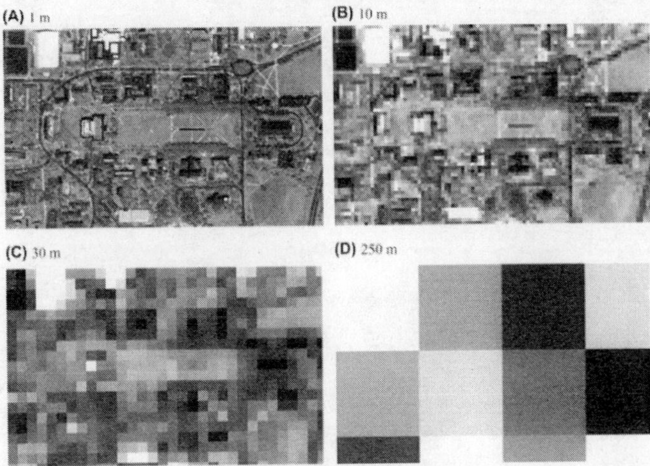

Fig. 5.13: Leslie, 2020

2. Spectral Resolution

The spectral resolution is the ability of a sensor to define the fine wavelength intervals in order to characterize different features of the Earth surface. The finer the spectral resolution, the narrower the wavelength range for a particular band. A satellite sensor, depending on the type, can capture the data in various spectral wavelength regions. In remote sensing, high spectral resolution is achieved by narrow the bandwidths which are collectively likely to provide more accurate spectral signatures of objects than the broad bandwidths, for example, hyperspectral sensors. Many remote sensing systems provide between three and eight bands data, called *multispectral images*, for example Landsat, SPOT and IRS. The number of bands and wavelength of bands, both are important while defining the spectral resolution.

Various sensors provide images at high spectral resolution (>100 bands), medium spectral resolution (3-15 bands), and low spectral resolution (3 bands). For example, IRS LISS-III provides 4 band images, and hyper-spectral sensors provide more than hundreds of very narrow spectral band images. With a higher spectral resolution, single objects can be identified and spectrally distinguished. But when several features/objects are to be identified such as vegetation type, built-up, water, rock classification, crop types, etc., multiple narrow band images are helpful than a single wide band. Figure 5.14 presents Landsat ETM images taken in 8 spectral regions. It is observed that some objects are easily identifiable in one spectral band, while other features are easily distinguishable on another spectral band. The three band images may also be used to create a colour composite.

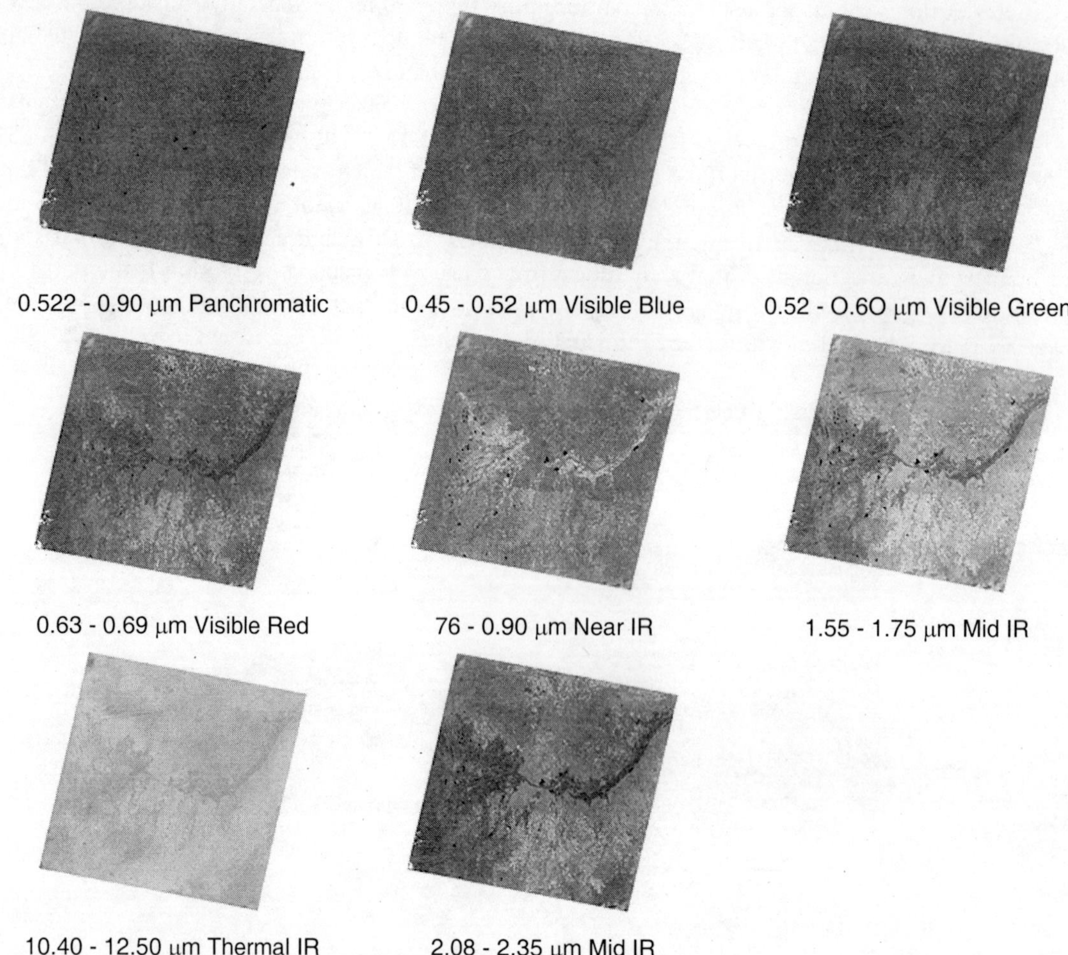

Fig. 5.14: Landsat ETM images at different spectral bands (Stevens et al., 2012)

3. Radiometric Resolution

Radiometric resolution is determined by the number of discrete levels into which reflected radiations may be divided (quantization) by a sensor. With a given spectral resolution, increasing the number of quantizing levels (radiometric resolution) will improve the clarity/identification of the objects. Radiometric resolution depends on the wavelengths and the type of the sensor used. If the radiometric resolution is higher, the small differences in reflected or emitted radiations can be measured accurately, but the volume of data storage will be larger. Most images used in remote sensing are 8 bit (i.e., 2^8= 256), so in many examples in remote sensing literature are related to 8 bit images. Generally, higher the bits, better is the image quality for interpretation. Example of various images at different bits include, Landsat-MSS [Landsat 1-3 provided 6 bits (64 grey values)], IRS-LISS I-III: 7 bits (128 grey values), Landsat-TM (from Landsat 4-5), SPOT-HRV: 8 bits (256 grey values), Landsat-ETM & ETM+ (from Landsat 6-7): 9 bits (only 8 bits are transmitted), IRS-LISS IV: 10 bits (only 7 bits are transmitted), and IKONOS and QuickBird: 11 bits images. Figure 5.15 shows four images of the same area taken by the sensor in 8 bit (256 levels), 4 bit (16 levels), 2 bit (4 levels) and 1 bit (2 levels).

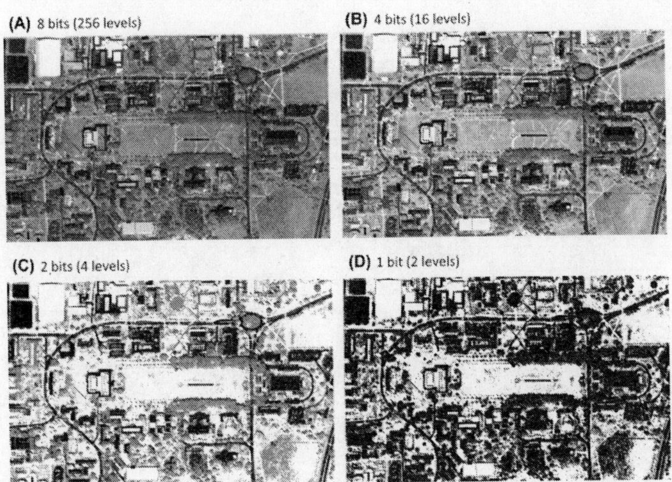

Fig. 5.15: Same image at different radiometric resolutions (Leslie, 2020)

4. Temporal Resolution

Temporal resolution is related to the repeat period between two successive visits of a satellite to a particular area. Smaller the revisit time, better is the temporal resolution of the sensor system. High temporal resolution may be <24 hours-3 days, medium temporal resolution between 4-16 days, and low temporal resolution >16 days. For example, the temporal resolution of IKONOS is 14 days, Landsat 7: 16 days, and SPOT: 26 days, while Meteorological satellites, such as METEOSAT 7 every half an hour and METEOSAT 8 with 15 min have extremely shorter repeat period. Temporal images are suitable for monitoring the dynamic surface features or processes, like the seasonal change of vegetation, growth and development of agricultural crops, floods, expansion of cities, deforestation, etc. Figure 5.16 present two image of urban area taken after a time interval of 4 years to study the changes in the urban area. It is evident that year 2008 image exhibits new buildings and infrastructure that have come up in last 4 years.

For monitoring the changes, temporal resolution is important when determining the resolution characteristics of a sensor system. Many times, images may have cloud cover present on the day of pass, in such cases, next clear image of the area/activity may be available after certain days. In order to reduce the temporal resolution to monitor such area/activity, temporal images of that area/activity can be acquired from various satellites or the off-nadir capabilities of the same satellite are used.

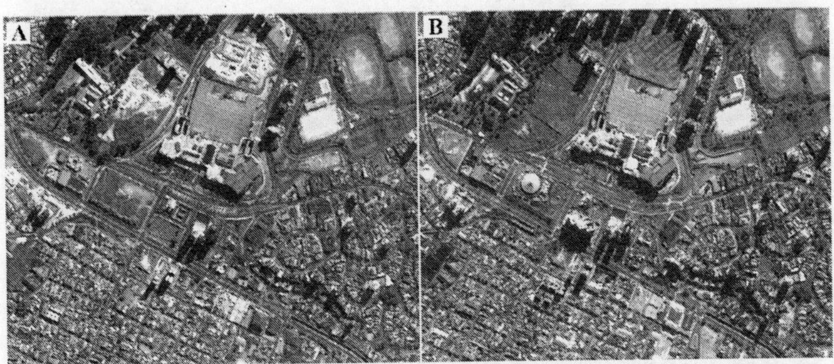

Fig. 5.16: Temporal images to study the changes (A) QuickBird (May 2004), and (B) WorldView-2 (June 2008) (Veettil and Zanardi, 2012)

Table 5.3 gives the spatial and temporal resolutions of some popular satellites.

Table 5.3: Spatial and temporal resolutions of some popular satellites (Garg, 2022)

Satellite	Sensors	Resolution		
		Spectral Bands (μm)	Spatial (m)	Temporal (days)
LANDSAT- 1, 2 and 3*	RBV	Blue-green (0.475-0.575)	80	18
		Orange-red (0.580-0.68) Red to Near-Infrared (0.69-83)	40*	
	MSS	Green (0.5 to 0.6) Red (0.6 to 0.7) NIR (0.7 to 0.8) NIR (0.8 to 1.1) TIR (10.4 to 12.6)*	80	
LANDSAT- 4, and 5	MSS	Same as LANDSAT -1 and 2	80	16
	TM	Blue (0.45-0.52) Green (0.52-0.60) Red (0.63-0.69) NIR (0.76-0.90)	30	
		NIR (1.55-1.75) Thermal (10.40-12.50) Mid-IR (2.08-2.35)	120	
LANDSAT 7	ETM+	Blue (0.45-0.52) Green (0.52-0.60) Red (0.63-0.69) NIR (0.77-0.90) NIR (1.55–1.75)	30	16
		Thermal (10.40-12.50) Mid-IR (2.08-2.35)	60	
		PAN (0.52-0.90)	15	
LANDSAT 8	Operational Land Imager (OLI)	(0.43–0.45) Blue (0.450–0.51) Green (0.53–0.59) Red (0.64–0.67) NIR (0.85–0.88) SWIR 1(1.57–1.65)	30	16
		SWIR 2 (2.11–2.29) PAN (0.50–0.68) Cirrus (1.36–1.38)	15	
	Thermal Infrared Sensor (TIRS)	TIRS 1 (10.6–11.19)	100	
		TIRS 2 (11.5–12.51)	100	

Satellite	Sensors	Resolution		
		Spectral Bands (μm)	Spatial (m)	Temporal (days)
SPOT-1, 2 and 3	HRV PAN	Green (0.50 – 0.59) Red (0.61 – 0.68) NIR (0.78 – 0.89) (0.50 – 0.73)	20 10	26 4-5 days by using steering capabilities
SPOT-4	HR-VIR	In addition to HRV and PAN MIR (1.58-1.75)	20	26 4-5 days by using steering capabilities
SPOT-5	2 HRG sensors	Green (0.50 – 0.59) Red (0.61 – 0.68) NIR (0.78 – 0.89) SWIR (1.58 – 1.75) Pan (0.51 – 0.73)	10 10 10 10 5	26 4-5 days by using steering capabilities
SPOT-6 and 7		Pan (0.45–0.75) Blue (0.45–0.53) Green (0.53–0.59) Red (0.62–0.69) NIR (0.76–0.89)	1.5 8.0	26 4-5 days by using steering capabilities
IRS-1A	LISS-I, and LISS-II A/B (3 sensors)	Blue (0.45-0.52) Green (0.52-0.59) Red (0.62-0.68) NIR (0.77-0.86)	72.5 m LISS-I 36 m LISS-II	22
IRS-1B	LISS-I and LISS-II	same as for IRS-1A		22
IRS-1C & 1D	LISS-III	Green (0.52-0.59) Red (0.62-0.68) NIR (0.77-0.86) NIR (1.55-1.70)	23.5 23.5 23.5 70	24
	PAN	(0.50-0.75)	5.8	24 (5)
	WiFS	Red (0.62-0.68) NIR (0.77-0.86)	188	5
IRS-P3	WiFS	Red (0.62-0.68) NIR (0.77-0.86) NIR (1.55-1.70)	188	5
	MOS-A MOS-B MOS-C	(0.75-0.77) (0.41-1.01) (1.595-1.605)	1500 520 550	Ocean surface

Satellite	Sensors	Resolution		
		Spectral Bands (μm)	Spatial (m)	Temporal (days)
IRS-P4 (OceanSat-1)	OCM MSMR	(0.40-0.90) 6.6, 10.65, 18, 21 GHz (frequencies)	360 x 236 105x68, 66x43, 40x26, 34x22 (km for frequency sequence)	2 2
IRS-P6 ResourceSat-1	LISS-IV	Green (0.52-0.59 Red (0.62-0.68 NIR (0.77-0.86	5.8 5.8 5.8	24 (5)
	LISS-III	Green (0.52-0.59) Red (0.62-0.68) NIR (0.77-0.86) NIR (1.55-1.70)	23.5 23.5 23.5 23.5	24
	AWiFS	Red (0.62-0.68) NIR (0.77-0.86) NIR (1.55-1.70)	70 70 70	5
IRS-P5 CartoSat-1	PAN-F PAN-A	(0.50-0.75) (0.50-0.75)	2.5 2.5	2-line stereo camera
CartoSat-2	PAN camera	(0.50-0.85)	< 1	
OceanSat-2	OCM SCAT ROSA	(0.40-0.90) 8 bands 13.515 GHz GPS occultation	360 x 236 25 km x 25 km	2

Table 5.4 presents the requirements of spatial and temporal resolutions for various applications (Briottet et.al., 2016), ranging from high to low. This table can guide in properly selecting the spatial and temporal resolution for a particular application.

Table 5.4: Requirements of spatial and temporal resolutions for various applications (Briottet et.al., 2016)

Broad application	Applications	Spatial resolution	Temporal resolution
Vegetation and Agriculture	Monitoring/status	High	High
	Monitoring/disease	High	High
	Classification	Medium/High	High
Geology and Soils	Mapping properties	Medium/High	Low
	Exploration	High	Low
Land use	Classification/change	Medium	Low
Urban	Classification/change	High	Low

Broad application	Applications	Spatial resolution	Temporal resolution
Water Resources	Quality assessment	Low	Low
	Bathymetry	Low	Low
	Classification of coastal ecosystems	Low	Low
	Component bloom	Medium	High
Disasters	Prevention	Medium	Low/High
	Monitoring	Medium/High	
	Post-crisis	Medium	Low/Medium

5.13 DIFFERENT TYPES OF SENSORS

Sensor is a device that captures the reflected/emitted radiations from the objects, and coverts these radiations (analog signals) into digital signals. Satellite sensors record the reflected and emitted radiations not only the visible spectrum, but also infrared, near infrared, and thermal infrared bands, and microwave bands. So, normally more than one sensor is deployed in a satellite. Figure 5.17 summarizes the types of sensors being used in remote sensing. There are two broad categories of sensors; Passive sensors and Active sensors, which are further sub-classified.

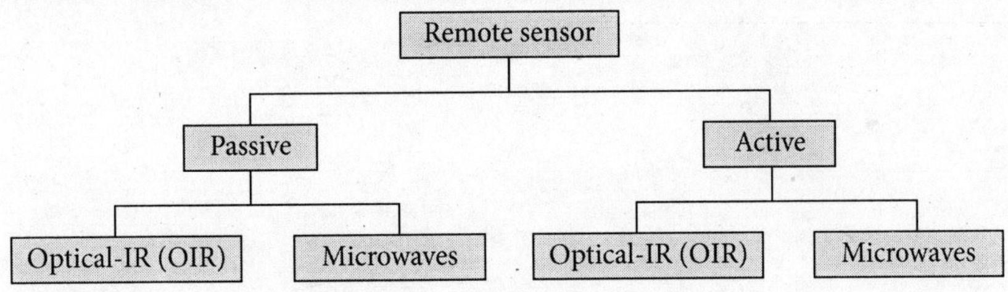

Fig. 5.17: Types of sensors

5.13.1 Based on the source of illumination

1. Passive Sensors

The Sun provides a useful and important source of natural energy for remote sensing applications. The sensors record the reflected Sun energy from the objects. A sensor that depends on an external (natural) source to illuminate the target to be sensed is called a *passive sensor*. In visible light, the Sun illuminates the target/object, and the reflected light from the target is detected by the sensor (Figure 5.18). Most sun-synchronous (Polar) satellites employ passive sensors with them. Landsat MSS is an example of passive sensor. Passive remote sensing images have been used frequently for natural resource mapping and monitoring.

2. Active Sensors

The Sun's energy is available only in the sunlit portion of Earth and not in the other half dark portion of Earth surface, at any given time. So, when taking images in dark or poor weather conditions, sensors carry their own source of energy to illuminate the objects and record reflected and emitted energy from these objects. A sensor that consists of both the source to illuminate the targets and sensor to record the reflected and emitted energy from the objects/targets is called *active sensor* (Figure 5.18). The majority of active sensors operate in microwave portion of the EMS, and are able to penetrate the atmosphere all the time under any conditions. These sensors have the advantage of obtaining data any time of day or season. The Synthetic Aperture Radar (SAR) is a good example of active sensor which transmits microwave pulses from its transmitter antenna to illuminate the target and receives the return signals by its receiver antenna. Active remote sensing images can be used for a variety of applications, such as soil moisture, agriculture, geology, vegetation, including marine, and search and rescue missions.

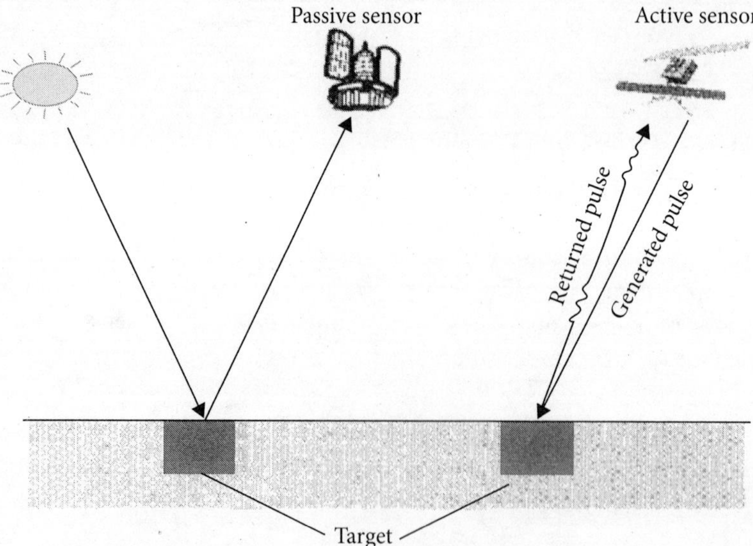

Fig. 5.18: Passive and active remote sensing sensors (Castleman, 1996)

Passive and active sensors both are further divided into *scanning* and *non-scanning systems*. A sensor classified as a combination of passive, non-scanning and non-imaging method is a type of profile recorder, for example a microwave radiometer. A sensor classified as passive, non-scanning and imaging method, is a camera, such as an aerial survey camera or a space camera. Sensors classified as a combination of passive, scanning and imaging are classified further into *image plane scanning sensors*, such as TV cameras and solid state scanners, and object plane scanning sensors, such as multispectral scanners (optical-mechanical scanner) and scanning microwave radiometers.

5.13.2 Based on internal geometry

The internal geometry of design of a space-borne multispectral sensor is quite different from an aerial camera. Figure 5.19 shows six types of remote sensing sensor systems; digital frame area array, scanning mirrors, linear pushbroom arrays, linear whiskbroom areas, and frame area arrays. A linear array, or pushbroom scanner is used in many space-borne platforms, such as SPOT, IRS, QuickBird, OrbView, and IKONOS. The geometric distortions in the images, such as skew caused by the rotation of the Earth, are required to be corrected before analysing the imagery. Several air-borne systems, like Leica ADS-40, ITRES CASI, SASI, and TABI also employ pushbroom technology where each line of imagery is captured at a time, corresponding to an instantaneous position and attitude of the aircraft.

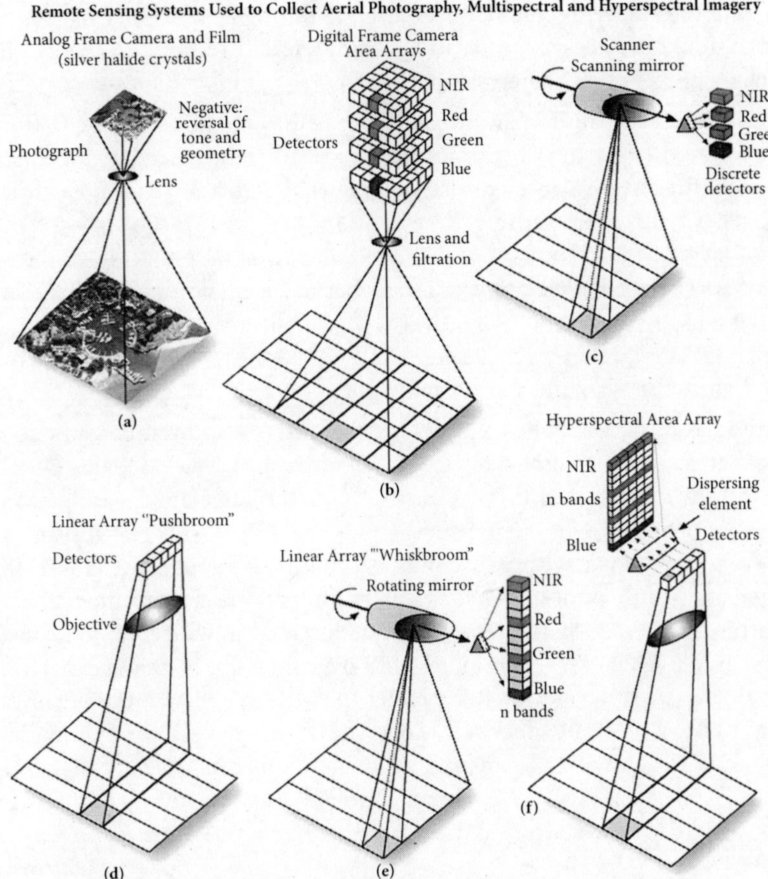

Fig. 5.19: Remote sensing sensor systems (Jensen, 2007)

The internal geometry of images captured by space-borne scanning systems is much more complex. Across-track scanning and whiskbroom systems are more similar to a LiDAR scanner than to a digital array sensor. Each pixel is captured at a unique time, and there is a stack of recording pixels, one for each spectral band, which are required to be precisely co-registered pixel-by-pixel to create an accurate multispectral image.

5.13.3 Based on the wavelength

1. Optical Sensors

The optical remote sensing devices operate in the visible, near infrared, middle infrared and short wave infrared portions of the EMS within a range 0.30 μm to 3.0 μm, e.g., bands of IRS P6 LISS IV sensor work in optical range of EMS.

2. Thermal Sensors

Thermal infrared energy is emitted as heat by all the objects, both natural and manmade, that have a temperature greater than absolute zero. Even in complete darkness and poor weather conditions, thermal imaging is able to detect small temperature differences. For example, water, rocks, soil, vegetation, and the atmosphere, all have the ability to conduct heat directly through them (thermal conductivity) onto another surface and to store heat (thermal capacity). Some materials respond to changes in temperature more rapidly or slowly than the others (thermal inertia).

Human eyes can't detect the thermal infrared energy as they are not sensitive to the infrared (0.7-3.0 μm) or thermal infrared (3-14 μm) regions. The 3 to 5 μm range is related to high temperature phenomenon, like forest fire, while 8 to 14 μm range is related with the general Earth features having lower temperatures. Thermal remote sensing is very useful to measure the surface temperature and thermal properties of targets, for example, fire detection and thermal pollution studies. For example, the last five bands of ASTER (Advanced Spaceborne Thermal Emission and Reflection Radiometer) and band 6 of Landsat ETM+ are thermal. The ASTER data are used to create detailed maps of surface temperature of land, emissivity, reflectance, geology and elevation. Useful reviews on thermal remote sensing are given by Kahle (1980), Sabins (1996) and Gupta (1991).

Thermal infrared sensor data may be collected by (i) across-track thermal scanners, and (ii) push-broom linear and area array charge-coupled device (CCD) detectors. The Daedalus DS-1260, DS-1268, and Airborne Multispectral Scanner (AMS) provide useful high spatial and spectral resolution thermal infrared data for monitoring the environment. Landsat TM 4 and 5 sensors collected 120 x 120 m thermal infrared data (10.4-12.5 μm) along with two bands of middle infrared data. The NOAA (National Oceanic and Atmospheric Administration) Geostationary Operational Environmental Satellite (GOES) collected thermal infrared data at a spatial resolution of 8 km x 8 km which could be used for weather prediction. Entire Earth is imaged every 30 minutes both day and night by the thermal infrared sensors. The NOAA Advanced Very High Resolution Radiometer (AVHRR) collected thermal infrared local area coverage (LAC) data at 1.1 x 1.1 km and global area coverage (GAC) at 4 x 4 km. Sensors, such as NOAA-AVHRR, ERS-ATSR and TERRA-MODIS can provide images at 3.8 μm wavelength that can be used for the detection of fire and hot spots. Figure 5.20 shows various sensors operating in thermal regions.

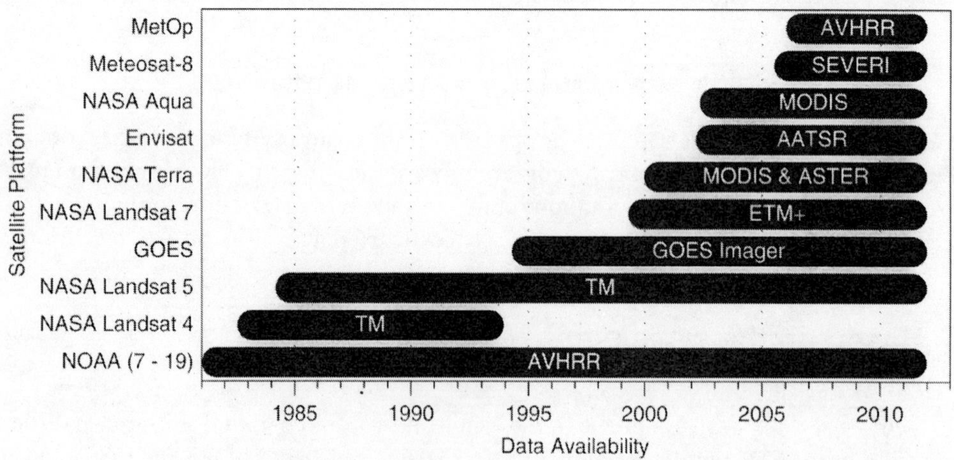

Fig. 5.20: Some thermal sensors (Kahle, 1998)

Thermal images have also been adopted by fire and rescue teams, security professionals, maintenance operations, coal fire team, etc. These sensors have been used for sea surface temperature measurement, detection of forest fires or other warm/hot objects, or monitoring volcanic activities, hydrology, coastal zones, seismology, environmental modelling, meteorology, medical sciences, intelligence/military applications, and heat loss from buildings.

3. Microwave Sensors

The microwave region falls between the IR and radio wavelengths, and has a long range extending from approximately 0.1 cm to 1 m. The microwave sensors, e.g., RADARSAT, having their own sources of energy, record the backscattered microwaves and operate independent of weather and solar energy. Microwave energy can pass through clouds, tree canopies, haze, dust and the rainfall, as these are not affected by atmospheric scattering. These sensors therefore have the capability to collect imagery day and night, under all weather conditions.

The microwave sensors could be passive or active type. The *passive microwave sensors* detect the naturally emitted microwave energy within its field of view. This emitted energy is related to the temperature and moisture properties of the emitting object or surface. Passive microwave sensors are typically spectro-radiometers or scanners, and they operate in the same manner as other radiometers/scanners, except that an antenna is used to detect and record the microwave energy. Since the field of view is large to detect reflected energy, most passive microwave sensors provide low spatial resolution data. The passive microwave sensors can be used to measure atmospheric profiles, determine water and ozone contents in the atmosphere, and measure soil moisture. Oceanographic applications of such sensors include mapping sea ice, currents, and surface winds as well as detection of pollutants, such as oil slicks.

The active microwave sensors have their own source of microwave radiations to illuminate the targets/objects. These sensors can be divided in to two distinct categories: imaging and non-imaging. An active, scanning and imaging sensor is RADAR (RAdio Detection And Ranging), for example SAR which can provide high resolution, imagery, day or night, even under the cloud cover. The working of a RADAR is shown in Figure 5.21. The SAR, whether used airborne or space-borne, emits microwave radiation in a series of pulses from the antenna. This backscattered microwave radiation is detected, and the time required for the energy to travel to the target and return back to the sensor determines the distance or range to the target. By recording the range and magnitude of the energy reflected from all the targets, a two-dimensional image of the surface can be produced.

The intensity in a SAR image would depend on the amount of microwave backscattered by the target and received by the SAR antenna. Since the physical mechanism responsible for this backscatter is different for microwave, the interpretation of SAR images requires the knowledge of how microwaves interact with the targets. Because of these differences, radar and optical data can be complementary to each other as they offer different information content. Other examples of active microwave sensors include ERS-1 launched in 1991, J-ERS satellite launched in 1992, ERS-2 in 1995, and Canada's advanced satellite Radarsat-1 launched in 1995 and Radarsat-2 launched in 2007, IRS-P4 (Oceansat-1), Oceansat-2, RISAT-1, RISAT-2, TerraSAR-X Radar satellite launched in 2007. The Shuttle Radar Topography Mission (SRTM) launched in 2000 uses InSAR which measures Earth's elevation with two antennas to create digital elevation models of Earth.

Non-imaging microwave sensors include altimeters, LiDAR and scatterometers. In most cases, these are profiling devices which take measurements in one linear dimension, as opposed to the two-dimensional representation of imaging sensors. Radar altimeters transmit short microwave pulses and measure the round trip time delay to targets to determine their distance (range) from the sensor. Radar altimetry is used on aircraft for altitude determination and on aircraft & satellites for topographic mapping and sea surface height estimation. The LiDAR is an active microwave sensor that measures ground height, as well as provides coordinates of the points. It consists of a sensor that uses a laser (light amplification by stimulated emission of radiation) to transmit a light pulse, and a receiver with detectors to measure the backscattered or reflected light (Figure 5.22). The distance to the object is determined by multiplying time of travel with the speed of light. Scatterometer is a high-frequency microwave radar designed specifically to measure the backscattered radiations. Measurements of backscattered radiations in the microwave spectral region can be used to derive maps of surface wind speed and direction over the ocean surfaces.

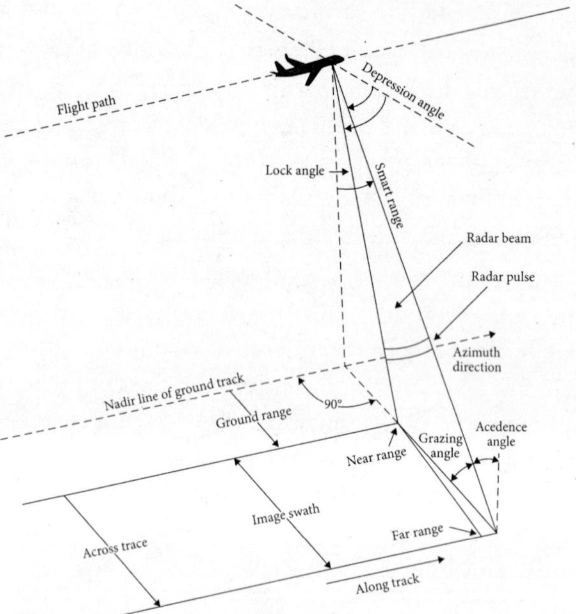

Fig. 5.21: Working principle of a RADAR (Gupta, 1991)

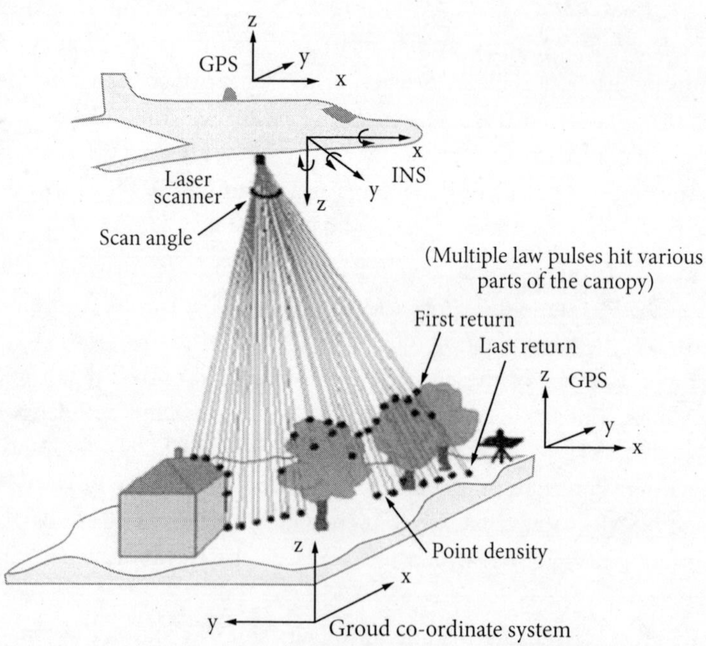

Fig. 5.22: LiDAR scanning (Holland et al., 2003)

Microwave sensors can be used for study of crops, agriculture, urban, land use and land cover, geology and hydrology, forest cover, snow and ice, soil moisture and soil types, snow studies, hydrocarbons, etc., (Calla, 2010). Its potential has been established in various sectors, like discrimination of crop types, crop condition monitoring, soil moisture retrieval, delineation of forest openings, fire scar mapping, monitoring wetlands and snow cover, coastal wind, wave slope measurement, ship detection, substrate mapping, and slick detection (Kasischke et al., 1997). Many studies have demonstrated the use of SAR remote sensing to retrieve biophysical characteristics from forest targets (Richards et al.,1993), and established useful relationships between the backscattering coefficients and the above-ground biomass (Imhoff; 1995). Microwave sensors are being used for planetary exploration. The planets, like Mars and Venus, and satellites, like Moon, have been explored to detect presence of frozen water on Moon (e.g., very successful Chandrayan Mission of India) and presence of buried channels under sand dunes on Mars (Calla, 2010).

4. Hyperspectral Imaging Systems

Multispectral images are usually taken in **3 to 10 bands**, where each band is obtained using a scanner/sensor/radiometer, whereas the hyperspectral images consist of much narrower bands (10-20 ηm) to record the images in more than hundred bands. In general, it comes from an imaging spectrometer. Hyperspectral sensors (also known as *Imaging Spectrometers*) collects images of a scene in tens to hundreds of narrow spectral bands, nearly simultaneously. The hyperspectral sensors have two key component technologies. One is the spectral filtering technique by which the observed scene radiance is divided into narrow distinct bands. The other key is the detector array technology which allows multiple spatial and/or spectral samples through one- or two-dimensional arrays.

The NASA successfully launched the Hyperion imaging spectrometer (part of the EO-1 satellite) which gives 30 m resolution images in 220 spectral bands (0.4-2.5 μm) with a 30 m resolution. The data collected are often termed an "image cube" where the two spatial dimensions are joined by the third spectral dimension (Figure 5.23). The Hyperion instrument provides a new class of observation data for improved characterization of Earth surface. The instrument can image a 7.5 km by 100 km land area per image, and provides detailed spectral mapping with high radiometric accuracy to map complex land eco-systems. Airborne Visible/Infrared Imaging Spectrometer (AVIRIS) is another example of NASA's hyperspectral airborne sensor. For example, AVIRIS collects data in 224 contiguous channels with wavelengths from 0.4-2.5 μm.

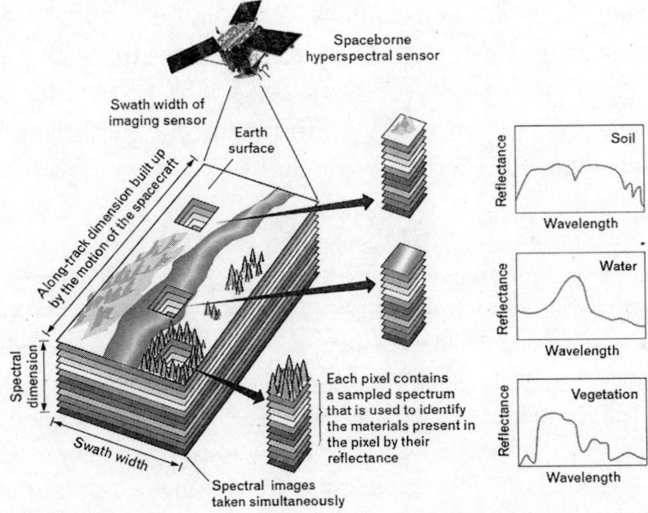

Fig. 5.23: Hyperion imaging spectrometer (Plaza et al., 2009)

Hyperspectral images are being used in vegetation studies (species identification, plant stress, productivity, leaf water content, and canopy chemistry), soil science (type mapping and erosion status), geology (mineral identification and mapping) and hydrology (snow grain size, liquid/solid water differentiation). Lake, river and ocean applications include biochemical studies (phytoplankton mapping, activity), and water quality (particulate and sediment mapping). Processing techniques generally identify the presence of materials through measurement of spectral absorption features. Often the hyperspectral data are post-processed to derive surface reflectance through the use of atmospheric radiative transfer models. One of the drawbacks with hyperspectral images is that it adds a level of complexity to reduce the redundancy from 200 narrow bands to work with. Hyperspectral images have many real-world applications, and give higher level of spectral detail and better capability to analyse minute information. For example, hyperspectral imagery has been used in mineral exploration.

5.14 SOME REMOTE SENSING SATELLITES AND SENSORS

Different satellites are designed and launched based on their intended uses. Satellite imagery employed in various application ranges from 1000 m to <1 m in spatial resolution.

1. Landsats

Landsat, launched by US in July 1972, is the longest running satellite program for acquisition of earth observation imagery. The Earth Resources Technology Satellite (ERTS) was renamed as Landsat in 1975. Since then, the Landsats have provided vast amount of images at medium resolution required to study and analyse the land use/land cover, vegetation and agricultural crops, urban, water, soils, geology and other Earth resources. Landsat has been one of the most important and demanding sources of medium resolution multispectral images globally.

The launch of Landsat-2 and Landsat-3 followed in 1975, and 1978, respectively. The first three satellites were identical and their payloads consisted of a Multispectral Scanner (MSS) and two video cameras, called *Return Beam Videcons* or (RBVs). One scene covered an area of 170 km x 185 km, as the satellites operated at an altitude between 907-915 km in sun-synchronous polar orbit with 103 minutes of orbital period and revisit time of 18 days.

Landsat-4 was launched in 1982. Then, Landsat-5 was launched in 1984, and it continued to deliver high quality, global data of Earth's surfaces for more than 28 years. Landsats-4 and 5 were equipped with two multispectral sensors, i.e., a MSS and a Thematic Mapper (TM). The altitude of the orbit was 705 km with 99 minutes' orbital period, and revisit time of 16 days. Other parameters were same as earlier satellites.

Landsat-6 was launched October 5, 1993, but was lost during launch. Thereafter, Landsat-7 was successfully launched in April 1999 which has 8 separate spectral bands with spatial resolutions ranging from 15-60 m, and a temporal resolution of 16 days (Figure 5.24). It was equipped with a multispectral sensor known as the *Enhanced Thematic Mapper Plus* (ETM+). The ETM+ provides data in one panchromatic band and 7 multispectral bands. Other parameters were same as Landsat-4 and -5 satellites.

Landsat-8 was launched in 2013, which continued to provide daily global data. It was launched in sun-synchronous orbit at an altitude of 705 km, circling the orbit every 98.9 minutes, covering the entire globe every 16 days. It consisted of two sensors. The Operational Land Imager (OLI) operates in nine spectral bands, including a pan band with resolution of 30 m in multispectral and 15 m in pan, while Thermal Infrared sensor (TIR) operates in two spectral bands: (10.6-11.19 μm) and (11.5-12.51 μm) with 100 m spatial resolution.

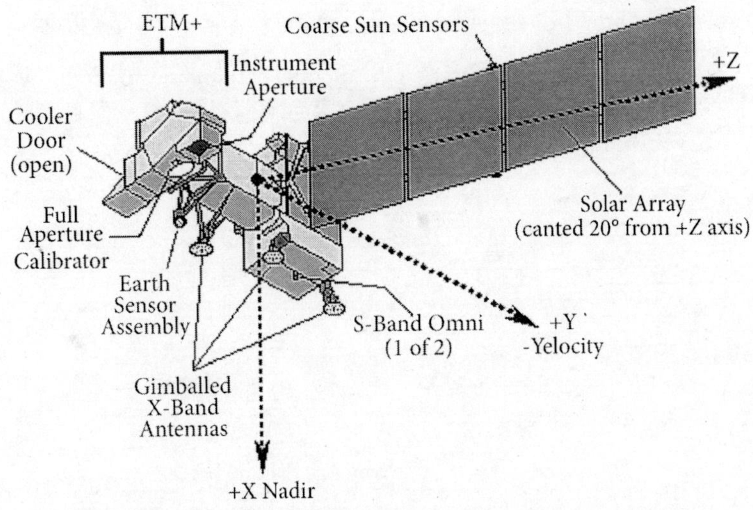

Fig. 5.24: A view of Landsat-7 *(https://landsat.usgs.gov/)*

Landsat-9 is launched on 27, September 2021. It carries the Operational Land Imager-2 (OLI–2), and the Thermal Infrared Sensor-2 (TIRS–2), The OLI–2 captures images of the Earth's surface in visible, near-infrared, and shortwave-infrared bands, and TIRS-2 measures thermal infrared radiation, or heat, emitted from the Earth's surface. Landsat 9 improvements include higher radiometric resolution for OLI-2 (14-bit quantization increased from 12-bits for Landsat 8) allowing sensors to detect more differences, especially over darker areas such as water or dense forests. With the higher radiometric resolution, Landsat 9 can differentiate 16,384 shades of a given wavelength. The TIRS-2 enables improved atmospheric correction and more accurate surface temperature measurements. The spatial resolution is same as of Landsat-8. The date-wise history of Landsats is given in Figure 5.25. For other details, refer to Table 5.4 and Table 5.5.

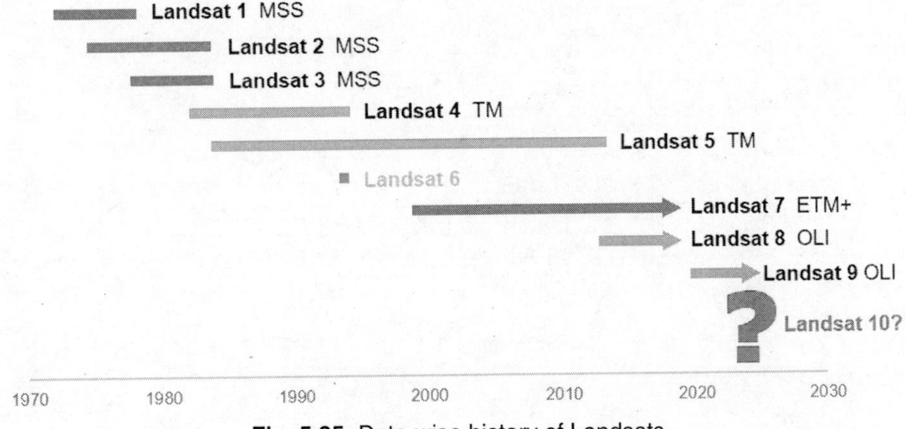

Fig. 5.25: Date-wise history of Landsats

Source: https://landsat.usgs.gov/landsat-missions-timeline

Table 5.5: Details of some satellites and their sensors (Garg, 2022)

Satellite	Launch date	Sensors	Spatial resolution (m)	Swath width (km)
Landsat-1	23. 07.1972	RBV MSS	80 80	185
Landsat-2	22.01.1975	RBV MSS	80 80	185
Landsat-3	05.03 1978	RBV MSS	40 80	185
Landsat-4	16.07.1982	MSS TM	80 30 and 120	185
Landsat-5	01.03.1984	MSS TM	80 30 and 120	185
Landsat-6	05.10.1993	ETM	could not achieve the orbit, and hence failed	
Landsat-7	15.04.1999	ETM+	30, 60 and 15	185
Landsat-8	11.02.2013	OLI TIRS	30 and 15 100	185
Landsat-9	27-09-2021	OLI TIRS	30 and 15 100	185
SPOT-1	21.02.1986	HRV PAN	20 10	60
SPOT- 2	22.01.1990	HRV PAN	20 10	60
SPOT-3	25.09.1993	HRV PAN	20 10	60
SPOT-4	24.03.1988	HR-VIR	20	60
SPOT-5	04.05.2002	HRG	10 and 5	60
SPOT-6	08 .09.2012	HRG	1.5	60
SPOT-7	30.06. 2014	HRG	8.0	60
IRS-1A	17.03.1988	LISS-I, LISS-II A/B	72.5 LISS-I 36.25 LISS-II	148 74 x 2 (148 km)
IRS-1B	29.08.1991	LISS-I and LISS-II A/B	72.5 LISS-I 36.25 LISS-II	148 74 x 2
IRS-P2	15.10.1994	LISS-II M	32 x 37	66 x 2 (131 km)
IRS-1C IRS-1D	28.12.1995 29.09.1997	LISS-III	23.5 and 70	142 and 148
		PAN	5.8	70
		WiFS	188	804

IRS-P3	21.03.1996	WiFS	188	804
		MOS-A MOS-B MOS-C	1500 520 550	195 200 192
IRS-P4 (Oceansat-1)	26.05.1999	OCM MSMR	360 x 236 105x68, 66x43, 40x26, 34x22 (km for frequency sequence)	1420 1360
IRS-P6 Resourcesat-1	17.10.2003	LISS-IV	5.8	70
		LISS-III	23.5	140
		AWiFS	70	740
IRS-P5 Cartosat-1	05.05.2005	PAN-F PAN-A	2.5 2.5	30 30
Cartosat-2	10.01.2007	PAN camera	< 1	9.6
Cartosat-3	27.11.2019	PAN camera Multispectral	0.25 4.0	16
Oceansat-2	23.09.2009	OCM SCAT ROSA	360 x 236 25 km x 25 km	1420 1400
RISAT	26.04. 2012	SAR instrument	< 2 to 50	100–600
RADARSAT-1	04.11.1995		9-100	50-100
ADEOS-1	17.08.1996	POLDER	6000-7000	1440-2200
Terra and Aqua	18.12.1999	MODIS	250, 500, 1000	2330
NOAA	1978	AVHRR	1100 at nadir	3000
	1975	GOES	8000	40
Sentinel-1	03.04.2014	C-Band (SAR)	5-40	80-400
Sentinel-2	23.06.2015	MSI	10, 20 and 60	290
Sentinel-5P	13.10.2017	Tropomi	3500, 7000	2600
Sentinel-3 A	16.02.2016	OLCI	300	1270
Sentinel-3 B	25.04.2018	SLSTR	500, 1000	1470
ERS-1	17.07.1991	ATSR-1	1000	80
ERS-2	21.04.1995	ATSR-2	1000	100
ENVISAT	01.03.2002	MERIS	300, 1200	1150
ALOS-2	24.05.2014	PALSAR-2	1-100	25-490

Various Landsats have employed various sensors, including Return Beam Videcon (RBV) Camera, Multispectral Scanner (MSS), Thematic Mapper (TM), Enhanced Thematic Mapper (ETM), ETM+, Operational Land Imager (OLI), etc. The details can be found in Garg, (2022). Landsat TM images taken in seven spectral regions are shown in Figure 5.26.

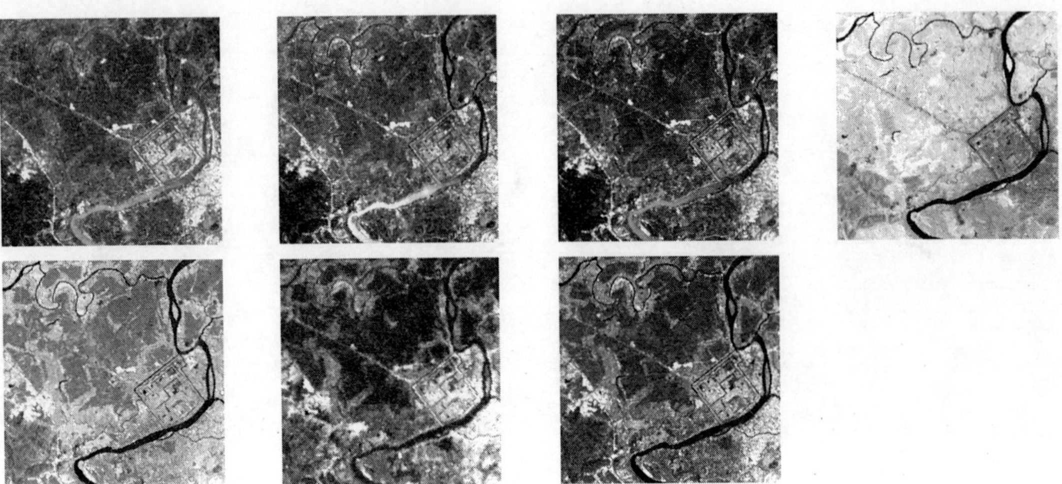

Fig. 5.26: Landsat TM images, April 21, 2003, (a) band 1 (0.45-0.52 μm), (b) band 2 (0.52-0.60 μm), (c) band 3 (0.63-0.69 μm), (d) band 4 (0.76-0.90 μm), (e) band 5 (1.55-1.75 μm), (f) band 6 (10.40-12.50 μm), (g) band 7 (2.08-2.35 μm) (Garg, 2019)

2. SPOT

The SPOT (Système Pour l'Observation de la Terre) (System for Earth Observation), a high-resolution optical imaging Earth observation satellite system, was initiated by the CNES (Centre National d'études Spatiales- the French Space Agency). The SPOT-1 was launched by Ariane-2 on 21 February 1986, in sun-synchronous orbit with an inclination of 98.7° from equator and an altitude of 832 km. The inclination of the orbital plane combined with the Earth's rotation around the polar axis allows a satellite to cover entire Earth every 26 days. It provides data in panchromatic at10 m and multispectral bands at 20 m spatial resolution. The satellite was designed to explore the Earth's resources, detecting and forecasting phenomena, such as climatology and oceanography, and monitoring human activities and natural phenomena. The SPOT satellites provided images with improved resolution over the Landsat images. The SPOT satellites also had the capabilities to collect stereo-images for 3D study of Earth. Refer to Table 5.4 and Table 5.5 for additional details.

The SPOT-2 was launched in 22 January 1990, and SPOT-3 on 25 September 1993 which has ceased operations, due to problems with stabilization. Sensors in SPOT 1-3 were identical, including two identical optical imaging instruments High Resolution Visible (HRV) which operated in 2 modes (panchromatic and multispectral), either simultaneously or individually. The panchromatic band has a resolution of 10 m, and the multispectral 3 bands, operating in Green, Red and NIR wavelengths, have a resolution of 20 m covering an area of 3600 km^2 in a single scene The satellites can offer an oblique viewing capability up to angles of ±27° from the satellite's vertical axis. In this way, the temporal resolution is shortened from 26 to 4-5 days for the temperate zones. Images of the same area are captured on successive days by the same satellite viewing off-nadir.

In March 1998, the SPOT-4 satellite was launched which carried two identical optical sensors; Visible & Infrared High-Resolution (HR-VIR) sensors. This sensor had all bands of HRV sensor plus one additional band in MIR region (1.58-1.75 μm) with 20 m ground resolution. The SPOT-4 also had onboard the first VEGETATION instrument, developed for observations of vegetation cover at global level.

The SPOT-5, launched in May 2002, has 2 High Resolution Geometrical (HRG) instruments offering a higher resolution of 2.5-5 m in panchromatic mode and 10 m in multispectral mode. It was designed to have shortwave infrared band (1.58-1.75μm, essential for VEGETATION data) at a resolution of 20 m, in addition to first three multispectral bands of SPOT-4 HR-VIR sensor. A view of SPOT-5 is shown in Figure 5.27. The SPOT-5 also features an imaging instrument HRS (High-Resolution Stereoscopic) operating in panchromatic mode, pointing forward and backward of the satellite so that it takes stereo-images near simultaneously to create 3D relief. Figure 5.28 presents a multispectral colour image taken by SPOT-5 sensor.

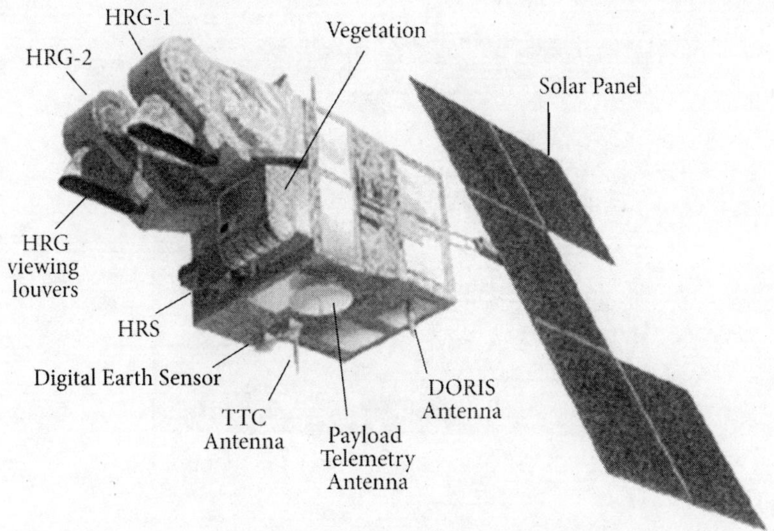

Fig. 5.27: A view of SPOT-5 satellite

https://www.eoportal.org/satellite-missions/spot-5

Fig. 5.28: Rome as seen from SPOT-5 image

https://www.esa.int/ESA_Multimedia/Images/2005/12/Rome_seen_by_France_s_Spot_5_satellite

The SPOT-6 and SPOT-7 were launched on 08 September 2012 and 30 June 2014, respectively, providing panchromatic (0.45–0.75) at 1.5 m resolution, and multispectral in Blue (0.45–0.53), Green

(0.53–0.59), Red (0.62–0.69), NIR (0.76–0.89) μm at 8 m resolution, covering an area 60 km x 60 km, as well as color merged products at 1.5 m, resolution. Figure 5.29 shows a panchromatic band image taken by SPOT-6. The SPOT-6 and SPOT-7 have identical sensors and operating characteristics, and are capable to provide a daily revisit everywhere on Earth. The off-nadir viewing capability is shown in Figure 5.30 where from 4 different times, the satellite captures the images of the same area. These images are very useful for creating 3D model of the area. The disadvantage is that those days the satellite will miss the images of areas down below it. The availability of stereo imagery is limited, as their collection requires special control on the satellite. Traditional photogrammetric techniques can be used to successfully create topographic 3D data of inaccessible areas of the world.

Fig. 5.29: SPOT-6 panchromatic image, August 2016

https://innoter.com/en/satellites/spot-6-7/

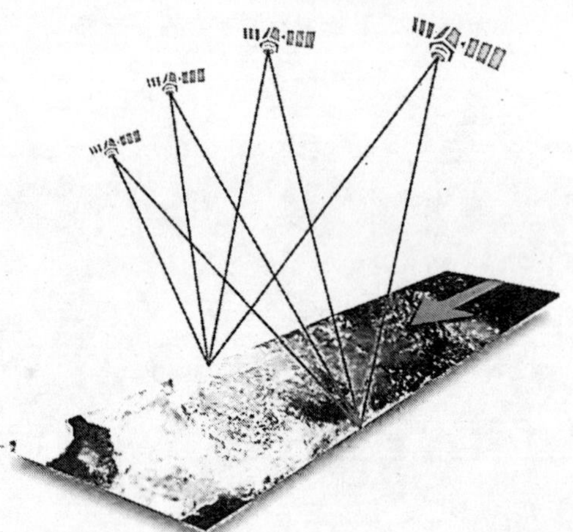

Fig. 5.30: Off-nadir capabilities of SPOT to collect images

https://crisp.nus.edu.sg/~research/tutorial/spot.htm

3. The Indian Remote Sensing (IRS) satellite

India's first civilian remote sensing satellite IRS-1A was launched in March 1988. The first generation satellites IRS-1A and -1B were designed, developed and launched successfully during 1988 and 1991 with multispectral cameras LISS-I (Linear Imaging and Self Scanning sensor) and LISS-II at spatial resolutions of 72.5 m and 36.25 m, respectively. It was launched in a sun-synchronous orbit, at a nominal altitude of 904 km. It had an orbital period of 103.2 minutes with a repeat cycle of 22 days, crossing the equator around 10:26 AM. The IRS-1B provided better repetivity. Figure 5.31 shows the IRS satellite. Refer to Table 5.4 and Table 5.5 for additional details.

Fig. 5.31: The IRS satellite

https://www.isro.gov.in/saga-of-indian-remote-sensing-satellite-system

The IRS-1A and -1B carried LISS-I and -II; each with a multispectral camera providing images in 4 bands (0.46-0.52 μm blue, 0.52-0.59 μm green, 0.62-0.68 μm red, and 0.77-0.86 μm NIR) at 7 bits radiometric resolution. The LISS-I camera provides 76 m resolution images with a swath of about 150 km, while the LISS-II provides images at 36.25 m resolution. Four LISS-II scenes cover the area of one LISS-I scene. Figure 5.32 shows a FCC of IRS-1B image.

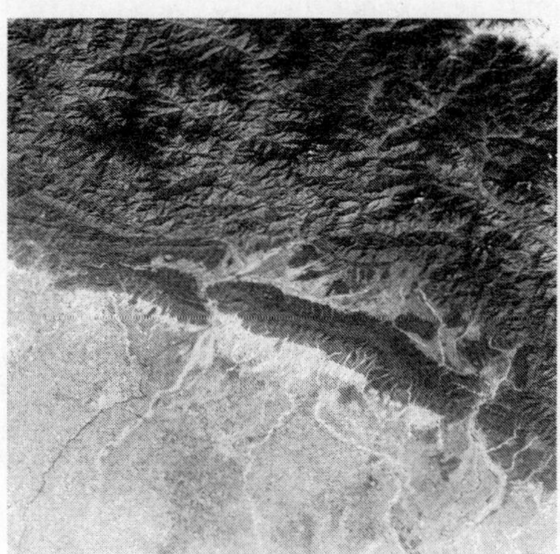

Fig. 5.32: The IRS-1B image of Himalaya and lower region (Pradhan et al., 2010)

The second generation satellites, IRS-1C and -1D with improved spatial resolutions have been successfully launched in 1995 and 1997, respectively. The PAN, LISS-III and WiFS (Wide Field Sensor) sensors on IRS-1C and -1D provided the increased coverage required for application areas, such as resources survey and management, forest studies and disaster monitoring, environmental studies, cartographic and town planning. The panchromatic band data was captured (0.5-0.75 μm) at 5.8 m resolution.

The IRS-1C and 1D employed LISS-III scanner, and provided 7 bits multispectral image with 23.5 m ground resolution in Green 0.52–0.59 μm, Red 0.62–0.68 μm, Near-Infrared 0.77– 0.86 μm, and Short-wave Infrared 1.55–1.7 μm band at 70.5 m resolution. It carried another sensor called, WiFS sensor, which provided 7 bits data at 188 m resolution in two spectral band (0.62-0.68 μm and 0.77- 0.86 μm). The third sensor provides Pan data in 6 bits at 5.8 m resolution in 0.5–0.75 μm wavelength region, covering 70 km swath. Figure 5.33 (left image) shows a PAN image of IRS-1D at 5.8 m resolution.

Resourcesat-1 (IRS-P6), the tenth satellite in IRS series, was launched in October 2003 to provide both multi-spectral and panchromatic imagery of the Earth's surface. It was placed in 820 km high polar Sun synchronous orbit. It has three sensors on-board: a LISS-III, an AWiFS (Advanced Wide Field Sensor) and a high resolution multi-spectral camera LISS-IV, working on 'pushbroom scanning' mechanism. Resourcesat-3, a medium resolution, planned to be launched in 2023, will carry Advanced Linear Imaging Scanning Sensor-3 (ALISS-3) payload consisting of VNIR and SWIR bands. It would also carry Atmospheric Correction Sensor (ACS) for quantitative interpretation and geophysical parameter retrieval and to improve the data products. The ground sampling distances is expected to be 20 m for VNIR and SWIR bands.

The IRS 1C and 1D employed a Wide Field Sensor (WiFS) having 188 m resolution and providing images in 0.62-0.68 μm (red), 0.77-0.86 μm (near IR) with a swath of 774 km. RESOURCESAT employed Advanced Wide Field Sensor (AWiFS) with 56-70 m ground resolution, providing four images in 0.52-0.59μm (green), 0.62-0.68 μm (red), 0.77-0.86 μm (near IR), and 1.55-1.70 μm (mid-IR) with 370-470 km swath. The AWiFS data with improved spatial and spectral resolution helped in better classification accuracy of agricultural related applications, flood inundation, vegetation stress etc.

The IRS-P6 carried LISS-IV scanner to provide data with 5.8 m resolution. This scanner can be operated in two modes: mono and multi-spectral. In the multispectral mode, data is collected in three spectral bands *viz*, 0.52 to 0.59 μm (Green band 2) 0.62 to 0.68 μm (Red band 3), and 0.76 to 0.86 μm (NIR band 4) with 23.5 km swath. In mono mode, the data is available corresponding to a swath of 70 km. The LISS-IV scanner an off-nadir viewing capability by ± 26°, thus providing a revisit of 5 days for any given ground area.

OCEANSAT-1 (IRS-P4) was launched in May 1999 to study physical and biological aspects of oceanography. It carried an Ocean Colour Monitor (OCM) and a Multifrequency Scanning Microwave Radiometer (MSMR). The OCM operates in 0.402-0.422, 0.433-0.453, 0.480-0.500, 0.500-0.520, 0.545-0.565, 0.660-0.689, 0.745-0.785 and 0.845-0.885 μm bands with 360 m spatial resolution and 1420 km swath. OCEANSAT-2, launched on Sept. 23, 2009, carried three payloads: OCM, Ku-band Pencil Beam Scatterometer (SCAT), and Radio Occultation Sounder for Atmosphere (ROSA). OCEANSAT-3, planned to be launched in September 2022, will carry Thermal IR sensor, 12 channel OCM, Scatterometer and Passive Microwave Radiometer for simultaneous measurement of ocean color and SST, improvements in signal to noise ratio. The IR sensor and OCM data would be used for the analysis of operational potential fishing zones.

CARTOSAT-1 carried two advanced Panchromatic (PAN) cameras with 30 km swath width and a spatial resolution of 2.5 m. The cameras are mounted on the satellite in such a way that near simultaneous imaging of the same area from two different angles is possible to generate DEM. The cameras are steerable across the direction of the satellite's movement to facilitate the imaging of an area more frequently. CARTOSAT-1 data provided enhanced inputs for large scale mapping applications, such as urban and rural development, land and water resources management, disaster assessment, relief planning and management, environment impact assessment and updating topographic maps.

CARTOSAT-2, launched in January 10, 2007, is capable of providing scene-specific images. It carried a single panchromatic camera providing better than 1 m spatial resolution image, with a swath of 9.6 km. Due to its along track and across track feature up to ±45°, it can take frequent images at revisit periods of 4 days and 1 day. CARTOSAT-2A, launched on 28 April 2008, can also provide scene specific PAN imagery with better than 1 m resolution. CARTOSAT-2B was launched on July 12, 2010, with a camera of spatial resolution of about 0.8m. CARTOSAT-3, launched on 27 Nov. 2019, provided images with 0.25 m resolution in Panchromatic and 1 m in 4 bands of multispectral modes with a swath of 16 km. Images from these satellites would be useful for cadastre, infrastructure mapping and analysis, disaster monitoring and damage assessment, cartographic applications, mapping urban and rural areas. Figure 5.33 (right image) shows a PAN image of CARTOSAT-3 at 0.25 m resolution.

Fig. 5.33: (Left image) IRS-1D panchromatic image at 5.8 m resolution

(http://www.swisstopo.ch/NPOC/Products/kosovo/irsp_sarajewo.html), and (right image) CARTOSAT-3 panchromatic image of Palm city area, Qatar, at 0.25 m resolution, dated 28-Dec-2019

(https://www.isro.gov.in/high-resolution-panchromatic-and-multi-spectral-images-observed-cartosat-3-calibration-validation-of)

National Remote Sensing Centre (NRSC), Hyderabad, a sole agency of Department of Space, Govt. of India, distributes the satellite data to its users in India. The NRSC has an Earth station at Shadnagar, about 55 km from Hyderabad, to receive data from almost all contemporary remote sensing satellites. The National Data Centre (NDC) is a one-stop-shop for procuring a range of data products with a wide choice of resolutions, processing levels, product media, area coverage, revisit, season and spectral bands. Data products can be supplied on a wide variety of media and formats.

4. IKONOS

GeoEye Inc. (formerly Orbital Imaging Corporation or ORBIMAGE) provides very high-resolution images from IKONOS, Orbview-2, Orbview-3, GeoEye-1, GeoEye-2, which are today an important source of large scale mapping. The IKONOS, derived its name from the Greek eikōn for image, is a commercial Earth observation satellite. It was launched on 24 September 1999 in an altitude of about 680 km to collect high-resolution multispectral imagery with 3375 pixels and panchromatic (PAN) imagery with 13500 pixels, providing a swath width of 11.3 km. It provides 1 m (0.82 m) panchromatic image, and 4 m multispectral (Blue, Green, Red, and NIR 0.45–0.90 µm) image and 1-m Pan-sharpened image. The revisit time for IKONOS is 3-5 days off-nadir and 144 days for true-nadir, with a swath width of 11 km × 11 km (single scene). The IKONOS-2 developed by the DigitalGlobe, was launched in September 1999 in a sun-synchronous orbit at an altitude of 681km. It provides multispectral data in B, G, R, and NIR bands with 4 m resolution and a panchromatic data with 1 m resolution. It provides 11 bits (2048 grey levels) radiometric resolution image, covering 11 km swath. Figure 5.34 shows an IKONOS image of the Rio de Janeiro Port, Brazil.

5. QuickBird

QuickBird-1, a very high resolution satellite, developed by DigitalGlobe, was launched in November 2000. It was lost due to launch vehicle failure. QuickBird 2, was launched on 18 October 2001. It was placed in Sun-synchronous polar orbit, at an altitude of 470 km, with 1-5 days' temporal resolution with up to 25° of viewing angle along-and cross-track covering 16.5 km swath. The satellite has 4 multispectral bands (B, G, R and NIR) with 2.4 m resolution and a panchromatic band with 0.61 m resolution with 11 bits radiometric resolution. Figure 5.34 shows a QuickBird image Houston Reliant Stadium.

6. WorldView-1

WorldView-1, was launched on September 18, 2007 by DigitalGlobe, providing the panchromatic images at 0.5 m resolution. It has an average revisit time of 1.7 days, and is capable of collecting in-track stereo. Operating at an altitude of 496 km, WorldView-1 can cover up to 750,000 km^2 per day. WorldView-2 was successfully launched on October 8, 2009, collecting up to 975,000 km^2 of imagery per day. WorldView-2 was the first very high resolution satellite to offer 8-band multispectral imagery with sub-metre resolution, along with increased agility, accuracy and stereo capability. With the additional four spectral bands, WorldView-2 offers unique opportunities for remote sensing analysis of vegetation, coastal environments, agriculture, geology, environment, tourism and many others.

WorldView-3 was launched on August 13, 2014, in a Sun-synchronous orbit, at an altitude of 617 km, with orbital period of 97 minutes. It provides 31 cm resolution in panchromatic, 1.24 m resolution in multispectral, 3.7 m resolution in SWIR, and 30 m resolution in CAVIS (Clouds, Aerosols, Vapors, Ice, and Snow). The CAVIS monitors the atmosphere and provides correction data to improve haze, soot, or dust. WorldView-3 has an average revisit time of <1 day and is capable of collecting up to 680,000 km^2 per day. WorldView-3 offers unique opportunities for remote sensing analysis of vegetation, coastal environments, agriculture, geology, environment, tourism and many others. It provides highly detailed imagery for precise map creation, and change detection studies in telecommunications, infrastructure planning, mapping/surveying, civil engineering, mining & exploration, oil & gas, and DEM generation. WorldView-4 (formerly GeoEye-2) was launched on November 11, 2016 in a Sun-synchronous orbit, at an altitude of 617 km, with 97 minutes' orbital period. It has an effective revisit time capability of ≤ 3 days. The satellite provides 31 cm resolution in panchromatic at nadir, 0.34 m at 20° off-nadir, 1m at 56° off-nadir, and 3.51 m at 65° off-nadir, as well as 1.24 m resolution in multispectral at nadir, 1.38 m at 20° off-nadir, 4 m, at 56° off-nadir, 14 m at 65° off-nadir in 4 bands (B, G, R and NIR) in 13.1 km wide swath with 2048 (11 bits) radiometric resolution. Figure 5.34 shows a WorldView-2 image of downtown Oakland, California.

Fig. 5.34: (left) IKONOS image of the Rio de Janeiro Port, Brasil

https://seos-project.eu/world-of-images/world-of-images-c04-p06.html), (middle) Quickbird image Houston Reliant Stadium. (http://www.usgsquads.com/digital-map-products/aerial-and-satellite-imagery/satellite-imagery/digitalglobe quickbird), and (right) WorldView-2 image of downtown Oakland, California (https://www.eoportal.org/satellite-missions/worldview-2)

Summary of some of the very high resolution satellites are given in Table 5.6.

Table 5.6: Characteristics of high resolution satellites (Garg, 2022)

Sensor	Resolution/Band*		Stereo-capability	Swath (km)
	Panchromatic	Multispectral		
IKONOS	0.82 m 445-900 nm	4 m (Blue, green, red, NIR) bands	Yes	11.3
IKONOS-2	1.0 m 445-900 nm	4 m (Blue, green, red, NIR) bands	Yes	11
QuickBird	Failed after launch			
QuickBird-2	61cm 450-900 nm	2.4m (Blue, green, red, NIR) bands	Along track	16.5
OrbView-3	1 m	4m (Blue, green, red, IR)	No	8
GeoEye-1	41 cm 450-800 nm	1.65 m Blue: 450-510 nm Green: 510-580 nm Red: 655-690 nm Near IR: 780-920 nm	Along track	15.2
WorldView-1	50 cm 400-900 nm	Not applicable	Along track	17.6
WorldView-2	50 cm 450-800 nm	2.0 m Coastal: 400-450 nm Blue: 450-510 nm Green: 510-580 nm Yellow: 585-625 nm Red: 630-690 nm Red Edge: 705-745 nm Near Infrared 1: 770-895 nm Near Infrared 2: 860-1040 nm	Along track	16.4
WorldView-3	31cm 450-800 nm	Multispectral bands (1.24 m resolution) Coastal: 400-450 nm Blue: 450-510 nm Green: 510-580 nm Yellow: 585-625 nm Red: 630-690 nm Red Edge: 705-745 nm	Yes	13.1

		Near Infrared 1: 770-895 nm Near Infrared 2: 860-1040 nm SWIR bands (3.7 m resolution) SWIR-1: 1195-1225 nm SWIR-2: 1550-1590 nm SWIR-3: 1640-1680 nm SWIR-4: 1710-1750 nm SWIR-5: 2145-2185 nm SWIR 6: 2185-2225 nm SWIR-7: 2235-2285 nm SWIR-8: 2295-2365 nm CAVIS bands (30 m resolution) Desert Clouds: 405-420 nm Aerosol-1: 459-509 nm Green: 525-585 nm Aerosol-2: 635-685 nm Water-1: 845-885 nm Water-2: 897-927 nm Water-3: 930-965 nm NDVI-SWIR: 1220-1252 nm Cirrus: 1365-1405 nm Snow: 1620-1680 nm Aerosol-3: 2105-2245 nm Aerosol-4: 2105-2245 nm		
WorldView-4 (formerly GeoEye-2)	31 cm 450-800 nm	1.24m Blue : 450–510 nm Green : 510–580 nm Red : 655–690 nm NIR : 780–920 nm	Yes	13.1
Pleiades	0.5 m	Panchromatic: 480-830 nm Blue: 430-550 nm Green: 490-610 nm Red: 600-720 nm Near Infrared: 750-950 nm	-	20 km
SkySat-C by Planet Labs	50 cm 0.81 cm	Panchromatic: 480-830 nm Blue: 430-550 nm Green: 490-610 nm Red: 600-720 nm Near Infrared: 750-950 nm	-	5.5 km

PlanetScope	m	Blue 440 – 510 nm Green 520–590 nm Red 630–685 nm Red Edge 690–730 nm NIR 760–850 nm	-	16 km x 24 km
Rapideye	5 m	Blue 440 – 510 nm Green 520 – 590 nm Red 630 – 685 nm NIR 760 – 850 nm	-	77 km
Cubasats	2-3 m	Blue 440 – 510 nm Green 520 – 590 nm Red 630 – 685 nm NIR 760 – 850 nm	-	147 m to 1.1 km

**Note: All data are 11 bit (2048 grey levels) radiometric resolution*

5.15 TYPES OF REMOTE SENSING IMAGES

Remote sensing images are available in two broad forms: hard copy or photographic products (B &W, colour), and soft copy (digital) product. The original product from the sensor is digital data acquired in individual band as B &W image. Human eyes have the limitations in identifying not more than 8-10 grey shades from B&W images, while up to 16 categories (colours) may be identified by the human eyes on a colour image. Therefore, colour images are preferred as compared to B&W images, which can be produced from three individual B&W images, taken in different spectral regions. Depending upon the range of wavelength used for capturing them, the images are classified into three main types.

1. B &W (Panchromatic) Images

A Panchromatic image, consists of a B&W image taken by the sensor a slightly larger range of visible part (refer to Figure 5.32). A panchromatic image may be interpreted and analysed in a similar way as B &W aerial photograph.

2. Multispectral Images

A multispectral image may consist of several bands of data (3-10 bands), such as from Landsat TM, where each band image has a specific utility. Human eyes can't appreciate the grey level variation of objects in each image, but there is variation which can be detected by the software. For visual colour display, each band of B&W image is displayed in one of the primary colours and superimposed. Thus, a colour composite image is much appreciated by human eyes for identifying many features/objects, as compared to individual B &W images. The interpretation of a colour composite image however will require the knowledge of the spectral signature of the objects present in the scene. A multispectral image in shown in Figure 5.35 (left image).

3. False Colour Composite (FCC) images

In displaying a colour composite image, basically three primary colours (red, green and blue) are used. When the three primary colours are mixed in various proportions, they can produce different colours in the visible spectrum. As we know that each spectral band in visible part is associated with a colour, so three images taken in three different wavelength regions are superimposed after passing them through three colour filters to obtain a colour composite image. If the colour of an object in the colour composite image may not have any resemblance to its actual colour, the output image is known as a *False Colour Composite* (FCC). There could be many possible combinations of producing a FCC image, however, a common scheme for displaying a multispectral image is NIR band in red, red band in green, and green band in blue colour (Figure 5.35, right image).

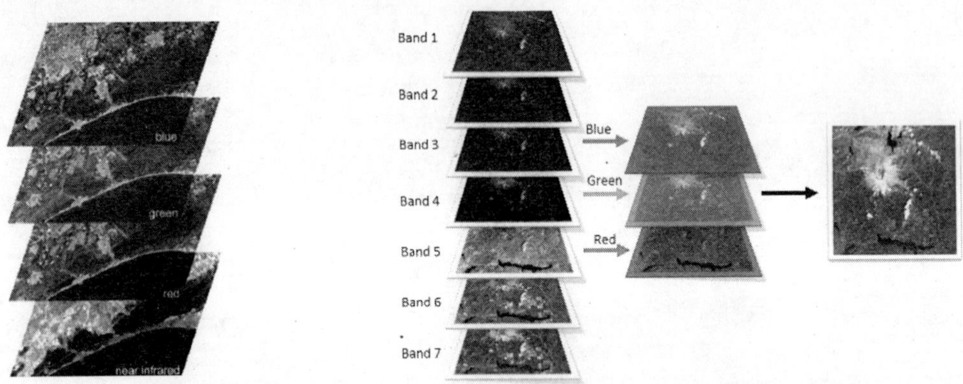

Fig. 5.35: (Left image) Multispectral images (Ose et al., 2016), and (Right image) False colour composite from multispectral image

https://gsp.humboldt.edu/olm/Courses/GSP_216/lessons/composites.html

In a FCC, vegetation would appear in different shades of red/orange colour, depending on the types and conditions of the vegetation. These shades of colours help identification of various vegetation types from FCC image. The vegetation types will have high reflectance in the NIR band as compared to the Red band. Clear water will appear as dark-bluish due to higher reflectance in green band. Bare soils, roads and buildings may appear in various shades of blue, yellow or grey, depending on their composition.

4. True Colour Composite (TCC) images

In a multispectral image, when the three primary colour bands are combined in a different way, the resultant image is called a true colour composite image (Figure 5.35, left image). For example, bands 3 (red band), 2 (green band) and 1 (blue band) of LANDSAT TM images are assigned R, G, and B colours, respectively for displaying the TCC. The TCC image resembles closely what is normally observed on Earth surface by the human eyes, e.g., vegetation would appear in different shades of green colour. Figure 5.36 shows a comparison of a TCC and a FCC.

Fig. 5.36: (Left) True colour composite, and (Right) False color composite (Kurnaz et al., 2020)

5. Hyperspectral images

Hyperspectral image consists of more than 100 of narrower bands (10-20 ηm) where images are recorded by an imaging spectrometer, such as AVIRIS. Hyperspectral remote sensing combines the imaging and spectroscopy in a single system which often includes large data sets, about 100 to 200 spectral bands of relatively narrow bandwidths. Hyperspectral imagery is typically collected (and represented) as a data cube with spatial information collected in the x-y plane, and spectral information represented in the z-direction, as shown in Figure 5.23. Hyperspectral remote sensing is used for detection and identification of minerals, vegetation, water vapor, cloud properties, aerosols in the atmosphere, chlorophyll, phytoplankton, dissolved organic materials, suspended sediments, agriculture and forest production, snow cover fraction, grain size, leaf water, pigments, etc. Some hyperspectral satellite sensors and platforms along with their year of operation is given in Figure 5.37.

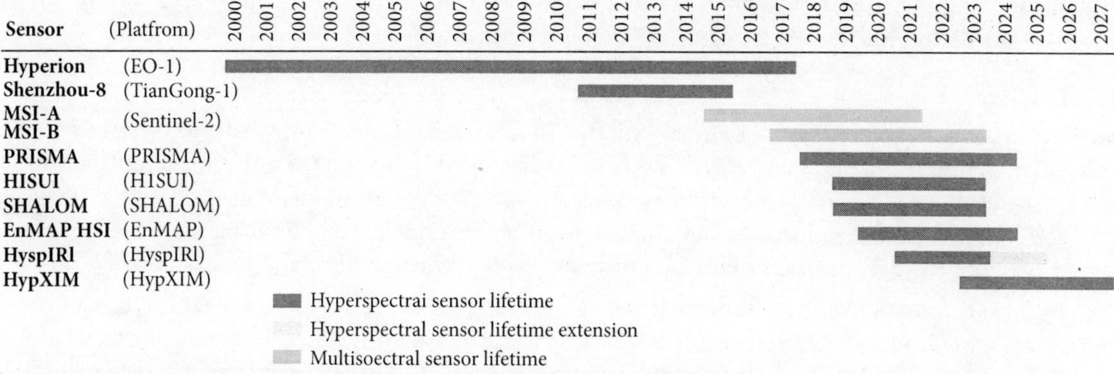

Fig. 5.37: Hyperspectral remote sensing sensors and satellites (Transon et al., 2018)

5.16 VISUAL INTERPRETATION METHODS

Interpretation and analysis of remote sensing imagery involves the identification and/or measurement of various objects present on an image. The visual interpretation (or *manual analysis*) of satellite images includes recognition of objects and ascertaining their properties. The recognition of objects includes, such as roads, forest, agricultural fields, rivers, etc., which would depend on resolution of images. In manual interpretation, the objects in an image may also be recognized with previous knowledge and experience of the area, which play an important role in the interpretation process. The recognition and interpretation of objects is a repetitive process, where both rely on each other. Thematic maps can thus be prepared by manual image analysis which are taken to the field for making any correction as well as verification of interpretation accuracy.

The accuracy of recognition of objects and subsequently determination of their properties would depend on the knowledge and experience of an image interpreter. As human eye cannot interpret all spectral differences in imagery, looking at the familiar areas on the imagery is the best way of gaining experience in visual image interpretation. However, full exploitation of the data characteristics can't be done by visual interpretation of images. Visual interpretation methods require very simple devices, that's why it is most common method used, particularly in developing countries, where highly sophisticated equipment and software are not available. The equipment used for visual interpretation includes, magnifying lens, light table, magnifying project table, etc. It is fairly easy to use with quick start. Normally, a piece of tracing paper is kept ion the image and details/objects identified are traced out manually on the tracing sheet.

The visual interpretation utilizes the visual ability and human brain as well as image characteristics to derive information from remote sensing data. The elements of visual image interpretation are colour (or tone for black and white photos), texture, pattern, shape, size, shadow and location. Tone is considered as the primary element, texture, size and shape are considered as the secondary elements, while pattern and shadow as tertiary and site and locations as lower order elements of interpretation. These elements are explained below, while Table 5.7 gives a summary of these interpretation elements.

1. Tone

It refers to the relative brightness or colour of objects on photographs/images (Figure 5.38). Generally, tone is the most important element required to distinguish between different objects or features, as without tone nothing would be visible. It is the variation in tone that allows the shape, texture, and pattern of different objects to be distinguished from each other.

2. Texture

It is the arrangement and frequency of tonal variation in an image (Figure 5.38). Texture is created by an aggregation of unit feature that may be too small to be discerned individually on the image, such as tree leaves. It determines the overall smoothness or coarseness of image features as visualised on the image. If the scale of the image is reduced, texture of any given object or area becomes progressively finer and ultimately disappears. Texture is combination of shape, size, pattern, shadow and tone.

Various features with similar reflectances can be distinguished based on their texture differences, such as the smooth texture of green grass as contrast with the rough texture of crowns of green trees. Smooth textures would have very little tonal variation, e.g., fields, asphalt, or grasslands, whereas the grey levels change abruptly in a small area, e.g., forest canopy, where rough texture is present. Sand has rough texture as compared to clay. Texture is also one of the most important elements for distinguishing features from Radar images.

3. Pattern

It relates to the spatial arrangement of visibly discernible objects (Figure 5.38). Typically, a repetition of similar tones and textures in a particular fashion will produce a distinct pattern which makes them distinct from each other. Many natural and man-made objects exhibit peculiar pattern, such as triangular, rectangular, square, pentagon, hexagon, circular or any other shape. Gardens with evenly spaced trees, and urban streets with regularly spaced houses are examples of pattern.

Table 5.7: Elements of image interpretation (Richards and Jia, 2013)

	What it is?	Example
Tone	Refers to the relative brightness or colour of objects in an image	Can distinguish between two crop fields due to different texture and shape of plants
Shape	Refers to the general form., structure, or outline of individual objects	Manmade objects tend to have straight edges, whereas natural objects have more irregular shapes
Size	Size of objects in an image is a function of scale	It is easv on an image to distinguish between a football stadium and a house

Pattern	Refers to the spatial arrangement of visibly discernible objects	In a housing estate, although individual houses cannot be made out, the recognisable pattern its produced
Texture	Refers to the arrangement and frequency of tonal variation in particular areas of an image	Different textures will appear differently on an image, for example, a sandy clearing within a forest will show up on the image
Shadow	May provide an idea of the profile and relative height of a target or targets which may make identification easier.	Shadow allows for identification of topographical landforms
Association	Association takes into account the relationship between other recognizable objects or features in proximity to the target of interest	Boats cannot be made out on their own within an image, however when in a marina, their proximity allows for identification

4. Shape

Shape can be a very distinctive clue for interpretation of various objects. It refers to the general form, configuration, or outline of individual objects (Figure 5.38). Shapes of some objects can be easily identified from stereo-photographs. Some objects are so distinctive that their images may be identified solely from their shapes. Shadow characteristics is also helpful to reveal the shape of the object. Straight edge shapes typically represent agricultural fields, while natural features, such as forest boundary, lakes, are generally more irregular in shape.

5. Size

Size of objects in an image is a function of scale of image. It is important to map the size of an object relative to other objects as well as its absolute size (Figure 5.38). The size of a feature will change on different scale images. A building may look like a point feature on a small scale image. For example, zones of land use, large buildings such as factories or warehouses would indicate commercial property, whereas small buildings would indicate residential use.

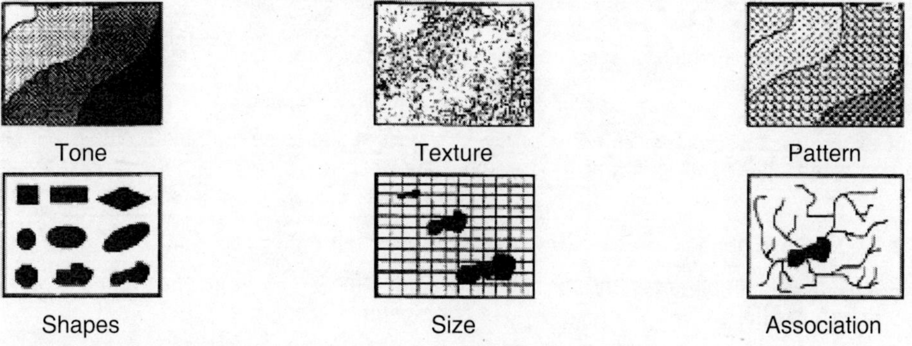

Fig. 5.38: Pictorial representation of various visual interpretation elements (Garg, 2019)

6. Shadow

Shadow is helpful in interpretation of objects from images in two respects. Firstly, the shape or outline of a shadow normally provides the height of objects, and secondly the objects on the ground can be identified with respect to their shape of the shadow. However, the areas under shadow would hide the information and create difficulty in interpretation. For example, the shadows casted by various cultural features (bridges, silos, towers, etc.) can aid in their identification on air-photos/images. The shadows resulting from variations in terrain elevations can aid in assessing the natural topographic variations or geological landforms, particularly in Radar imagery.

7. Site/Association

Site refers to the topographic or geographic location of objects. It is an important clue in the identification of certain types of features (Figure 5.38). For example, certain tree species would occur on well-drained upland sites, whereas other tree species would occur on poorly drained lowland sites.

Association refers to the presence of certain features in relation to others. It takes into account the relationship between the recognizable objects/features in proximity to the object to be identified. For example, the snow cover is related to higher elevation zones, commercial properties are associated with major transportation routes.

5.17 DIGITAL IMAGE INTERPRETATION METHODS

This section deals with the interpretation of optical digital remote sensing images. Digital remote sensing images consist of pixels, and have a square grid structure, where one grid represents a *pixel* (i.e., the smallest element in a digital image). Each pixel is associated with a DN in a particular wavelength. Identification and separation of objects/features with respect to their DN values is called *digital image processing*. Pixels with similar DN values are grouped into various classes. In digital image processing requires a digital image and a software along with the skilled manpower for the interpretation, analysis and mapping. A comparison between visual and digital methods of interpretation is given in Table 5.8.

Table 5.8: Comparison of visual interpretation and digital interpretation methods (Garg, 2022)

S.No.	Manual Interpretation	Digital Interpretation
1	Hard copy data are required	Digital data are required
2	Requires simple and economical equipment or no equipment	Requires specialized, and often expensive equipment and software
3	Limited to analyse a single image at a time	Simultaneous interpretations of multispectral images
4	It involves subjectivity. i.e., the results may vary to some extent with different interpreters	It is an objective process which is based on the spectral signature of objects, and the images are analysed through a computer, so results are almost consistent
5	It is a qualitative method	It is a quantitative method
6	Cost-effective for small areas and for one-time interpretation	Cost-effective for large geographic areas, and for repetitive analysis.
7	No algorithm is used	Complex interpretation algorithms are required
8	Time-consuming and laborious	Speed may be an advantage
9	Difficult to change the scale of mapping	Easy to change the scale of mapping
10	Output product is in hard copy format so not compatible with other data	Compatible with other digital data

The digital images give us the flexibility to pre-process the digital pixel values in an image. The entire digital image processing consists of four basic operations: image pre-processing, image enhancement, image transformation, and image classification (Eastman, 1999). The details of these processes can be found in Garg (2022).

5.17.1 Image pre-processing

It involves the initial processing of raw image data to apply corrections for geometric distortions, calibrate the data radiometrically, and remove the noise present in the data, if any.

a. Geometric Corrections:

It has two basic steps, as explained below-

i. ***Georeferencing***

Raw remote sensing data is without any geographic coordinates, and has distortions, mainly caused by the sensor geometry. Therefore, these can't be used as such for any quantitative measurements on them. Georeferencing is the conversion of image coordinates to ground coordinates by removing the distortions caused by the sensor geometry. Georeferencing is important to deal with various images, create mosaicking and compare various scenes (e.g., change assessment). It is a process of locating an entity/object in real world coordinates, also called *geo-rectification* or *geo-registration.*

The direction of satellite motion in the orbit and on-board sensors while taking images is not exactly north-south or west-east, respectively. In addition, there is a rotation of the Earth about its own axis while taking the images, so images are not perfect square but they have somewhat skewed shape. Georeferencing re-orients the image to a coordinate system representing the Earth, and making its geometry same as the Earth. Georeferenced images can be viewed, compared, and analysed with other geographic data.

To do georeferencing, the exact locations of several known points, called *Ground Control Points (GCPs)*, are required. These GCPs are normally selected as prominent objects whose geographical locations can be accurately determined either from the topographic maps or GPS survey. A minimum of four control points are required for georeferencing, however, additional control points would help increasing the accuracy of georeferencing. These GCPs are also identified on the image to be georeferenced. With these two sets of coordinates, polynomial is fitted amongst the GCPs, and *rms* error is minimized to ± 1 pixel size. After georeferencing, each point on the image has real-world coordinates associated. The accuracy of the georeferencing would depend on the number, accuracy, and distribution of the control points and the choice of transformation polynomial. Normally, 2nd or 3rd order polynomial is used.

ii. ***Resampling***

After the georeferencing process, we may find that the pixels have been oriented differently than the way they were present in the original image coordinate system. Resampling is the process of interpolation the new DN values of the displaced pixels (new pixel location) in the new coordinate system. Three methods of resampling are commonly used, as given below (Figure 5.39).

a. Nearest Neighbour:

In this method, the attribute value of the original pixel nearest to a pixel in the output image is assigned to the corresponding cell.

b. Bilinear Interpolation:

It assigns the value to a pixel in the output image by taking weighted average of the surrounding four pixels in the original grid nearest to it.

c. Cubic Convolution:

It assigns the value to a pixel in the output image by taking weighted average of the surrounding sixteen pixels in the original grid nearest to it.

Among the three methods, nearest neighbour is a preferred method as it doesn't alter the values of the original grid cells assigned to the resampled grid cells but it produces a blocky image. The cubic convolution on the other side does change the values but is more accurate. It generates a smoother image.

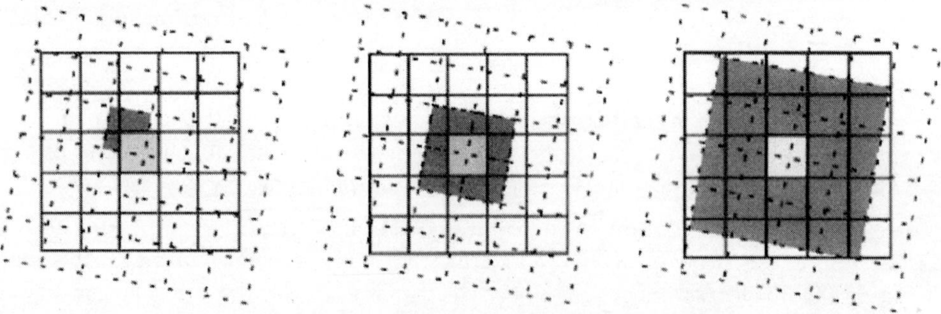

Fig. 5.39: Resampling methods (Nady, 2020)

d. Atmospheric Correction

Atmospheric correction is done to modify the DN values to remove noise, i.e., contributions to the DN due to intervening atmosphere. Lower wavelengths of visible spectrum are more subject to haze, which falsely increases the DN values. The simplest approach for its correction is known as *dark object subtraction method*, which assumes that if there was no haze present then the dark pixel will have zero DN value, e.g., deep water in near infrared will have complete absorption, and therefore those pixels will have zero DN values. But in reality, we won't find any dark pixel with zero values. For such image, the lowest DN value present in the image is determined, and is subtracted from all the DN values of the image, so that the dark (water) pixels are now zero.

5.17.2 Image enhancement

Image enhancement is mainly carried out to improve the quality of images. Often, the images do not have the optimum contrast so objects are not clearly identified visually. Enhancement is made to make the visual appearance of image better. It is the modification of images to improve the interpretability through human vision. The main purpose of image enhancement is to increase the contrast in a low contrast image. It does not add any information to the original image but it enhances the visual appearances of already captured features. Enhancement of an image can be implemented by using various methods. They improve the image quality so that the enhanced image is better than the original image for a specific application. Before image enhancement is done it is necessary to understand the image characteristics through its *histogram*.

a. Image Histogram

A histogram is a graphical representation of the DN values (i.e., 0-255) in an image that are displayed along x-axis, while the frequency of occurrence of these values is plotted on y-axis. Image histogram is a way to portray the information present on an image. In raw imagery, the useful data often occupies only a small portion of the available range of DN values (256 levels in an 8 bit image). In an 8-bit image, in a histogram, the x-axis will contain 256 DN values and the y-axis will display how many of each intensity value is present.

A digital image can be represented by three effective ways (Figure 5.40): (i) Pictorially, in the form of image (ii) Numerically, in the form of DN values arranged in the matrix, and (iii) Graphically, through its histogram. A single peak bell-shaped histogram is considered as the best shape of a histogram of image data. It conveys about the homogeneity and well distribution of grey levels in the image. Two peaks (bi-model) histogram indicates that there could be two pre-dominant classes (e.g., water and vegetation) present on the image.

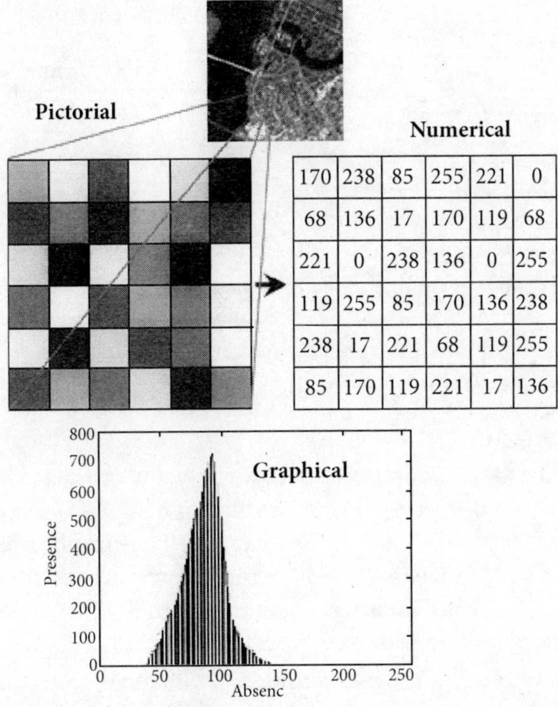

Fig. 5.40: Representation of an image (pictorially, numerically and graphically) (Garg, 2019)

In reality, the shape of the histogram of an image is quite different that the ideal shape of histogram, as shown in Fig. 5.41. The DN values are normally skewed either towards the lower values or towards the higher values, therefore, they indicate the presence or absence of features with higher or lower reflectances. Histograms alone can provide a lot of information about images to an interpreter even without looking at the images, such as likelihood of presence or absence of type of features, distribution etc. These help evaluate images statistically, e.g., normal, skewed, bimodal distribution, etc. The histograms are then used in individual image enhancement, image segmentation and image classification. Histograms also help matching of images across time or space.

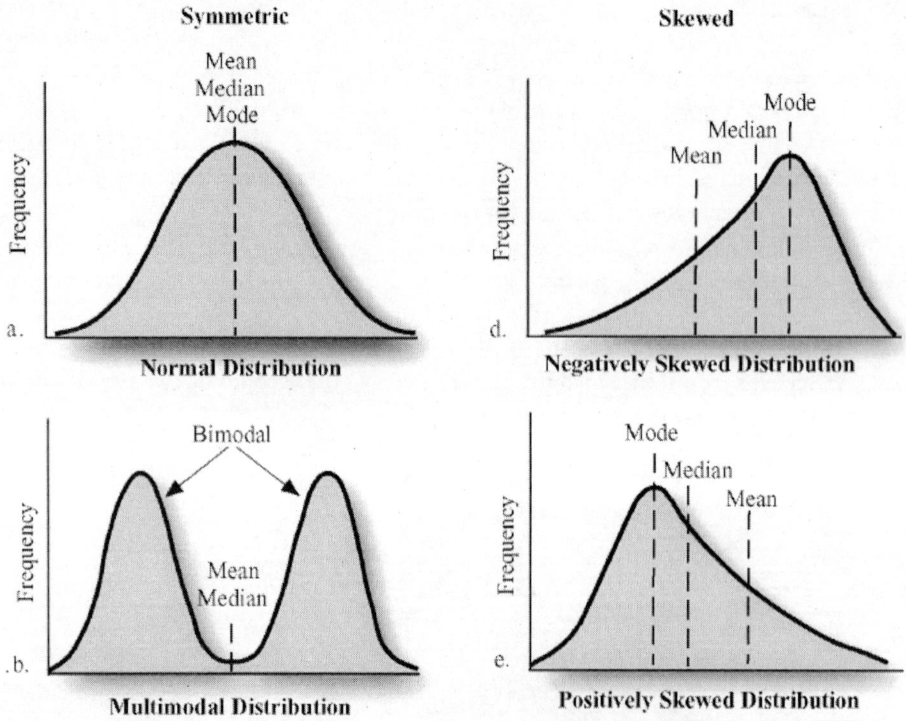

Fig. 5.41: Various shapes of histogram of an image (Jenson, 1986)

Various enhancement techniques interpolate the range of DN values in an image, which can be represented graphically by its histogram.

b. Contrast Enhancement

The contrast is defined as the maximum difference in color or intensity between two objects in an image. If the contrast is poor (low), it is impossible to distinguish between various objects and they are perceived as the same object. Contrast enhancement involves changing the original DN values so that more available range is utilised, thereby increasing the contrast between the objects and their background. It basically improves the interpretability for human viewing, and provides enhanced input to be used for image processing.

There are many different techniques and methods of enhancing the contrast, and details can be found in Garg (2022). The *linear contrast enhancement* is the most popular technique used for image enhancement. It involves identifying the minimum and maximum DN values in the image, and applying a linear transformation to stretch the present range to occupy the full range (e.g., 0-255 in an 8-bit image). In Figure 5.42, for example, minimum DN value (occupied by actual data) in the histogram is 84 and the maximum DN value is 153, so these 70 grey levels (153- 84 =70) occupy less than one-third of the full 256 levels available. A linear stretch will uniformly expand this small range to cover almost the full range of values from 0 to 255. It would enhance the contrast in the image with light toned areas appearing lighter and dark areas appearing darker, and thus making the identification of objects/features much accurate and effective.

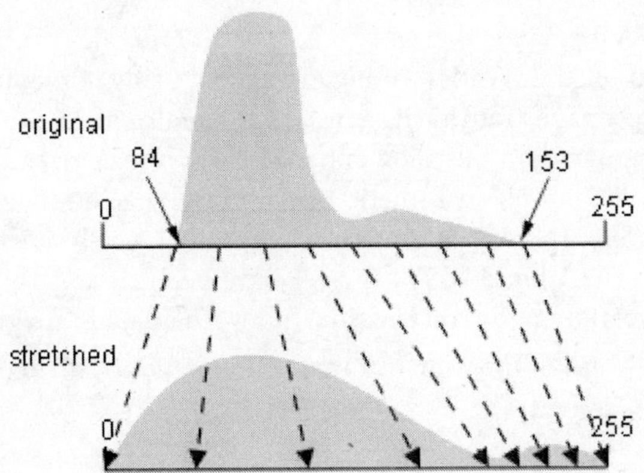

Fig. 5.42: Original histogram and after linear contrast enhancement (Garg, 2022)

The linear contrast enhancement is mathematically represented as-

$DN_{out} = ((DN_{in}-DN_{min})/(DN_{max-}DN_{min}))*$ No. of grey levels (5.8)

where,

DN_{out} represents the output values in the image

DN_{in} represents the DN value at that pixel location from input image.

DN_{min} and DN_{max} are the minimum and maximum DN value, respectively, in the input image.

No. of grey levels are the total number of intensity values that can be assigned to a pixel. For example, in 8 bit images, the maximum number of grey levels is 255.

Figure 5.43 shows the results of linear contrast enhancement on Landsat ETM+ images. The main purpose of image enhancement is to increase the contrast in a low contrast image. It does not add any information to the original image but it enhances the visual appearances of already captured features.

a. Before Enhancement b. After Enhancement

Fig. 5.43: (Left) original satellite image, and (Right) after linear contrast enhancement

(https://www.gisoutlook.com/2019/08/digital-image-processing-image.html)

c. Image Transformations

The image transformation is the creation of new image by using some mathematical function on the original images. The image transformation will normally yield synthetic images which are very useful for specific applications, as they enhance certain features of interest. Some examples of transformations include; simple arithmetic operations, Vegetation Indices (VI), Normalised Difference Vegetation Index (NDVI), Principal Component Analysis (PCA) and Tasselled Cap Transformations (TCT). The VI and NDVI images have been frequently used world-wide for the study of forest cover, vegetation and crop classifications. The details are given in Garg (2022).

The VI is obtained as the ratio of the near-infrared (NIR) band to the Red band-

VI = NIR band / Red band (5.9)

If both the Red and NIR bands (or the VIS and NIR) have similar reflectance, the value of ratio is 1 or close to 1. Ratio values for bare soils generally are near 1; as the amount of green vegetation increases, the ratio increases. The value of ratio can increase far beyond 1 up to 30.

Most popular and commonly used vegetation index is the NDVI. The NDVI is a measure of the vegetative cover on the land surface, but it is also used to identify the water and soil. Vegetation differs from other land features because it tends to absorb strongly red wavelengths of sunlight and reflect in the near-infrared wavelengths. It is a measure of the difference in reflectance between these wavelength ranges. The NDVI is computed as-

$$NDVI = (NIR - Red) / (NIR + Red) \quad (5.10)$$

The NDVI values range between -1 and 1; value 0.5 indicates dense vegetation and values < 0 indicate no vegetation. Since vegetation has high NIR reflectance but low red reflectance, vegetated areas will have higher values as compared to non-vegetated areas. The NDVI has been used world-wide to monitor vegetation condition, vegetation health, cover and phenology over large areas, and therefore can provide early warning on droughts and famines.

5.17.3 Digital image classification

The digital classification of optical images and microwave images required different approaches, and software. In this section, optical image classification and associated algorithms have been discussed. Digital image classification is a software-based classification technique used for information extraction from optical images based on their DN values. It requires identification of spectral signature of various objects in an image, and subsequently tagging the pixels with similar spectral signatures. Each pixel (or groups of pixels) of an image may be assigned to a land use or land cover class. So, the image classification will categorise each pixel into at least one of the classes. The statistical decision rules are used which allow grouping of the pixels in different classes.

In optical remote sensing, there are broadly two classification techniques; *supervised* and *unsupervised classification*. Both the approaches of classification have their own strengths and weaknesses associated with the classification process and results of the analysis. These are briefly explained below.

a. Supervised Classification

Supervised classification consists of three distinct stages; *training, allocation* and *testing*, as shown in Figure 5.44. Training is the first stage where the identification of a sample of pixels of known classes is done with the help of reference data, such as field visits, existing maps and aerial photographs. The DN values of these known classes are determined to check the homogeneity. In supervised classification, an analyst uses previously acquired knowledge of an area, or a priori knowledge, to locate specific areas, or training sites, which represent homogeneous samples of known land use and/or land cover types. These classes are interactively marked on the digital image in the form of polygons. Based on statistics of these training sites, each pixel in an image is then assigned to a user-defined land use type (e.g., residential, industrial, agriculture, etc.) or land cover type (e.g., forest, grassland, snow cover, etc.).

In the second stage, the training pixels are used by the software to derive various statistics for each class, and are correspondingly assigned signatures. These samples are referred to as training areas or samples. In supervised classification, the analyst identifies the homogeneous representative samples of different land use and land cover types (information classes) of interest on the imagery. Information classes imply that the actual land cover of the area under consideration which the analyst wants to classify, like vegetation cover, agriculture, urban or water bodies. A histogram for each band of the training areas/ samples can be drawn. The normal histogram with a single peak would indicate a single class but a bimodal response would indicate two class present in the training pixels. In case, some training pixels are to be dropped or new added, it can also be done here itself. Thus, the classification of image data may be improved if each of the class in training sample has one single peak in the histogram.

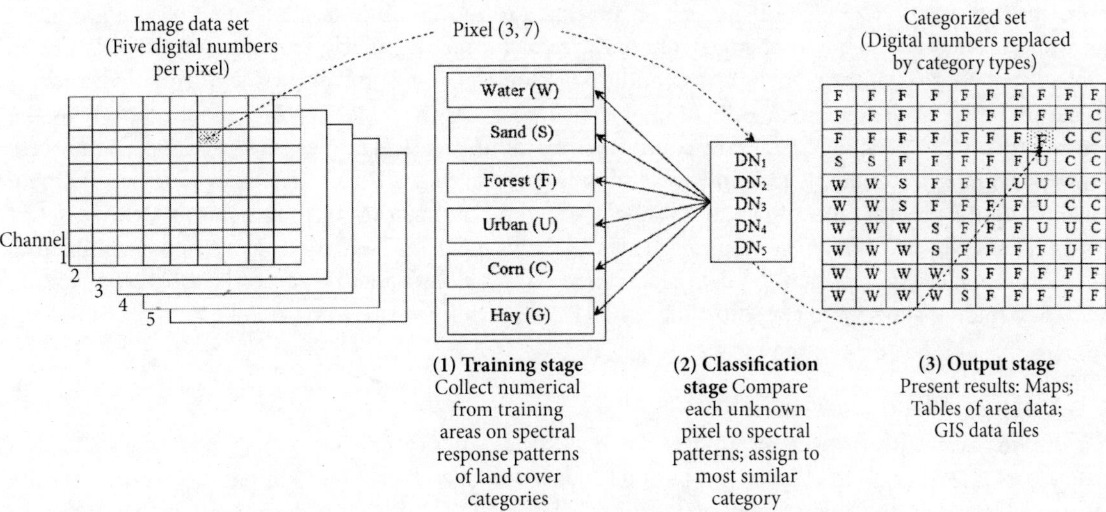

Fig. 5.44: Steps in supervised classification (Alakhouri, 2014)

Training process helps in collection of statistical parameters which describe the spectral response pattern for the information classes that are being considered for image classification. Thus, a sufficient number of homogeneous training samples are required for each class to represent the tonal variation present within each class in the image. The variation in spectral reflectances of a training set per class in multispectral bands is used to derive the statistics. It is important that the training sites are well distributed throughout the image, as far possible, as they are the representative samples for the entire image. Training sets also increase the chance to incorporate the site specific variations of that class.

One of the important points in supervised classification is to avoid overlap in the training spectra for each class through the spectral plot. It is therefore important to identify the combination of bands that might be best for discrimination of classes through scatter plots between two spectral bands. The scatter plot is a graphical representation of DN values of two bands data; one on x-axis and another on y-axis. The scatter plots provide information related with the spectral separability. The training sample size is also important in classification which might vary from a minimum of 10N -100N per class, where N is the number of bands. The selection of accurate training areas is based on the analyst's familiarity and knowledge with the geographical area and actual surface cover types obtained through site surveys and present in the image. For each class, several training samples are identified from the images.

In the third stage, the remaining pixels of the image are allocated to the same class with which they show greatest similarity based on the established signature files in the second stage. Once the statistical characteristics are computed for each information class, the image is classified using any method, like *Parallelepiped classifier, Minimum distance* or *Maximum likelihood classifier*. For details of these techniques, refer Garg (2022). For the training samples, digital values (spectral signature) and their statistics in all spectral bands are used to "train" the software, and to recognize spectrally similar areas for each class. The known spectral signature for each class is then compared by the software to the matching signatures of unknown pixels and labeled the class as soon as it closely resembles the signature. This process is repeated for each unknown/unclassified pixel in the image. Figure 5.45 summarizes the supervised classification procedure in the form of a flow diagram.

It is important to note that training sites developed in one scene may or may not be replicable to entire study area due to variation in ground objects, ground conditions, illumination conditions, or atmospheric effects, which may change from one area to another. Similarly, training samples may not be usable directly across time due to lighting conditions, cloud cover response and due to growth of various vegetation types. At the end of classification, if a particular class has not been picked up properly by the software, the image is again classified by refining the training areas/samples for that class. The process allows interpreter to refine the training areas/samples several times, till a satisfactory result is obtained. It is an iterative process, and the analyst "supervises" the classification of images into a defined set of classes. That's why it is known as supervised classification method. Thus, in a supervised classification, the information classes are identified which are then used to determine the spectral classes representing them. Accuracy of supervised classification results would depend entirely on the collection of a sufficient number of training sites, and purity of samples.

b. Unsupervised classification

This method is almost opposite to the supervised classification process. The unsupervised classification is very useful technique to classify the remotely sensed image where the field data/reference data are not available. Here, the image is classified purely based on the spectral variations of the classes to identify the major classes that already existed in the image. Unsupervised classification techniques do not require training sample signatures, prior to analysis of the scene. Statistical algorithms group DN values with similar pixels into various spectral classes, and later analyst will identify or combine these spectral classes into information classes (Jensen 2005). Spectral classes are grouped, solely based on DN values in the data, and then these DNs are matched by the analyst to information classes. Several **clustering algorithms** are used to determine the natural (statistical) groupings in the image data. The basic assumption in unsupervised classification is that a particular land cover in a scene would form a single cluster. Thus, the algorithm classifies the pixel data based on similar properties of the data itself. Figure 5.45 shows various steps involved in the unsupervised classification procedure.

In unsupervised classification, only major land classes are separated as clusters, while for smaller classes it may be not be possible. The decision for the number of clusters may be based on histogram analysis of the reflectance values. Usually, the analyst specifies the number of groups or clusters (classes) to be identified from the scene. The analyst also specifies the parameters related to the separation distance among the clusters, number of iterations required, and the variation (like standard deviation) within each cluster, as input to software. Often, the number of peaks as seen in the histogram can also be considered as the number of clusters present in the scene. The iterative clustering process may result into some clusters that the analyst wants to subsequently combine, or clusters that should be split further, based on ancillary/reference data available for the site. Alternatively, the complete process can be re-started by changing the input parameters into the software, till a satisfactory result is obtained. Thus, unsupervised classification is faster and less dependent on human intervention.

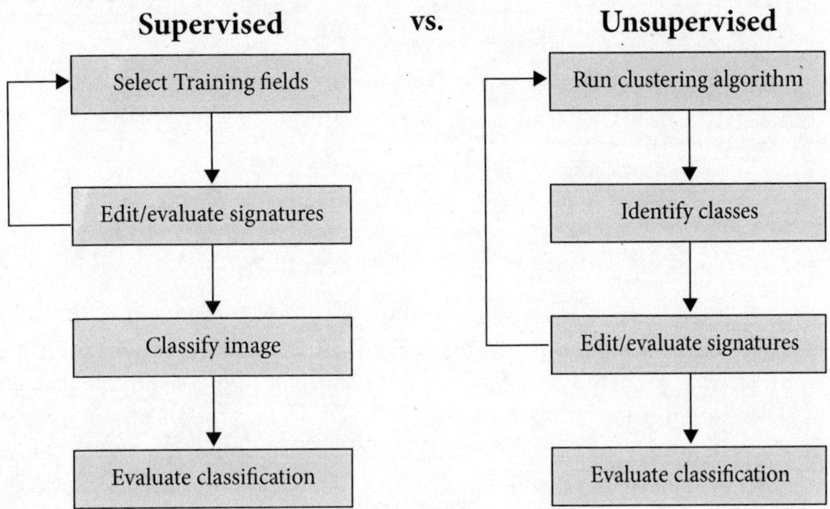

Fig. 5.45: Broad steps involved in supervised classification and unsupervised classification procedures

There are two most popular clustering methods used for unsupervised classification; K-means and Iterative Self-Organizing Data Analysis Technique (ISODATA). These two methods rely purely on spectrally pixel-based statistics and require no prior knowledge of the characteristics of the themes being studied. In K-means approach, classes are determined statistically by assigning the pixels to the nearest cluster mean based on all available bands. The result of the K-means clustering could be influenced by the number of cluster centers specified, the choice of the initial cluster centre, the sampling nature, properties of the data, and clustering parameters. The ISODATA method uses the minimum spectral distance to assign a cluster to each pixel. The method requires a specified number of arbitrary cluster means or the means of existing signatures. It then processes the data repetitively, so that those means shift to the means of the clusters in the data. The input to ISODATA is number of clusters: 10 to 15 per desired land cover class, convergence threshold: percentage of pixels whose class values should not change between iterations; generally, set to 95%, and the maximum number of iterations: ideally, the convergence threshold should be reached. It should set "reasonable" parameters so that convergence is reached before iterations run out. In the iterations, pixels assigned to clusters with closest spectral mean; mean recalculated; pixels reassigned. The process continues until maximum iterations or convergence threshold reached. A graphical example is shown in Figure 5.46, where left graph shows the results of clustering by ISODATA after first iteration, and right graph shows the results after the second iteration. One may see the changes in the size of clusters. This way iteration continues till the specified threshold value is matched.

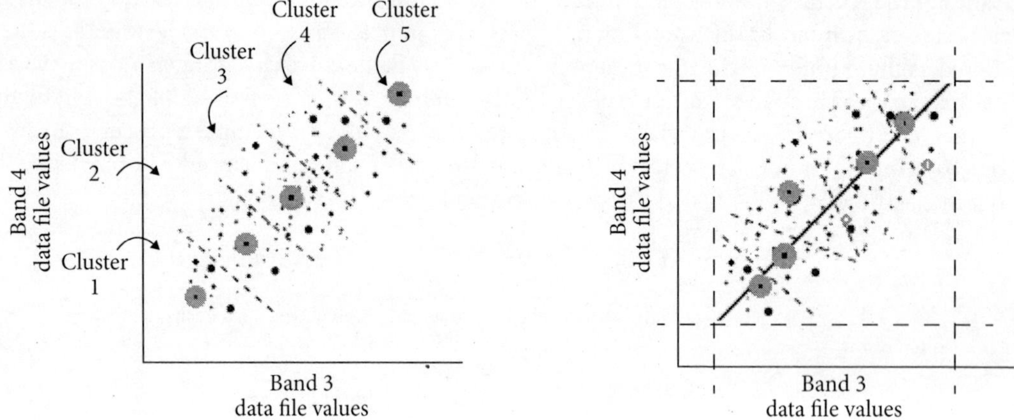

Fig. 5.46: ISODATA Clustering techniques, result after (left) first iteration, and (right) after second iteration (Richards, 2013)

The supervised technique has some advantage over the unsupervised approach, as in supervised approach, information categories are distinct first, and then their spectral separability is examined while in the unsupervised approach, the software determines spectrally separable classes based, and then defines their information values (Lillesand and Keifer 1994). Unsupervised classification is easy to apply, and does not require analyst specified training samples. It automatically converts raw image data into useful information as long as there is higher classification accuracy (Langley et al., 2001). But, the disadvantage is that the classification process has to be repeated, if new classes are added. Figure 5.47 presents the results of supervised and unsupervised classification.

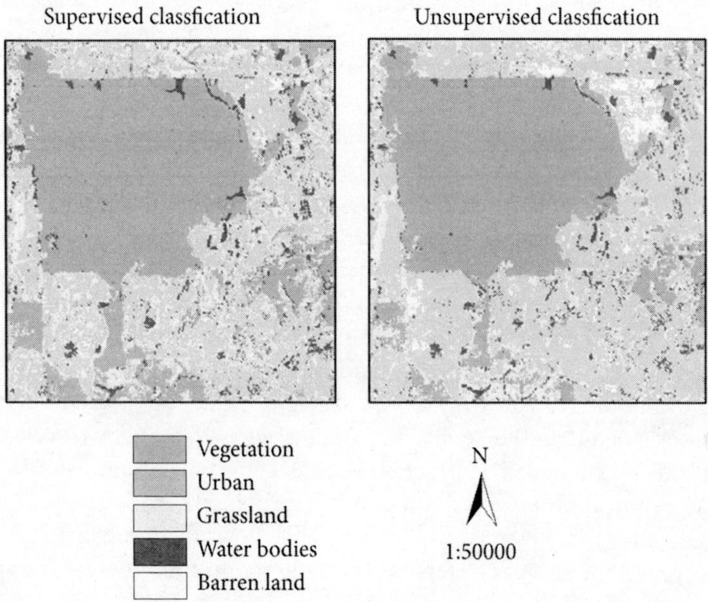

Fig. 5.47: (a) Supervised, and (b) unsupervised classification of SPOT 5 image of the area (Ismail et.al., 2009)

While using high resolution images, it is still difficult to obtain satisfactory results by using supervised and unsupervised methods alone. Researchers have therefore developed advanced classification procedures to get high speed and better accuracy. These are summarised in Table 5.9.

Table 5.9: Various image classification techniques along with their salient characteristics (Garg, 2022)

Methods	Examples	Characteristics
Supervised	Maximum Likelihood, Minimum Distance, and Parallelepiped classification etc.	Analyst identifies the training sites to represent classes and each pixel is classified based on the statistical analysis
Unsupervised	ISODATA and K-means etc.	Prior ground information is not required. Pixels with similar spectral characteristics are grouped according to specific statistical criteria
Parametric	Maximum Likelihood classification and Unsupervised classification, etc.	Data are normally distributed, Prior knowledge of class density functions
Non-parametric	Nearest-neighbor classification, Fuzzy classification, Neural networks and support Vector machines, etc.	No prior assumptions are made
Non-metric	Rule-based Decision tree classification	Can operate on both real-valued data and nominal scaled data statistical analysis
Hard (parametric)	Supervised and Unsupervised classifications	Classification using discrete categories
Soft (non-parametric)	Fuzzy Set classification logic	Considers the heterogeneous nature of real-world, Each pixel is assigned a proportion of the in land cover type found within the pixel
Pre-pixel		Classification of the image pixel by pixel
Object-oriented		Image regenerated into homogenous objects, Classification preformed on each object and pixel
Hybrid approaches		Includes expert systems and artificial intelligence

5.17.4 Accuracy assessment

The increased use of remote sensing data and techniques has made analysis faster and more powerful, but the spectral and spatial complexity in the images have also created increased possibilities for errors. Thematic maps generated from remotely sensed image may contain some errors because of; (i) geometric error, (ii) atmospheric error, (iii) clusters incorrectly labeled after unsupervised classification, (iv) training sites incorrectly labeled for supervised classification, and (v) separability of classes.

Image classification is considered to be incomplete without estimating the accuracy of classification. It measures the agreement between a reference (assumed as 100% accurate) and a classified image of unknown accuracy. The accuracy assessment or error analysis is the quantitative comparison to validate the classified map with the actual reference data/image.

Accuracy assessment of land use or land cover classification obtained by remote sensing images is necessary to evaluate the quality of classification. Accuracy assessment is a critical step in the use of the results of analyses from remotely sensed data. A typical approach for validation and accuracy assessment is to use a statistically sound sampling design to select a sample of ground locations (number of pixels) in the scene. The land use or land cover classification assigned at these ground locations is actually compared with the true classification to ascertain the accuracy. An *error matrix* or *confusion matrix* is thus generated, as shown in Figure 5.48. An error matrix is a common tool to assess the accuracy of classification. Error matrix compares the pixels or polygons in a classified image against the ground reference data (Jensen 2005). A confusion matrix (or error matrix) is usually a quantitative method of characterising the image classification accuracy. It shows a correspondence between the classification results with respect to a reference image. The error matrix is created with the help of reference data which includes thematic map, aerial photos, ground surveys, GPS surveys and a high resolution satellite data used for classification.

To select the ground samples, five important sampling techniques are often used. These are:

1. ***Simple Random Sampling***: observations are randomly placed, and no rules are used. It is done using a completely random process.
2. ***Systematic Sampling***: observations are placed at equal intervals according to a planned strategy.
3. ***Stratified Random Sampling***: a minimum number of observations are randomly placed in each class. **Sampling** points are generated proportionate to the distribution of classes in the image
4. ***Stratified Systematic Un-aligned Sampling***: a grid is laid out which provides even distribution of randomly placed observations.
5. ***Cluster Sampling***: randomly placed "centroids" used as a base of several nearby observations. The nearby observations can be randomly selected, systematically selected, etc.

		Reference image			
		A	B	C	Total
Classified image	a	37	3	7	∑a = 47
	b	9	25	5	∑b = 39
	c	11	2	43	∑c = 56
	Total	∑A = 57	∑B = 30	∑C = 55	N = 142

Fig. 5.48: Error matrix or Confusion matrix

It is not practically possible to test each and every pixel in the classification image, and therefore a representative sample of reference points in the image with known class values is used. Ground reference pixels earlier used to train the classification algorithm are not used now for the assessment of classification accuracy. Normally half of the training samples are used for classification, and remaining half for estimating the accuracy. The accuracy of a classification can be defined as; (i) Overall accuracy, (ii) Producer's accuracy, (iii) User's accuracy, and (iv) Kappa coefficient.

The overall accuracy of the classification map is determined by dividing the total correct pixels (sum of the major diagonal) by the total number of pixels in the error matrix. The total number of correct pixels in a category is divided by the total number of pixels of that category as derived from the reference data (i.e., the column total). It indicates the probability of a reference pixel being correctly classified, and is a measure of *omission error*. This statistics is called the *producer's accuracy* because the producer (the analyst) of the classification is interested in how well a certain area can be classified. Thus, the errors in classification can further be obtained individually for each class. The classification accuracy for each class indicates if there is any over-estimation or under-estimation in classification based on the input from reference data.

Columns of the table in Figure 5.48 are the reference (ground truth) classes, while rows are the classes of classified image whose accuracy is to be assessed. Various cells of the Table show number of pixels for all the possible correlations between the ground truth and the classified image. The diagonal elements of the matrix are highlighted which contains the number of correctly identified pixels. Classification overall accuracy is obtained by dividing the sum of these diagonal pixels by the total number of pixels. It is computed as:

$$\{(37 + 25 + 43)/142\}\ 100 \approx 0.74 \tag{5.11}$$

The overall accuracy measures the accuracy of the entire image without reference to the individual categories. It is sensitive to differences in sample size, and biased towards classes with larger samples. In addition, the accuracy of individual class needs to be assessed. The non-diagonal cells in the matrix contain classification errors, i.e., the number of pixels in reference image and the classified image don't match. There are two types of errors: under-estimation (omission errors) and over-estimation (commission errors).

If the total number of correct pixels in a class is divided by the total number of pixels that were actually classified in that category, the result is a measure of *commission error*. This measure, called the *user's accuracy* or *reliability*, is the probability that a pixel classified on the map actually represents that category on the ground. For any class, error of commission occurs when a classification procedure assigns pixels to a certain class that don't belong to it in reality. Number of pixels incorrectly allocated to a class is found in column cells of the class above and below the main diagonal. The sum of these yellow is divided by the total number of class pixels to get the commission error for class A:

$$\{(9 + 11)/57\}\ 100 \approx 35\% \tag{5.12}$$

The Producer's accuracy is the number of correctly identified pixels divided by the total number of pixels in the reference image. For class A, it is:

$$(37/57)\ 100 \approx 65\% \tag{5.13}$$

The *Commission error* quantifies the Producer's accuracy. The sum of Error of Commission and Producers Accuracy for class A is 100%.

The probability a reference pixel is being properly classified measures the "Omission error" (reference pixels improperly classified are being omitted from the proper class). For any class, error of omission occurs when pixels that in fact belong to one class, are included into other classes. In confusion matrix, number of such pixels is found in the row cells to the left and to the right from the main diagonal. The sum of these orange cells is divided by the sum by the total number of pixels in the classified image to compute omission error:

$$\{(3 + 7) / 47\}\ 100 \approx 21\% \tag{5.14}$$

User's accuracy is the number of the correctly identified pixels of a class, divided by the total number of pixels in that class in the classified image. For class A, it is computed as:

$$(37 / 47)\ 100 \approx 79\% \tag{5.15}$$

The Omission error quantifies the User's accuracy. The sum of Error of Omission and User's Accuracy for class A is 100%.

The overall accuracy of classification can sometimes be misleading, as it does not reveal if error was evenly distributed between the classes or if some classes were really badly classified and some really good. Therefore, it is always better to compute the values of commission and omission errors. It is likely that the overall accuracy of an image classification might be quite high, whereas an individual class may have a low accuracy. If the accuracy of individual classes is most important to users, the classification results can't be accepted as such, even if the overall accuracy is coming out be higher. Additionally, the producer's and user's accuracy can't be separately used as an indicator of the classification accuracy, as these values do not provide the complete details.

Unit Summary

This unit discusses various remote sensing data products and their utilisation. It focusses mainly on the use of optical satellite images for carious purposes. The technical terms used in remote sensing are defined. Various comonents of remote sensing and interaction of EMR with the atmosphere are discussed. Laws governing black body are described. Resolutions are very important while dealing with the satellite images, so these have been detailed out in the unit. Large number of sensors and their chracteristics is described. Details of various optical satellites and high resolution satellites are given. Two methods of interpretation of images; manual interpretation and digital interpretation are described. While carrying out digital interpretation, the characteristics of digital images can be changed, e.g., contrast enhancement, vegetation indices, and used for classification. Two important classification methods; supervised classification and unsupervised classification have been described. At the end, accuracy assessment method of thematic map created from the satellite image has been discussed.

Exercises for Practice

a. Short questions

5.1 Define remote sensing. What is its utility?

5.2 List the advantages and disadvantages of remote sensing.

5.3 Draw visible part of electro-magnetic spectrum. Show various divisions.

5.4 In remote sensing, why most of the time visible wavelength bands are used.

5.5 State the relationship between wavelength and frequencies involving the velocity of EMS.

5.6 Which part of the EMR spectrum is used in (i) thermal and (ii) microwave remote sensing?

5.7 What is a black body? How is emissivity related to black body radiation?

5.8 Define Scattering. Why do you see sky as blue during day?

5.9 What is the effect of scattering on remote sensing images? Why do you see colour of the cloud as white?

5.10 Draw a standard spectral signature curve of vegetation, water, and soil.

5.11 Define (i) Atmospheric window, and (ii) Spectral reflectance of an object.

5.12 Explain why chlorophyll in green vegetation absorbs mainly in the visible region of the EMR spectrum.
5.13 Draw orbits to show Geostationary orbit and Polar sun synchronous orbit.
5.14 Define the following terms with neat sketches; (i) Swath, and (ii) Path-row number.
5.15 Define the; Pixel, (iv) Digital Number and (v) Spectral band.
5.16 What do you mean by the (i) histogram of an image, and (ii) resolution of an image?
5.17 Define the IFOV. How is this related to resolution or pixel?
5.18 What is difference between a true color, and false color image?
5.19 What is a sensor? Differentiate between active sensors and the passive sensors.
5.20 Write any four thermal sensors and their salient features.
5.21 What is RADAR? What are the advantages of microwave sensors over the optical sensors?
5.22 Write the wavelength of various bands in TM sensor.
5.23 What are various LISS sensors used in IRS?
5.24 What is the major difference between visible, thermal infrared and microwave remote sensing?
5.25 Write any four sensors used for collecting stereo-pair images.
5.26 Define interpretation of images. What are the two methods of image interpretation?
5.27 Draw a pyramid diagram to show primary, secondary and tertiary elements of visual image interpretation.
5.28 What is a digital remote sensing image? What is the objective of digital image processing?
5.29 What are the causes of geometric errors in satellite images?
5.30 Define georeferencing of images. Why georeferencing is required?
5.31 Draw the diagrams to show images involved in all the three resampling processes.
5.32 Define (i) image enhancement and (ii) image transformation.
5.33 What is vegetation Index? Write its significance.
1.34 Write the advantages and disadvantages of supervised and unsupervised classification methods.

b. Long questions

5.35 Describe various parts of an ideal remote sensing system.
5.36 Show major components of electromagnetic spectrum (EMS), along with their wavelength and frequency.
5.37 Discuss the interaction of electromagnetic radiation with the Earth's atmosphere.
5.38 Explain (i) Planck's Law (ii) Stefen-Boltzmann law, and (iii) Wein's Displacement law.
5.39 Explain four types of interactions between the EMR and atmosphere.
5.40 What do you mean by atmospheric scattering? Explain Mie Scattering?
5.41 Draw a diagram to explain the transmittance of various part of EMR.
5.42 How atmospheric window regions are useful in remote sensing, explain?
5.43 What is spectral signature curve? How is spectral signature useful in identifying the object?
5.44 Explain the use of field spectro-radiometer.
5.45 Differentiate between Geostationary orbit and Polar sun synchronous orbit. Give examples.
5.46 How do you take help of a histogram in image analysis?
5.47 Discuss (i) Spatial resolution, and (ii) Radiometric resolution, including their importance.

5.48 Describe (i) Spectral resolution, and (ii) Temporal resolution, including their importance.

5.49 Discuss different types of remote sensing data products available?

5.50 What are the different types of sensors based on energy source? Write the advantages of active sensors?

5.51 Draw a neat diagram for the classification of imaging sensor systems.

5.52 Define thermal infrared sensors. Discuss most important applications of thermal infrared sensors.

5.53 Discuss various microwave sensors.

5.54 Enumerate various applications of microwave sensors.

5.55 Explain any three platforms used in remote sensing.

5.56 Describe in detail the characteristics of Landsats.

5.57 Discuss the salient features of SPOT satellites.

5.58 Describe the characteristics of IRS satellites.

5.59 Discuss the approach used in collecting PAN stereo-images?

5.60 Write a short note on the Indian remote sensing programme.

5.61 Describe high/very high resolution satellites/sensors. What are their main applications?

5.62 Discuss the characteristics of IKONOS image.

5.63 Describe the advantages of hyperspectral over multispectral remote sensing.

5.64 Write the merits and demerits of two broad methods of image interpretation.

5.65 Explain various image interpretation elements with suitable examples.

5.66 Discuss various steps involved in pre-processing of remotely sensed data.

5.67 Explain geo-referencing approach. What is the role of ground control points?

5.68 Describe the re-sampling methods used in image processing.

5.69 Explain the procedure to apply radiometric correction to satellite images.

5.70 Discuss the broad two approaches used for image enhancement.

5.71 Define (i) Ratio and (ii) NDVI. Why NDVI is most frequently and popular approach in image transformation.

5.72 Describe the two broad digital image classification techniques used. List the advantages and disadvantages of each.

5.73 Explain the role of training samples. How do you obtain these samples?

5.74 Discuss various supervised image classification techniques.

5.75 Explain the maximum likelihood classification technique.

5.76 Discuss various unsupervised classification methods.

5.77 Explain the following terms (i) Error of omission, (ii) Error of commission, (iii) User's accuracy, and (iv) Producer's accuracy.

Unsolved Numerical Questions

5.78 The wavelength of an electro-magnetic wave is 100 mm. What is the frequency of energy of that wave?

(Ans: 3 Ghz)

5.79 If the frequency is given as 67 Hz, what will be the wavelength of radiation?

*(Ans: 0.044*108 m)*

5,80 If the wavelength of EM energy is given as 43 m, what is the value of energy?

*(Ans: 0.46*10-26 J)*

5.81 If the energy produced is 36 J, determine the wavelength of EM energy.

*(Ans: 0.55*10-26 m)*

5.82 If the absolute temperature of a blackbody is 300 K, what is the spectral existence of the black body?

*(Ans: 1698*10-11 W/sq.m)*

5.83 If spectral existence of the body is 3.55 * $10^{-1°}$ W/sq.m, what is the absolute temperature of the body as per the Stefan–Boltzmann law?

(Ans: 1.58 K)

5.84 As per the Wien's displacement law, if wavelength is given as 0.05 m, what is the absolute temperature of the body?

*(Ans: 57.8*10-3 K)*

5.85 If the temperature of the body is given as 560 K, determine the value of wavelength (λ).

*(Ans: 5.16*10-14 m)*

5.86 If the reflected and incident energies are given as 43 J and 87 J, respectively, what will be the value of reflectance?

(Ans: 0.49)

1.87 Determine the area covered by an image of 512 x 512 size with 10 m pixel size.

(Ans: 26.22 km2)

1.88 How many grey levels are there in 10 bit and 12 bit radiometric resolution image?

(Ans: 21° = 1024 grey levels and 212 = 4096 grey levels)

References and Suggested readings

1. Alkhouri, Saher. (2014). Developing spatial information database for Jericho and Al Auja, West Bank, Palestine. 10.13140/RG.2.1.2426.1606.
2. Annual Report (2017-2018) ISRO, Bangalore, Department of Space, Govt. of India.
3. Birk, R.J. and T.B. McCord, (1994), Airborne hyperspectral sensor systems, IEEE AES Systems Magazine, October 1994, pp. 26-33.
4. Briottet, X.; Chehata, N.; Oltra-Carrio, R.; Bris, A.L. and Weber, C., (2016), Optical remote sensing in urban environments, In Land Surface Remote Sensing in Urban and Coastal Areas, edited by N. Baghdadi and M. Zribi, 1–62, Elsevier.

5. Calla, O.P., (2010), Applications of microwaves in remote sensing IP Multimedia Communications, In Special Issue of IJCA -www.ijcaonline.org.
6. Castleman, K. R., (1996), Digital Image Processing, Prentice-Hall, London
7. Chavez, Jr., P.S., Marino, C.M. and Schowengerdt, R.A., (1994) (Eds) Recent Advances in Remote Sensing and Hyperspectral Remote Sensing, SPIE Vol. 2318.
8. Colwell, R. N. (1983) Manual of Remote Sensing, Published by American Society of Photogrammetry, USA.
9. Congalton, R. (1991) A review of assessing the accuracy of classifications of Remotely Sensed Data, Remote Sensing of Environment, 37: 35, 46.
10. Congalton, R. G. and Green, K., (1999), Assessing the Accuracy of remotely sensed data: Principles and Practices, Lewis Publishers, Boca Raton.
11. Crist, Eric P. and Cicone, Richard C., (1984) A physically-based transformation of thematic mapper data-the TM tasseled cap, IEEE Transactions on Geoscience and Remote Sensing, Volume: GE-22, Issue: 3, May 1984.
12. Dong, Han; Tao, Li; Jiabing, Leng; Lingyan, Kong and Gang, Bai, (2017), GCN: GPU-Based Cube CNN Framework for hyperspectral image classification, 46th International Conference on Parallel Processing (ICPP) (2017): 41-49.
13. Eastman, J R. (1999), IDRISI 32- Guide to GIS and Image Processing, Vol. 1 & II, Worcester, MA, Clark University, USA.
14. Fung, T. and LeDrew, E. (1987), Application of principal components analysis to change detection, ISPRS Journal of Photogrammetry and Remote Sensing, Vol. 53, No. 12, 1987, pp. 1649-1658.
15. Garg, P K (2019), Principles and Theory of Geoinformatics, Khanna Book Publishers Ltd., New Delhi.
16. Garg, P K., (2022), Remote Sensing: Theory and Applications, New Age India International Publishers, New Delhi.
17. Gibson P.J. (2000), Introductory Remote Sensing–Principles and Concepts, Routledge, London.
18. Gupta, R.P. (1991), Remote Sensing Geology, Springer Verlag, Berlin.
19. Holland, D., Guilford, R. and Murray, K. (Eds), 2003. Topographic Mapping from High Resolution Space Sensors, OEEPE Official Publication No. 44.
20. Imhoff, M.L., (1995) Radar backscatter and biomass saturation: ramifications for global biomass inventory, IEEE Trans. Geosci. Remote Sensing 33, 511–518.
21. Ismail, Mohd Hasmadi, Pakhriazad, HZ and Shahrin, MF, (2009), Evaluating supervised and unsupervised techniques for land cover mapping using remote sensing data. Geografia: Malaysian Journal of Society and Space.
22. Jain, A.K., (1989), Fundamentals of Digital Image Processing, Prentice-Hall, London.
23. Jensen, J. R., (1986), Introductory Digital Image Processing. A Remote Sensing Perspective, Englewood Cliffs: Prentice-Hall, London.
24. Jensen, J. R., (2007), Remote Sensing of the Environment: An Earth Resource Perspective. 2nd Edition, Pearson Prentice Hall, Upper Saddle River.
25. Kahle, B., (1998), A temperature and emissivity separation algorithm for advanced spaceborne thermal emission and reflection radiometer (ASTER) images. IEEE Transactions on Geosciences and Remote Sensing, 1998. 36(4): p. 1113-1126. 2.

26. Kasischke, E; Melack, J. and Dobson, M., (1997), The use of imaging radars for ecological applications–a review, Rem. Sen. of Environment, 59 (2), 141-156.
27. Kamavisdar, Pooja, Saluja, Sonam and Agrawal, Sonu (2013) A survey on image classification approaches and techniques, International Journal of Advanced Research in Computer and Communication Engineering, Vol. 2, Issue 1, January 2013
28. Kauth, R. J. and Thomas, G.S., (1976), The tasseled cap-a graphic description of the spectral-temporal development of agricultural crops as seen by LANDSAT, Symposium on Machine Processing of Remotely Sensed Data, Purdue University of West Lafayette, Indiana.
29. Kurnaz, Bahadır; Bayik, Caglar and Abdikan, Saygin, (2020), Forest fire area detection by using Landsat-8 and Sentinel-2 satellite images: a case study in Mugla, Turkey. 10.21203/rs.3.rs-26787/v1.
30. Lechner, Alex; Foody, Giles and Boyd, Doreen, (2020), Applications in remote sensing to forest ecology and management. One Earth. 2. 405-412. 10.1016/j.oneear.2020.05.001.
31. Leslie, R V (2020), A systematic view of remote sensing, Advanced Remote Sensing, Academic Press, https://doi.org/10.1016/B978-0-12-815826-5.00001-5
32. Lillesand, T.M. and Kiefer, R.W., (1994), Remote Sensing and Image Interpretation, 3rd Ed. John Wiley & Sons, New York.
33. Lillesand, T. M., Kiefer, R.W. and Chipman, J. W., (2004), *Remote Sensing and Image Interpretation*, Fifth Edition, Wiley, Chichester. pp 35-38
34. Lunetta, R., Congalton, R., Fenstermaker, L., Jensen, J., McGwire, K. and Tinney, L., (1991), Remote sensing and geographic information system data integration: error sources and research issues. Photogrammetric Engineering and Remote Sensing, 57(6):677{687.
35. Nady, Beshoy. (2020). Road network extraction from satellite images in Egypt, 10.13140/RG.2.2.16038.27207.
36. Ose, Kenji; Corpetti, Thomas and Demagistri, Laurent, (2016), Multispectral satellite image processing, Editor(s): Nicolas Baghdadi, Mehrez Zribi, Optical Remote Sensing of Land Surface, Elsevier, https://doi.org/10.1016/B978-1-78548-102-4.50002-8. and (Right image) False colour composite from multispectral image.
37. Plaza, Javier; Plaza, Antonio; Valencia, David and Paz, Abel, (2009). Massively parallel processing of remotely sensed hyperspectral images. Proceedings of SPIE- The International Society for Optical Engineering. 7455. 10.1117/12.825455.
38. Poon, J. and Fraser, C.S., (2006) High-resolution satellite imagery in remote regions, a case study in Bhutan. Australian Remote Sensing and Photogrammetry Conference, ARSPC-2006, Canberra, 20-24 Nov, 5pages
39. Pradhan, Biswajeet; Saied, Pirasteh and Renuganth, Varatharajoo, (2010), Enhancement of semi-automated lineament extraction from IRS-1B satellite images for part of himalayan region. International Journal of Geoinformatics. 6. 41.
40. Prasad, S.V.S. Satya Savithri, T. and Murali Krishna, I.V., (2015), Techniques in image classification; a survey, Global Journal of Researches in Engineering: Electrical and Electronics Engineering, Volume 15, Issue 6, Version 1.0.
41. Prolite (2021), Field-portable spectroradiometers for remote sensing, Prolite, https://www.pro-lite.co.uk/File/Field_portable_Spectroradiometers.php
42. Richards, J. A. and Jia, X., (2013), Remote Sensing Digital Image Analysis, Springer-Verlag, Berlin.
43. Richards, Kenneth; Moulton, Robert and Birdsey, Richard. (1993). Costs of creating carbon sinks

in the US. Energy Conversion and Management- Energ Conv Manage. 34. 905-912. 10.1016/0196-8904(93)90035-9.

44. Sabins, F.F., (1996), Remote Sensing Principles and Interpretation. 3rd Edition, W.H. Freeman and Company, New York.

45. Stevens, Joshua; Smith, Jennifer M. and Bianchetti, Raechel A., (2012), Survey of multispectral imagery types and their applications, GEOG-160 Mapping Our Changing World, Editors: Alan M. MacEachren and Donna J. Peuquet, University Park, PA: Department of Geography, The Pennsylvania State University.

46. Transon, Julie; d'Andrimont, Raphaël d'Andrimont; Maugnard, Alexandre; and Defourny, Pierre; (2018) Survey of hyperspectral earth observation applications from space in the Sentinel-2 context, Remote Sens, 10, 157; doi:10.3390/rs10020157

47. Veettil, K and Zanardi, Rafael, (2012), A comparative study of various urban change detection techniques using high spatial resolution commercial satellite images: Quickbird and Worldview-2. International Journal of Advances in Remote Sensing and GIS. 1. 76-84.

Web resources:

http://what-when-how.com/remote-sensing-from-air-and-space/atmospheric-absorption-scattering-and-turbulence-visible-imagery-remote-sensing/

https://crisp.nus.edu.sg/~research/tutorial/spot.htm

https://www.isro.gov.in/saga-of-indian-remote-sensing-satellite-system

http://www.nrcan.gc.ca/node/9295

http://labs.cas.usf.edu/geodesy/sar.html

http://www.markelowitz.com/Hyperspectral.html

https://www.e-education.psu.edu/geog160/node/1958

https://crisp.nus.edu.sg/~research/tutorial/spot.htm

http://web.sonoma.edu/users/f/freidel/techniques/exer/rem_sens/remsen_e.html)

http://hyperphysics.phy-astr.gsu.edu/hbasees/quantum/radfrac.html

http://www.harrisgeospatial.com/Support/SelfHelpTools/HelpArticles/HelpArticles-Detail/TabId/2718/ArtMID/10220/ArticleID/17333/Atmospheric-Windows-and-Optical-Sensors.aspx

https://www.nrcan.gc.ca/earth-sciences/geomatics/satellite-imagery-air-photos/satellite-imagery-products/educational-resources/9389

http://www.sc.chula.ac.th/courseware/2309507/Lecture/remote18.htm

https://www.researchgate.net/publication/290899564_Satellite_Hyperspectral_Earth_Observation_Missions_-_A_Review.

https://www.geoimage.com.au/satellite/geoeye-1

https://www.geoimage.com.au/satellites/very-high-resolution-satellites

https://www.itc.nl/Pub/sensordb/AllSensors.aspx

http://instruct.uwo.ca/earth-sci/300b-001/remote.htm

https://crisp.nus.edu.sg/~research/tutorial/opt_int.htm

https://www.spectravista.com/ger1500/

http://gisknowledge.net/topic/remote_sensing_and_dia/tutorial_true_colour_composite.htm

http://www.euroargo-edu.org/argoeu_4.php.

http://forsys.sefs.uw.edu/fusion.html

https://www.satimagingcorp.com/services/resources/eoedu.belspo.be/en/satellites/spot. htm).

https://landsat.usgs.gov/

http://digipiczone.com/editing-techniques-for-contrast-enhancement

https://www.codeproject.com/Articles/1203383/A-Fast-Simple-and-Powerful-Contrast-Enhancement- http://www.eotec.com/images/IRS_-_Current_and_Future_-_Web.pdf

https://calval.cr.usgs.gov/JACIE_files/.../Karnik_Deviprasad_JACIE2010FINAL .pdf

https://www.eoportal.org/satellite-missions/spot-5

https://innoter.com/en/satellites/spot-6-7/

https://www.esa.int/ESA_Multimedia/Images/2005/12/Rome_seen_by_France_s_Spot_5_satellite

http://www.swisstopo.ch/NPOC/Products/kosovo/irsp_sarajewo.html

https://www.isro.gov.in/high-resolution-panchromatic-and-multi-spectral-images-observed-cartosat-3-calibration-validation-of

https://seos-project.eu/world-of-images/world-of-images-c04-p06.html

http://www.usgsquads.com/digital-map-products/aerial-and-satellite-imagery/satellite-imagery/digitalglobe-quickbird

https://www.eoportal.org/satellite-missions/worldview-2

https://gsp.humboldt.edu/olm/Courses/GSP_216/lessons/composites.html

https://www.gisoutlook.com/2019/08/digital-image-processing-image.html

CO AND PO ATTAINMENT TABLE

Course outcomes (COs) for this course can be mapped with the programme outcomes (POs) after the completion of the course and a correlation can be made for the attainment of POs to analyse the gap. After proper analysis of the gap in the attainment of POs necessary measures can be taken to overcome the gaps.

Table for CO and PO attainment

Course outcomes	Attainment of Programme Outcomes *(1- Weak correlation; 2- Medium correlation; 3- Strong correlation)*											
	PO-1	PO-2	PO-3	PO-4	PO-5	PO-6	PO-7	PO-8	PO-9	PO-10	PO-11	PO-12
CO-1												
CO-2												
CO-3												
CO-4												
CO-5												
CO-6												

INDEX

U

V